Membrane Contactor Technology

Membrane Contactor Technology

Water Treatment, Food Processing, Gas Separation, and Carbon Capture

Edited by Mohammad Younas and Mashallah Rezakazemi

WILEY-VCH

Editors

Professor Mohammad Younas
University of Engineering & Technology
Department of Chemical Engineering
No. 2 University Rd
25120 Peshawar
Pakistan

Associate Professor Mashallah Rezakazemi
Shahrood University of Technology
Chemical and Materials Engineering
Shahrood
Iran

Cover Image: © amixstudio/Getty Pixabay

Library of Congress Card No.: applied for

British Library Cataloguing-in-Publication Data
A catalogue record for this book is available from the British Library.

Bibliographic information published by the Deutsche Nationalbibliothek
The Deutsche Nationalbibliothek lists this publication in the Deutsche Nationalbibliografie; detailed bibliographic data are available on the Internet at <http://dnb.d-nb.de>.

Print ISBN: 978-3-527-34861-9
ePDF ISBN: 978-3-527-83102-9
ePub ISBN: 978-3-527-83104-3
oBook ISBN: 978-3-527-83103-6

Typesetting Straive, Chennai, India
Printing and Binding CPI Group (UK) Ltd, Croydon, CR0 4YY

Printed on acid-free paper

C110177_221121

Dedicated to my whole family, teachers, and friends whom I owe a lot.

Contents

Preface

Membrane technology has been emerging as a sustainable technology in separation science and engineering for the last two decades. Membranes have become the center point in water treatment and desalination, CO_2 capture technology, and food and pharmaceutical industry. Many parts of the world are now in critical need of clean water, energy, and food security. Among the membrane technology, the membrane contactor technology is gaining increasing importance in water treatment and desalination, gas separation, and juice concentration because it is working at low or atmospheric pressure and taking other natural forces as driving forces. Membrane contactors have a bright future in diverse application areas. Tremendous work has been carried out on membrane contactor technology and is still in progress in different world regions to further improve the technology than its current status. However, if observed closely, one can realize that research and development activities have mostly focused on membrane materials synthesis and development rather than focusing on membrane contactor module design and development. It appears there is a timely need for a book that illustrates the basic theory, applications, and practices of membrane contactor technology in water treatment, gas separation, and juice concentration.

To undertake the task, we are composed of young and enthusiastic engineers and scientists to help us in drafting the contents of the book that can cover up the theory, applications, and practice of membrane contactors in a succinct but easy-to-understand way for a novice working in membrane contactor technology. Together, we prepared 11 chapters starting from the presentation of the membrane technology in general and the illustration of the need for membrane contactor technology, followed by the detailed description of theory and processes used in membrane contactor technology. Chapters 2, 3, and 5 cover the theoretical aspects and working principles of membrane contactors used as liquid–liquid, gas–liquid, and gas–gas contactors. Design aspects of membrane contactor were described in Chapter 4. Chapters 6–9 focused on illustrating important applications of membrane contactors in water treatment, juice concentration, and pre- and post-combustion CO_2 capture. Chapters 6–9 indeed cover a vast field of applications succinctly and briefly. Market aspects, conclusions, and perspective of this technology were presented at the end of the chapters.

We believe that the unique feature of this book is the presentation of theory, design, practice, and applications of membrane contactor technology very briefly in one book. This is, of course, a relative comparison, in general, with the other published books on membrane contactors. Another distinctive feature of the book is that a very strong coverage of all the separation processes using membrane contactors from micro- to macrolevel is presented for the reasons described above.

We are truly thankful for the strong cooperation from all the contributing authors. As a matter of fact, we still have quite many ideas that have been part of the book but have not finished due to the fact that we always bound ourselves to be very brief and to address the readers who are new to this field. Meanwhile, we are indeed delighted to have this book published for beginners in membrane contactor technology. We also wish to thank all the reviewers and chapter contributors.

July 2021 | *Mohammad Younas*
Peshawar, Pakistan | *Mashallah Rezakezemi*

About the Authors

Dr. Mohammad Younas obtained the Master (II) by research and doctorate degrees from the Université de Montpellier 2, France, during 2006–2011. He also worked in the University of Technology, Eindhoven, Netherlands, during 2011–2012 as a part of European applied research project FP7 (DemoCLOCK). Currently, Dr. Younas is working as professor and leader of the "Membranes and Process Intensification" group in the Department of Chemical Engineering at University of Engineering and Technology, Peshawar, Pakistan. He is also engaged with State Key Laboratory of Separation Membranes and Membrane Processes, School of Materials Science and Engineering, Tianjin Polytechnic University, Tianjin 300387, China, and the University of Toledo, OH, USA, as visiting research scholar.

His research focus is in the general area of the water–energy–food nexus, CO_2 capture, gas separation, and desalination, to the service of the broad areas of learning and training. Specifically, his research in engineered and natural environmental systems involves membrane-based processes for energy-efficient desalination, environmental applications and implications of nanomaterials, water treatment, and juice concentration. He has coauthored in more than 70 highly cited journal publications, conference proceedings, and book chapters.

Dr. Mashallah Rezekazemi received his BSc and MSc degrees in 2009 and 2011, respectively, both in Chemical Engineering from the Iran University of Science and Technology (IUST) and his PhD from the University of Tehran (UT) in 2015. In his first appointment, Rezakazemi has been served as professor of the Faculty of Chemical and Materials Engineering at the Shahrood University of Technology since 2016. He has also received his degree promotion to associate professor in 2019.

Dr. Rezakazemi's research is in the general area of the water–energy nexus, CO_2 capture, gas separation, and desalination, to the service of the broad areas of learning and training. Specifically, his research in engineered and natural environmental systems involves (i) membrane-based processes for energy-efficient desalination, CO_2 capture, gas separation, and wastewater reuse; (ii) sustainable production of gas stream, water, and energy generation with the engineered membrane; (iii) environmental applications and implications of nanomaterials; and (iv) water and sanitation in developing countries.

He has coauthored in more than 100 highly cited journal publications, conference articles, and book chapters. He has received major awards and grants from various funding agencies in recognition of his research. Notable among these is the Khwarizmi Youth Award from the Iranian Research Organization for Science and Technology (IROST) in 2012 and Outstanding Young Researcher Award in Chemical Engineering from Academy of Sciences of Iran. He is being named as top 1 most Highly Cited Researcher by Web of Science (ESI).

1

Introduction to Membrane Technology

Mohammad Younas[1] and Mashallah Rezakazemi[2]

[1] *University of Engineering and Technology, Department of Chemical Engineering, Peshawar 25120, Pakistan*
[2] *Shahrood University of Technology, Faculty of Chemical and Materials Engineering, Shahrood, Iran*

1.1 Overview of Membrane Technology

Membrane technology is a general term used for a range of different separation processes. Membrane separation processes have been proven to be well-established technologies in a wide range of water, energy, food, and environmental applications throughout the production, purification, and formulation of useful products [1–4]. Thus, the membrane separation processes have become the leading separation technology over the past two decades. The membrane is defined as a selective thin layer of a semipermeable material that acts as a selective barrier and separates undesired species from a feed solution based on their sizes or affinity by exerting a potential gradient, such as pressure, temperature, electrical, or concentration difference (Figure 1.1). Separation is accomplished if one species of a mixture moves through the membrane faster than another species in the mixture. The main advantage of membrane technology, which differentiates it from traditional separation, purification, and formulation processes, is that it produces stable products without adding chemicals with a relatively low energy consumption with a remarkable potential for an environmental impact. Other benefits include modular and easy to scale-up, well-arranged, compact, and straightforward process in concept and operation, decreased capital and operational cost of technology applications using membrane, and environment friendly.

In general, membranes are classified based on their average pore size, driving force, morphology, and materials. The pore size of the membrane material or surface is a paramount factor in its first differentiation. Nevertheless, membrane materials can be organic and inorganic. All of the membrane separation processes are effective methods of treating the feed mixture, e.g. water, gas, and food that hardly is treated using conventional separation methods.

Membrane Contactor Technology: Water Treatment, Food Processing, Gas Separation, and Carbon Capture, First Edition. Edited by Mohammad Younas and Mashallah Rezakazemi.

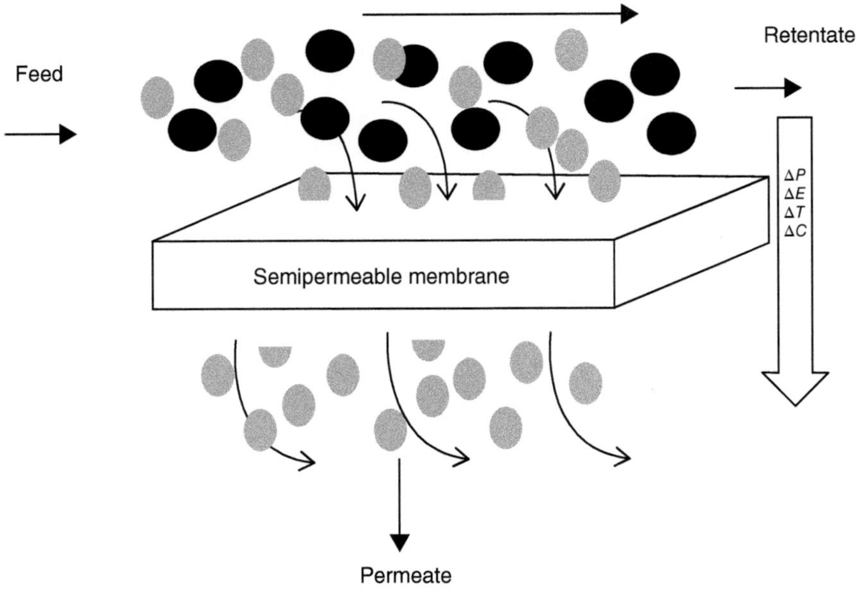

Figure 1.1 Typical membrane separation process.

1.2 Conventional Membrane Separation Processes

1.2.1 Microfiltration (MF)

Microfiltration (MF) is the first classification of membrane separation techniques based on pore size. The MF membrane was first developed to analyze the bacteria in the water. In the 1960s, the first commercial MF membrane was also developed in biological and pharmaceutical applications. Since then, MF membranes have been widely applied in wastewater treatment and juice technology to remove microorganisms, clarify cider and other juices, and sterilize beer and wine. The separation mechanism in MF membranes is governed by the sieving effect or size exclusion technique. Thus, the species are separated according to their size. Large pores of MF remove suspended solids, while even proteins can pass through the MF membrane easily. The MF membranes can also be used to separate sand, clays, algae, and some bacteria from aqueous feed streams. They are recommended to separate species with a diameter larger than 0.1 μm. The applied pressure in MF is low (usually <2 bar), while this is the lowest applied pressure in other pressure-driven membrane separation processes [5, 6].

1.2.2 Ultrafiltration (UF)

Ultrafiltration (UF) is also included in size exclusion-based pressure-driven membrane separation processes. The pore size of UF membranes is around 0.01 μm. These membranes can prevent species in the molecular weight range of 300–500 000 Da to pass through. UF rejects protein and suspended solids. However, dissolved substances could not be removed by UF unless they are first pretreated in an adsorption column like with activated carbon or coagulated with alum or iron salts. Similarly, UF membranes cannot retain the mono- and disaccharides, salts, amino acids, organics, inorganic acids, or sodium hydroxide. They exhibit small osmotic pressure differentials due to their inability to reject salts, as compared with reverse osmosis (RO). UF processes operate at 2–10 bars. Separation efficiency will further be augmented if the difference in the sizes of the species is high enough. UF is considered nowadays to be the dominant part of membrane separation processes due to its diverse applications in water, energy, food, and the environment. UF processes are considered the most used membrane separation process next to dialysis and MF [7].

1.2.3 Nanofiltration (NF)

Nanofiltration (NF) is another pressure-driven membrane process between RO and UF pore size of around 0.001 μm. NF membranes remove most organic molecules, viruses, and a range of salts. These membranes are often applied to soften the hard water by removing divalent ions. NF membranes possess a negative charge on the surface. It demonstrates the anion repulsion, which mainly causes the species rejection. Low rejection is witnessed for salts with monovalent anion and nonionized organics with a molecular weight below 150. However, high rejection can be observed for salts with di- and multivalent anions and organics with a molecular weight above 300. NF is advantageous over RO in different aspects, such as being operated at low pressure, giving high permeate flux, retention of multivalent salt and organic solutes, and having low investment and operation and maintenance costs.

NF membrane is more suitable for ions with more than one negative charge in single charged ions pass, such as sulfate or phosphate. However, NF membranes also reject uncharged and positively charged ions according to the molecule's size and shape. For example, the same rejection of calcium chloride and sodium chloride can be observed while the rejection of sodium sulfate is the same for magnesium sulfate. Instead, the rejection of di- and multivalent anions is high compared with that for monovalent ions. The species rejection decreases with increasing concentration. The Donnan exclusion model can explain this phenomenon. The higher the species concentration, the more cations available to shield the negative charges on the membrane surface, making it easier for the anions to pass through the membrane pores. On the other hand, the charge density of ions also plays an important role in its rejection. For example, the sulfate ion has a higher charge density than the chloride ion and is almost completely repelled by the NF membrane even in a high ionic strength solution such as seawater [8].

1.2.4 Reverse Osmosis (RO)

RO demonstrates, in principle, the least possible pore structure among the membranes. Water is the only species that can pass through the RO membrane;

essentially, all dissolved and suspended species are rejected. RO membranes have a pore size of around 0.0001 μm. The permeate is essentially the pure water because RO also removes most healthy minerals such as calcium, zinc, magnesium, etc. that are present in the water and are useful in a certain quantity for drinking water especially for people with inadequate diets and people living in hot climates. The water can be made healthy bypassing the RO water through calcium and magnesium beds. RO removes monovalent ions to desalinate the saline water. Both NF and RO are also termed as dense membrane separation processes because separation relies to some extent on physicochemical interactions between the permeate (species) and the membrane material. In wastewater treatment and reclamation, RO systems are typically used as the last step for removing total organic carbon (TOC). RO has been proven to remove dissolved species effectively, microbes, and neutral base compounds [9, 10].

To understand the working principle of RO, it is helpful to understand first osmosis. Osmosis refers to the migration of water from a weaker solution to the stronger solution when a semipermeable membrane separates two salt solutions of different concentrations. The migration of salts continues until the two solutions reach the same concentrations, achieving the osmotic equilibrium. The semipermeable membrane allows the water species to pass through naturally, but not the salt. In RO, the two solutions are still separated by a semipermeable membrane, but the pressure is applied to reverse the water's natural flow. This forces the water species to move from the more concentrated solution to the weaker. Thus, the solute aggregate on one side of the semipermeable membrane and the pure water pass through the membrane on the other side. The concept of osmosis and RO is described schematically in Figure 1.2 where (a) and (b) illustrate the process of osmosis and (c) represents the RO. If a certain pressure (ΔP) applied to the concentrated solution equals the osmotic pressure difference between the two solutions ($\Delta\pi$), the system reaches the osmotic equilibrium, and water flow stops. If the applied pressure exceeds osmotic pressure ($\Delta P > \Delta\pi$), water flows from the concentrated solution to the dilute solution. A summary of pressure-driven processes is outlined in Tables 1.1 and 1.2.

1.2.5 Electrodialysis (ED)

Electrodialysis (ED) refers to an electrically driven membrane separation process in which charged ions are separated from a feed solution through selectively

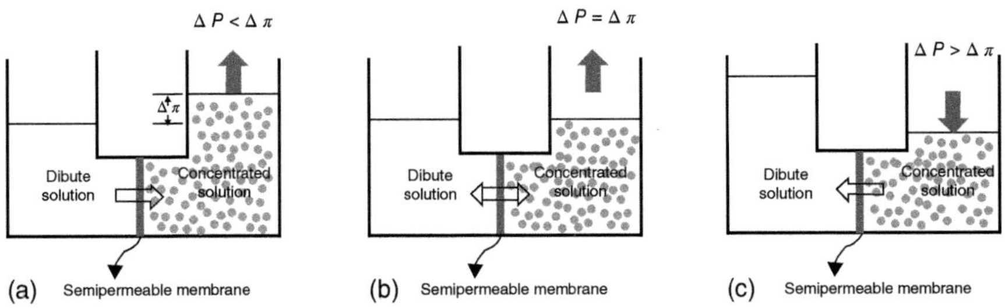

Figure 1.2 Osmotic phenomena: (a) osmosis, (b) equilibrium, and (c) reverse osmosis.

Table 1.1 Pressure driven size-based membrane processes for the removal of typical pollutants.

	Membrane separation process			
Feed component	**Microfiltration (MF)**	**Ultrafiltration (UF)**	**Nanofiltration (NF)**	**Reverse osmosis (RO)**
Water	↘	↘	↘	↘
Monovalent ions	↘	↘	↘	↘↗
Multivalent ions	↘	↘	↘↗	↘↗
Dissolved substances	↘	↘	↘	↘↗
Viruses	↘	↘↗	↘↗	↘↗
Bacteria, protozoa	↘↗	↘↗	↘↗	↘↗
Suspended solids	↘↗	↘↗	↘↗	↘↗

ion-permeable membranes. In an ED process configuration, cationic and anionic membranes are alternately arranged between an anode and a cathode plate. By applying an electrical potential, the ions migrate toward the anode and cathode, and consequently, the water molecule is deionized. A typical ED cell consists of electrodes and ion-permeable membranes, as shown in Figure 1.3. When an electric field across the membranes is applied, the cations move toward the cathode, and the anions migrate toward the anode. The cations pass through the cation-selective membrane, while anions pass through the anion-selective membrane. Thus, the feed became diluted in one side and concentrated in the electrolyte on the other side. Best performance in ED membranes could be achieved by selecting the highly permselective, physically strong, and low electrical resistance membranes [5].

1.2.6 Pervaporation (PV)

Pervaporation (PV) is a membrane separation process used to recover more volatile components in liquid mixture through a dense membrane. The PV is governed by a partial pressure difference across the membrane as the driving force by applying a vacuum at the permeate side [11–14]. The solution–diffusion model generally describes the transport of species across nonporous membranes in PV. Because of the negative pressure on permeate side, the osmotic pressure is not a limiting factor, as is the case for RO. The partial pressure difference at feed and permeate sides causes the more volatile liquid to vaporize within the membrane. The vapor passes through the membrane and finally condenses at the permeate side (Figure 1.4). PV

Table 1.2 Comparative analysis of conventional membrane processes.

	Microfiltration (MF)	Ultrafiltration (UF)	Nanofiltration (NF)	Reverse osmosis (RO)	Electrodialysis (ED)	Membrane pervaporation (MPV)
F/P	L/L	L/L, G/L	L/L	L/L	L/L	L/G
Membrane material	Polymeric	Polymeric/ ceramic	Polymeric/ ceramic/ mixed matrix	Polymeric	Polymeric	Polymeric; polyvinyl alcohol composites, silicones, cellulose acetates
Membrane structure	Symmetrical/ asymmetrical	Asymmetrical	Asymmetrical	Asymmetrical	Asymmetrical	Asymmetrical
Membrane morphology/ thickness	Porous	Porous	Porous/dense	Dense	Dense	Dense
Support layer	10–150 μm	150–250 μm	150 μm	150 μm		
Thin film	1 μm	1 μm	1 μm	1 μm		
Pore size	0.05–10 μm	0.001–0.05 μm	0.5–2 nm	<0.002 μm	MW < 200 Da	Nonporous
Driving force	ΔP	ΔP, activity difference, concentration difference, temperature difference	ΔP	ΔP	ΔE	ΔP vacuum, chemical potential gradient
Separation principle	Sieving mechanism	Sieving mechanism	Donnan exclusion/ solution–diffusion/ capillary flow	Solution–diffusion	Solution–diffusion/ion migration	Donnan exclusion/solution–diffusion
Operating pressure	<2 bar	2–5 bar	5–15 bar	15–100 bar	Electrical potential	Partial pressure difference
Membrane module type	Tubular, hollow fiber	Plate and frame, spiral wound, tubular, hollow fiber	Plate and frame, spiral wound, tubular	Plate and frame, spiral wound, tubular		Plate and frame, spiral wound, tubular, hollow fiber

MWCO		300–500 000 Da	200–1000 Da	<500 Da		
Applications	Separation of macromolecular to cellular size particles (bacteria, fat, proteins, whey industry)	Fruit juice clarification and concentration, milk separation, food, beverage, and dairy, biotechnology, medical applications	Water softening, removal of color, hardness, TOC, sulfate from water, concentration of organics with molecular weight of 300–1000 in the food and pharmaceutical	SWRO, BWRO desalination	Separation of ions mostly in desalination of water	Dehydration of liquid organic, ethanol, isopropyl alcohol, ethylene glycol
Permeate flux		150 l/m^2/h	43 I/m^2/h/bar			
Solute rejection (type)	Particles, clay, bacteria	Macromolecules, proteins, polysaccharides, sugars, biomolecules, polymers, colloidal particles	HMWC, mono-, di-, and oligosaccharides, polyvalent ions (−ive), $MgSO_4$, glucose, sucrose	HMWC, LMWC, sodium chloride, glucose, amino acids		
Solute rejection (%)		>90%	Up to 100%	>90%		
Issues and problems		Membrane fouling and concentration polarization				

F/P, feed/permeate; TMP, transmembrane pressure difference; HMWC, high-molecular-weight compounds; LMWC, low-molecular-weight compounds; SWRO, seawater reverse osmosis; BWRO, brackish water reverse osmosis; MWCO: molecular weight cutoff.

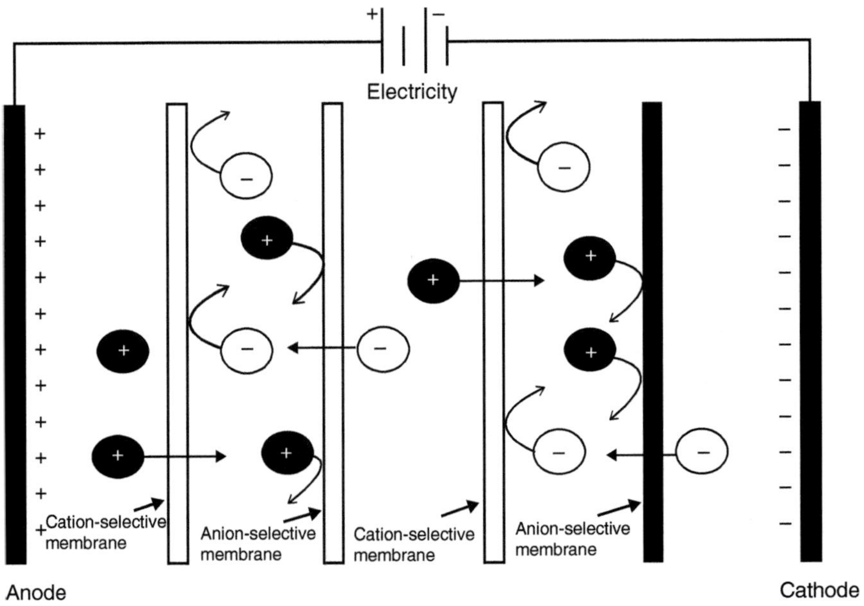

Figure 1.3 A basic electrodialysis system.

is characterized by the imposition of a barrier layer between two phases. Mass transfer occurs selectively across the membrane from one side to the other side of the membrane. The unique phenomenon of PV is the phase change required of the one phase (feed) diffusing across the membrane [17–19]. Since different species present in the feed mixture permeate through the membrane at different rates, a low concentration component in the feed mixture can be highly enriched in the permeate. Thus, the membrane's selectivity becomes the defining factor in the relative flow of the different species. PV has gained more attention from the chemical industry in the past decade due to the effective separation process for recovering volatile components in liquid mixtures. It is currently considered more effective for dehydration of liquid hydrocarbons to yield high-purity organics, most notably ethanol, isopropyl alcohol, and ethylene glycol. PV, due to its simplicity and easy installation, is used as an integrated process with distillation [20, 21].

1.3 Molecular Weight Cutoff (MWCO)

Molecular weight cutoff (MWCO) is a useful tool for characterizing filtration membranes. In early development, UF membranes were used to purify macromolecules in bioseparation processes such as to retain the proteins. Since their molecular weight characterizes macromolecules, the membranes are also characterized by whether the macromolecules up to certain molecular weights are retained. It depends on the size of the pore of the membranes. MWCO is indicated in Dalton that refers to the MWCO of species or solute with 90% rejection. MWCO 500 describes

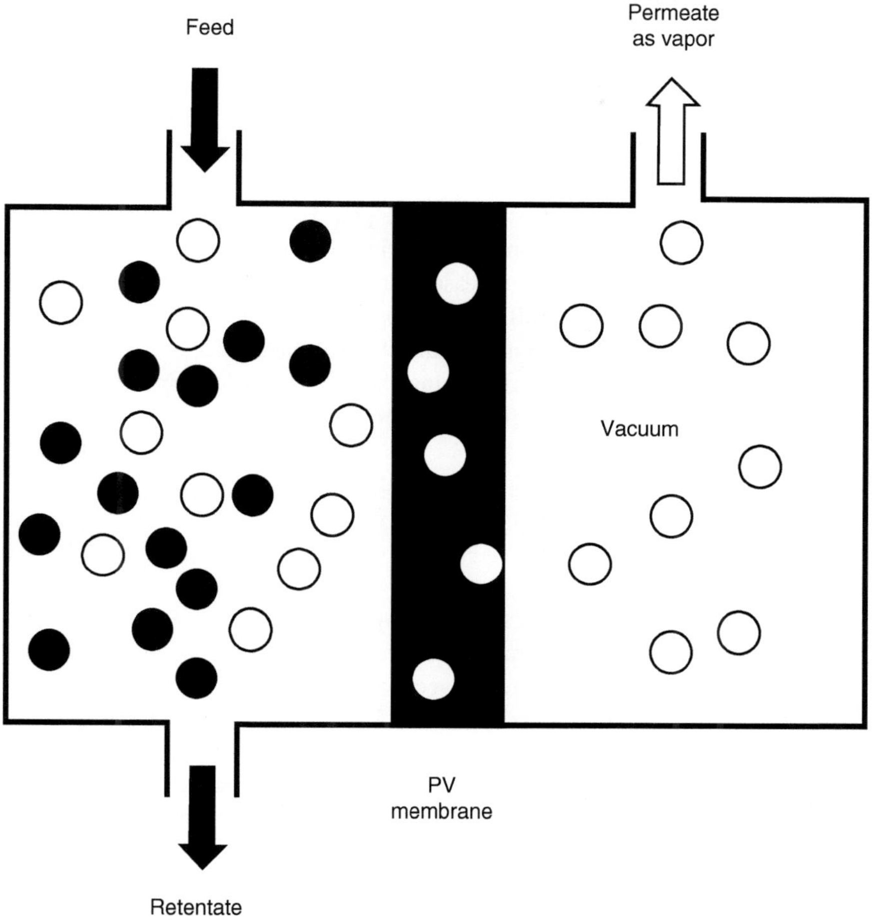

Figure 1.4 Membrane pervaporation process. Source: Based on Winzeler and Belfort [15] and Shirazi et al. [16].

that the molecules with molecular weight (MW) above 500 are rejected, and those with below 500 are passed through the membrane. The MWCO of any membrane can be altered from the chemistry of the solute with membrane interaction, their molecular orientation and configuration, and the operating conditions [22].

1.4 Concentration Polarization

Concentration polarization in membrane filtration is one of the significant problems that hinders the solvent flux and solute rejection. Concentration polarization is an important feature in membrane separation processes. Species rejected by the membrane accumulate at the membrane surface. This accumulation of species on the surface of the membrane is called concentration polarization. It produces a concentration gradient in the zone where the species accumulate. There remains a balance between species brought to the membrane surface by convective flow of the solvent and back-diffuses to the bulk. At times, however, the balance in species concentration at the membrane surface diminishes. It reaches its solubility limit, which is lower than that predicted by the fluid hydrodynamics of the system.

Consequently, the membrane effectively experiences a higher feed side concentration at its interface, resulting in reduced flux and reduced apparent solute rejection. Often the severity of concentration polarization can be controlled by operating conditions, module geometry, and fluid hydrodynamics. Concentration polarization practices to a smaller increase in transmembrane solvent flux with the rise in operating feed pressure until a gel layer is formed at the membrane's surface. It can also be lessened by increasing the fluid share at the surface of the membrane or producing the turbulence by introducing the channel spacers in the modules. Thus, the transmembrane solvent flux shows no further increase with the pressure and is termed as limiting flux [15, 23].

1.5 Membrane Fouling

Fouling refers to the deposition of solute or any other species in feed on the membrane surface or inside the membrane pores. For example, if the balance in species concentration at the surface due to convective flow and feed bulk concentration reaches the point where species precipitates or forms a thixotropic gel, the situation is termed as fouling. The formed gel layer causes an additional mass transfer resistance in conjunction with the membrane itself. In such cases, increased applied feed pressure may not improve the transmembrane flux; rather it will increase or densify the gel layer [24].

Fouling may be caused by the pore geometry/tortuosity or species–pore wall interactions. Consequently, the pores are blocked entirely or be marginally reduced in diameter, causing a decline in transmembrane flux while the rejection may be either constant or may increase. Proper and scheduled membrane/module cleaning may reverse the fouling; however, irreversible fouling may also occur over time, permanently deteriorating the membrane surface and pores. In such cases, the membrane's replacement becomes indispensable to regain the actual transmembrane flux and the species rejection. Although both concentration polarization and fouling reduce transmembrane solvent flux, they have opposite effects on species rejection. For example, concentration polarization is a function of operating parameters like pressure, temperature, feed concentration, and velocity but is independent of time. In contrast to that, fouling is partially dependent on operating parameters, particularly feed concentration, but is also time dependent. These phenomena have been described schematically in Figure 1.5 [15, 16].

In UF, feed-side mass transfer resistance and resistance due to the gel/cake layer formation on the membrane surface because of fouling play an essential role in transmembrane flux and species rejection. Usually, the proper membrane process and material selection are chosen to decrease the fouling tendencies of the membrane surface. The base polymer surface chemistry can be modified to increase

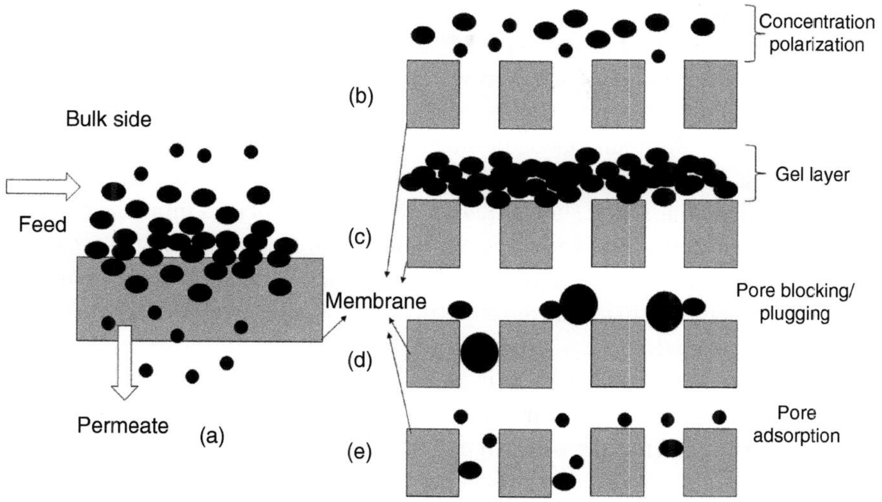

Figure 1.5 Membrane fouling and concentration polarization. (a) Membrane separation process. (b) Concentration polarization. (c) Gel/cake layer. (d) Pore blocking/plugging. (e) Pore adsorption.

hydrophilicity of the membrane with the contacting fluid to increase the flux and reduce the fouling in most aqueous applications. Fouling, scaling, or chemical interaction greatly affects the NF and RO systems while MF and UF are mildly affected in their effective operation. Thus, extensive pretreatment is occasionally made mandatory before NF and RO to avoid conditions leading to fouling, scaling, or chemical interaction [16, 23].

1.6 Diafiltration

In some cases, during filtration, recovery of species may be hindered due to reduced flux, high feed viscosity, solubility limits of nonpermeating solutes, etc. For such cases, the concentrate is diluted by solvent (water) during continuous filtration until satisfactory recovery of the permeable species, which is termed diafiltration.

1.7 Historical Perspective

The history of membrane separation technology dates back to 1748 when the French Abbe Nollet published his observations on osmotic phenomena [25]. The study of UF has been closely associated with that of dialysis. Dialysis experiments through artificial membranes of collodion were recorded by Fick [26]. Similar observations on dialysis were made by Hoppe-Seyler [27] and Schumacher [5]. The pioneering study UF process was reported by Schmidt [28], who investigated the filtration of a solution of protein or gum Arabic through an animal membrane.

In the third decade of twentieth century, membranes were regarded as mechanical sieves, and permeability was considered as the sole dependent on particle and pore dimensions. The concept of semipermeability of the membrane and the theory of partial solubility was also introduced in this decade, which describes membrane's permeability as solvent dissolving in the membrane from one side to the other. For the first time, the membrane was used in seawater desalination to produce sources for freshwater, put forth by Hassler [29]. Later, Reid and Breton introduced the RO membranes when they developed the cellulose diacetate film showing salt rejection up to 96% [30]. However, the breakthrough was achieved when Loeb and Sourirajan developed a cellulose diacetate asymmetric membrane and successfully tested for high flux and salt rejection.

DuPont developed a hollow fiber capillary membrane from aromatic polyamide. However, in 1985, Cadotte prepared high-performance membranes using in situ interfacial polycondensation between poly/monomeric amine and poly/monomeric functional acid halide [31, 32]. This opened a new era for the researchers to explore the polyamide films by crosslinking, which gave high permeate flux that cellulose acetate (CA) membranes [9].

FilmTec introduced the two-layer design membrane modules for water desalination at the industrial scale, which is still dominating the desalination industry. Undoubtedly, the membrane material and the modules have been improved over the years, but the basic concept adopted by FilmTec is still widely accepted. Desalination Systems, Inc. (DSI) introduced three-layer composite membranes for NF and RO. In the twenty-first century, membrane separation processes such as UF, NF, and RO emerged as reliable technology in water, food, environment, and food.

1.8 Concluding Remarks and Future Challenges

Membrane technology has become an important entity of our daily life routine work. Membranes have a potential in the future. Water scarcity and water stress, carbon capture, food security, energy constraints, environmental regulations, and nanotechnology are key drivers to boost the membrane technology further. However, the development in membrane technology would rise exponentially if "engineering aspects" in all membrane separation processes with key attention to "industrialists" and "entrepreneurs" are correctly addressed. Membranes perform the specific task for which they are designed. Each type of membrane filtration, e.g. MF, UF, NF, and RO, has its own role depending upon the pores' size, driving force, operating conditions, membrane material properties, and physicochemical interaction with feed components. However, if a specific membrane was chosen for the particular application and process, it performs well and achieves the required objectives. Two facts should be kept firmly in mind before deciding any membrane process: "Membranes do not lie." The statement describes that membranes do exactly what they can do under the given circumstances. For example, if the membrane material is not compatible with the feed solution or cannot withstand the operating parameters, the membrane will not perform to expectations. In such cases, this will not be the deficiency of the membrane. The other fact is: "Membranes are designed to reject dissolved solids." This means that if the feed mixture contains substantial and diverse undesired components like suspended solids, then the membrane systems will perform very poorly. So for each membrane separation process, the feed characteristics have also been described, and their protocol should be obeyed. Otherwise, pretreatment of feed should be ensured that the feed solution is free of species that may precipitate or degrade the membrane pores and surface due to their aggregation during the process.

In the twenty-first century, the world is facing more severe challenges than ever toward sustainable development in terms of water quality and sources in developed and developing countries, meeting increasing energy demands, securing the food shortages, and controlling the adverse effects of global warming. Therefore, the demand for the use of novel membranes, innovative processes, and compact modular designs to address these issues in various applications will continue to increase. The conventional membrane separation processes have already emerged as a promising technology in different water food and environment sector applications. Still, there remained a gap to develop the membrane technology to be driven by higher productivity, lower cost of production, and increased development speed. It was learned that several membrane characteristics could determine a membrane's suitability for a specific separation application. These include (i) porosity, (ii) morphology, (iii) surface properties, (iv) mechanical strength, (v) chemical resistance, (vi) selectivity, and (vii) driving force. These characteristics depend on the proper choice of membrane material and the synthesis technique. Further to that, module design is also essential to a great extent to achieve these properties. These characteristics are interrelated; for example, a highly porous membrane structure can be maintained only if the polymer has adequate mechanical strength or the membrane should be operated at low or atmospheric pressure. Surface properties and pore morphology are linked to fouling properties, flux through the membrane, and solute separation. There is a need to reduce or even remove the gas between scientists and industrialists. For example, scientists and engineers' major challenges are as follows: (i) membrane designs should be manufacturer specific, and (ii) application-specific membranes should be developed targeting the specific industry. Membrane system costs and applications are currently materially limited, whereas membrane performance is measured as solvent flux and selectivity which are the limiting factors for scientists and engineers. However, for an efficient and economically feasible industrial application, membranes need to keep their whole lifetime integrity. Unfortunately, the integrity and flux or selectivity is often in the opposite trend. Less integrity will lessen the membrane life and thus is meant for higher replacement costs of the membrane. It is also noted that membrane technology has its own disadvantages. For example, high pressure as a driving force causes high energy consumption and pollution to the environment and uses a range of chemical solvent that could be very harmful to the environment. Thus, the future development of membrane technology and its applications could conform with the sustainable development goals (SDG). Theoretically, 0.7 kWh/m^3 should be the minimum energy required to convert seawater to pure water [33]. Membrane separation technology is currently considered among the best available technologies (BAT) in the nexus of many processes and applications like food, water, energy, and the environment. However, with the current choice of materials, modules, and technology, the energy consumption still stands between 2 and 5 kWh/m^3 [34]. Thus, the research is focused on increasing the separation efficiency, reducing energy consumption, and making it more environment friendly and fouling resistant. The gap between scientists and industrialists should be removed. Such objectives could be achieved by adopting the membrane contactor technology and switching over to concentration difference as a driving force instead of using pressure difference as the driving force. The successful design and operation of membrane systems lie in a deeper understanding of principles, engineering, and practical aspects such as interfacial phenomena, rheology, material science, and module design of membrane separation processes. The research and development ("R&D") efforts should be focused rather on "engineering applications" such as water, energy, food, and the environment. This would also lead to clear the approach that any membrane module will result in the expected separation.

References

1 Sheikh, M., Pazirofteh, M., Dehghani, M. et al. (2020). Application of ZnO nanostructures in ceramic and polymeric membranes for water and wastewater technologies: a review. *Chemical Engineering Journal* 391: 123475.

2 Mirqasemi, M.S., Homayoonfal, M., and Rezakazemi, M. (2020). Zeolitic imidazolate framework membranes for gas and water purification. *Environmental Chemistry Letters* 18 (1): 1–52.

3 Rezakazemi, M., Sadrzadeh, M., and Matsuura, T. (2018). Thermally stable polymers for advanced high-performance gas separation membranes. *Progress in Energy and Combustion Science* 66: 1–41.

4 Rezakazemi, M., Dashti, A., Riasat Harami, H. et al. (2018). Fouling-resistant membranes for water reuse. *Environmental Chemistry Letters* 16 (3): 715–763.

5 Lonsdale, H.K. (1982). The growth of membrane technology. *Journal of Membrane Science* 10 (2–3): 81–181.

6 Fane, A.T., Wang, R., and Jia, Y. (2011). Membrane technology: past, present and future. In: *Membrane and Desalination Technologies* (eds. L.K. Wang, J.P. Chen, Y.-T. Hung and N.K. Shammas), 1–45. Springer.

7 Zeman, L.J. and Zydney, A. (2017). *Microfiltration and Ultrafiltration: Principles and Applications*. CRC Press.

8 Jye, L.W. and Ismail, A.F. (2016). *Nanofiltration Membranes: Synthesis, Characterization, and Applications*. CRC Press.

9 Uemura, T. and Henmi, M. (2008). Thin-film composite membranes for reverse osmosis. In: *Advanced Membrane Technology and Applications* (eds. N.N. Li, A.G. Fane, W.S.W. Ho and T. Matsuura), 1–19. Wiley.

10 Amjad, Z. (1993). *Reverse Osmosis: Membrane Technology, Water Chemistry, and Industrial Applications*. New York: Van Nostrand Reinhold.

11 Rezakazemi, M., Sadrzadeh, M., and Mohammadi, T. (2018). Separation via pervaporation techniques through polymeric membranes. In: *Transport Properties of Polymeric Membranes*, Chapter 13 (eds. S. Thomas, R. Wilson, S.C. George, et al.), 243–263. Elsevier.

12 Rezakazemi, M., Marjani, A., and Shirazian, S. (2018). Organic solvent removal by pervaporation membrane technology: experimental and simulation. *Environmental Science and Pollution Research* 25 (20): 19818–19825.

13 Rezakazemi, M., Shahidi, K., and Mohammadi, T. (2015). Synthetic PDMS composite membranes for pervaporation dehydration of ethanol. *Desalination and Water Treatment* 54 (6): 1542–1549.

14 Shahverdi, M., Baheri, B., Rezakazemi, M. et al. (2012). Pervaporation study of ethylene glycol dehydration through synthesized (PVA–4A)/polypropylene mixed matrix composite membranes. *Polymer Engineering & Science* 53 (7): 1487–1493.

15 Winzeler, H.B. and Belfort, G. (1993). Enhanced performance for pressure-driven membrane processes: the argument for fluid instabilities. *Journal of Membrane Science* 80 (1): 35–47.

16 Shirazi, S., Lin, C.-J., and Chen, D. (2010). Inorganic fouling of pressure-driven membrane processes – a critical review. *Desalination* 250 (1): 236–248.

17 Rezakazemi, M., Iravaninia, M., Shirazian, S. et al. (2012). Transient computational fluid dynamics modeling of pervaporation separation of aromatic/aliphatic hydrocarbon mixtures using polymer composite membrane. *Polymer Engineering & Science* 53 (7): 1494–1501.

18 Rezakazemi, M., Shahverdi, M., Shirazian, S. et al. (2011). CFD simulation of water removal from water/ethylene glycol mixtures by pervaporation. *Chemical Engineering Journal* 168 (1): 60–67.

19 Rezakazemi, M., Razavi, S., Mohammadi, T. et al. (2011). Simulation and determination of optimum conditions of pervaporative dehydration of isopropanol process using synthesized PVA–APTEOS/TEOS nanocomposite membranes by means of expert systems. *Journal of Membrane Science* 379 (1): 224–232.

20 Crespo, J. and Brazinha, C. (2015). Fundamentals of pervaporation. In: *Pervaporation, Vapour Permeation and Membrane Distillation* (eds. A. Basile, A. Figoli and M. Khayet), 3–17. Elsevier.

21 Figoli, A., Santoro, S., Galiano, F. et al. (2015). Pervaporation membranes: preparation, characterization, and application. In: *Pervaporation, Vapour Permeation and Membrane Distillation* (eds. A. Basile, A. Figoli and M. Khayet), 19–63. Elsevier.

22 Meireles, M., Bessieres, A., Rogissart, I. et al. (1995). An appropriate molecular size parameter for porous membranes calibration. *Journal of Membrane Science* 103 (1–2): 105–115.

23 Cui, Z., Jiang, Y., and Field, R. (2010). Fundamentals of pressure-driven membrane separation processes. In: *Membrane Technology* (eds. Z.F. Cui and H.S. Muralidhara), 1–18. Elsevier.

24 Vilker, V.L., Colton, C.K., and Smith, K.A. (1981). et al, The osmotic pressure of concentrated protein solutions: effect of concentration and pH in saline solutions of bovine serum albumin. *Journal of Colloid and Interface Science* 79 (2): 548–566.

25 Van Lente, H. and Rip, A. (1998). The rise of membrane technology: from rhetorics to social reality. *Social Studies of Science* 28 (2): 221–254.

26 Fick, A. (1855). Uber diffusion. *Annalen der Physik* 94: 59–86.

27 Ferry, R.J. (1923). Studies in the chemistry of hemoglobin I. The preparation of hemoglobin. *Journal of Biological Chemistry* 57 (3): 819–828.

28 Ferry, J.D. (1936). Ultrafilter membranes and ultrafiltration. *Chemical Reviews* 18 (3): 373–455.

29 Glater, J. (1998). The early history of reverse osmosis membrane development. *Desalination* 117 (1–3): 297–309.

30 Reid, C. and Breton, E.J. (1959). Water and ion flow across cellulosic membranes. *Journal of Applied Polymer Science* 1 (2): 133–143.

31 Shenvi, S.S., Isloor, A.M., and Ismail, A.F. (2015). A review on RO membrane technology: developments and challenges. *Desalination* 368: 10–26.

32 Cadotte, J.E. (1985). Evolution of composite reverse osmosis membranes. *ACS symposium series* 269: 273–294.

33 Schiffler, M. (2004). Perspectives and challenges for desalination in the 21st century. *Desalination* 165: 1–9.

34 Judd, S.J. (2017). Membrane technology costs and me. *Water Research* 122: 1–9.

2

Introduction to Membrane Contactor Technology

Mohammad Younas[1] *and Mashallah Rezakazemi*[2]

[1] *University of Engineering and Technology, Department of Chemical Engineering, Peshawar 25012, Pakistan*
[2] *Shahrood University of Technology, Faculty of Chemical and Materials Engineering, Shahrood, Iran*

2.1 Membrane Contactor Separation Processes

2.1.1 Membrane Contactors

A membrane is defined as a semipermeable material that acts as a barrier between two phases. If one component in the mixture passes from one stage to the other phase through the membrane, a separation is accomplished. The classical membrane separation techniques discussed in Chapter 1 are usually termed as pressure-driven operations, which work on the feed–retentate–permeate system. Membrane contactors differ from classical membrane separation processes because in these processes, membrane only provides the surface of contact for the two phases and does not give selectivity to the separation processes. The applied pressure is no more the driving force. Instead, the pressure difference is used to stabilize the membrane and prevent both phases from mixing or disperse with each other. Membrane contactors are defined as the devices in which the semipermeable membrane provides only the surface of contact for the two phases and allows both phases to contact and permits the transfer of selected specie at the mouths of membrane pores to pass through from one phase to the other in cross-flow direction. These two phases are generally termed as feed phase and receiving phase. A simple schematics of membrane contactors is shown in Figure 2.1.

Thus, the feed and receiving phases are in contact only through the membrane pores. As in comparison with the other membrane processes such as microfiltration (MF), ultrafiltration (UF), nanofiltration (NF), and reverse osmosis (RO) in which pressure gradient is used as the separation force here, the concentration gradient between the fluids is the driving force for membrane contactors, and pressure is only applied to pass the fluids through the tubes and shell. The main differences between the classical membrane separation processes and membrane contactors are listed in Table 2.1.

For many of the previous decades, the traditional fluid/fluid contactors like packed columns or absorption towers have been extensively used in industries, but they

Membrane Contactor Technology: Water Treatment, Food Processing, Gas Separation, and Carbon Capture, First Edition. Edited by Mohammad Younas and Mashallah Rezakazemi.

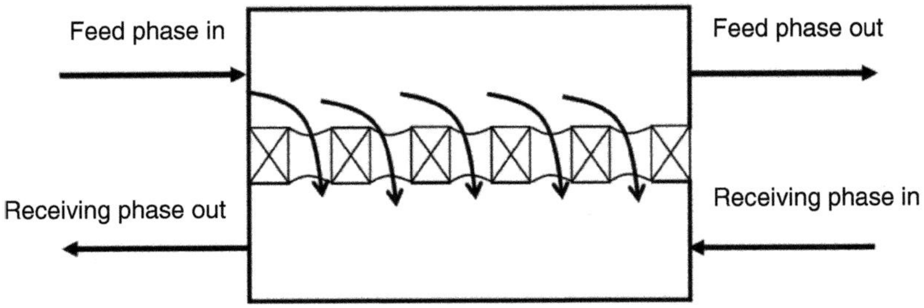

Figure 2.1 General schematics of membrane contactor separation process.

Table 2.1 Difference between classical membrane separation processes and membrane contactors.

Classical membrane separation processes such as MF, UF, NF, and RO	Membrane contactors
• Pressure is applied to generate the driving force for separation	• Pressure is applied on either side to stabilize the surface of membrane
• Concentration, temperature, and activity differences are the factors affecting the separation or efficiency	• Concentration, temperature, and activity difference are the driving force
• Works as dead-end or cross-flow	• Works as cross-flow
• Usually works in three ports like feed, retentate, and permeate	• Often works as four port like phase 1 in (feed-in), phase 1 out (feed-out), phase 2 in, phase 2 out
• Not in contact with permeate	• In contact with permeate at membrane interface by nondispersive mixing
• Devices are in the shape of a spiral wound	• Devices are in the shape of tubular, capillary, or hollow fiber

face some inherited operational issues such as flooding, entrainment, foaming, and emulsion formation that limit its operational flexibility. Although the proper selection of packing material and uniform distribution of fluids in the packed column increase its interfacial area, membrane contactors provide a substantial contact area greater than the traditional fluid/fluid contactors. The fluids flowing on the opposite side of the membrane contact at the mouth of each pore, and an interface between the phases is thus created.

The membrane acts as a barrier between two phases, and absorption/desorption of certain component(s) in the feed phase with the receiving phase provides selectivity. There are many advantages of the membrane contactors over the conventional column system. In membrane contactor processes, the main target is to increase the interfacial area and thus the mass transfer of certain component(s) from feed across the membrane to the other side of the membrane in the receiving phase. Similarly, membrane contactors also possess some disadvantages. A description of the merits and demerits of membrane contactors has been provided in Table 2.2.

Table 2.2 Advantages and disadvantages of membrane contactors.

Advantages	Disadvantages
• In contrast to the absorption column, the membrane contactors provide a fixed contact area between the two phases • Unlike traditional column extractors, the interfacial area is independent of the flow rates of either of the fluids • Membrane contactors can be operated with a variety of fluid densities unlike the absorption column, which can only be brought into use if the contacting fluids have a considerable density difference • In membrane contactors, fluids do not disperse into each other, so there is no chance of emulsion formation • Since membrane contactors are available in the modules, scaling up of the process in membrane contactors is easy and linear. On the other hand, in conventional processes, the scaling up is not as easy and linear • The modular arrangement also gives us the freedom of using membrane contactor for the required capacity. If the capacity required is high, more numbers of modules are used and vice versa • Because the contact area is known for each module in membrane contactor and scaling up can be done according to the required contact area • Membrane contactors provide 30 times more contact area as compared with the conventional absorption column, thus decreasing the height of the transfer unit (HTU) • In case membrane contactors are used as reactor and separator, and wherein equilibrium limited chemical reactions in which reactivity stops when the reaction reaches equilibrium conditions, membrane contactors can be used to withdraw the product by diffusing it through the membrane, and hence the rate of product formation can be increased • In comparison with the mechanical separation equipment, membrane contactors have no moving part • Unlike traditional processes, the fluids do not disperse into each other in membrane contactors, so there is no solvent loss • The opportunity for back mixing of the dispersed phase is reduced • Capable of coupling the unit operations in one step like absorption and stripping, extraction and regeneration, and reaction and separation • Reaction and separation can occur in one step • Consecutive phase separation step (e.g. decantation and centrifugation) is no longer needed • Lower operating temperatures for distillation or evaporation processes • Azeotropic mixtures can be easily separated than in conventional units	• In membrane contactors, the membrane itself offers resistance to mass transfer, which is not available in the conventional absorption process • Membranes suffer from the problem of fouling during its operation, although the problem is more dominant in the pressure-driven processes as compared with the concentration driven such as membrane contactors, but still it exists in membrane contactors • Membranes have a limited lifetime and often after a certain period of use it need to be replaced • Wettability is another issue that is encountered in membrane contactor. The solvent entraps in the pores and thus causes resistance to mass transfer. Wettability causes the mass transfer to decrease six times as compared with the non-wetted pores • Normally membrane contactor modules are formed by combining of bundle of fibers in a single module using epoxy as an adhesive material. However, these potting materials are often susceptible to chemical and thermal attack • For mass transfer to occur, membrane contactors are dependent upon the equilibrium stages, but sometimes, equilibrium stages are limited by pressure drop constraints along the length of the membrane contactor module, thus caused additional pressure or loss of efficiency • Transmembrane pressure is low in membrane contactors, which can limit its capacity in a single membrane contactor module. Applied pressure cannot be more than the critical breakthrough pressure, which is different for different membrane material, its pore size, and the wetting liquid • Unlike conventional extractors or absorbers, membrane contactor can only be fabricated by selected materials that are generally polymeric materials • Unlike conventional extractors or absorbers, membrane contactors cannot handle the solid particle in separation processes • Channeling and bypassing of fluids • Pretreatment before membrane contactor separation • Stability of solvent in liquid membrane contactors

Source: Based on Gabelman et al. [1], Li and Chen [2], Mansourizadeh and Ismail [3], Noble and Stern [4], and Wang et al. [5].

Apart from all the limitations and disadvantages of the membrane contactor described in Table 2.2, the advantages outweigh its disadvantages, and hence it is under extensive research by the researchers. Membrane contactors can be used as an alternative method for many processes such as pharmaceutical, food processing, biomedical, petrochemical, galvanic, and chemical industries. Moreover, they can be used both for product purification and for product recovery. Some applications use membrane contactors as devices. Still, they become so specific and popular that those processes are not generally called as membrane contactors, but it is named according to its application such blood oxygenator, hemodialyzer, membrane degasifier, membrane deaerator, and membrane distillation [6–8]. Further to that, a new horizon for the membrane contactor technology has opened with the proposals of utilizing it to recycle wastewater in space missions [9, 10] and using it as an artificial lung for contagious lung disease patients [9]. Nondispersive gas absorption or stripping was used for blood oxygenation for the first time by Tsuji et al. The oxygen was flowing on one side of the membrane while the blood was flowing on the other side. The oxygen diffused through the membrane into the blood, while in another module, the CO_2 was diffused out of the blood using a solvent [10]. Using nondispersive solvent extraction mode, Alonso et al. worked on the removal of Cr(VI) from the galvanic process wastewater [11].

Dispersion-free mode of separation processes using membrane contactor with solid porous membranes operates only the extraction step in a single module while further stripping or regeneration of the receiving phase can take place in another module. On the other hand, liquid membranes (LMs) can operate the extraction (absorption) and regeneration (stripping) in one step using one single membrane contactor module [11]. The use of membrane contactors in various applications is described in Table 2.3.

Membrane contactors are used in various separation processes as per the required application [13–17]. Though membrane contactors only provide the surface of contact for both the feed and receiving phases, these processes may be classified based on the nature of membrane material, hydrophobicity, separation mechanism, and application [18–21]. In membrane contactors, the membranes act as a medium of contact, which separates the phases, which is also viewed as semipermeable phase separators [22–24]. The medium could either be solid, liquid, or gas, depending upon the concept; it provides the additional mass transfer and remains as a stagnant medium. Membrane contactor classification has been described in Table 2.4 with examples in various applications while various recognized technologies using membrane contactors are listed in Table 2.5. The differences between membrane contactors using solid or liquid membrane as a medium of contact are listed in Table 2.6.

2.1.2 History and Background of Membrane Contactors

The inception of membrane contactors was dated back to 1861–1866 when Graham developed dialysis and gas permeation theory. Kolff first introduced hemodialysis in 1944, and first commercial hemodialyzer was made available in 1950 as an

Table 2.3 Diverse applications of membrane contactors.

Industry	Applications
Metal extraction	• Extraction of (from aqueous streams) metals as cations and anions like Cu, Zn, Ni, Co, Al, Cd, As, Cr, Hg
Organic pollutant extraction	• Extraction of (from aqueous streams) phenol, toluene, 2-chlorophenol, benzene, chloroform, 1,1,2-trichloroethane, trichloroethylene, carbon tetrachloride, tetrachloroethylene, nitrobenzene, trichloromethane, tetrachloromethane, and acrylonitrile
Aromatics extraction	• Aroma extraction from aqueous streams like dimethyltrisulfide, ethyl butyrate, benzyl aldehyde
Pharmaceutical extraction	• Extraction of 4-methylthiazole, 4-cyanothiazole, and mevinolinic acid • Production of diltiazem by selectively removing the undesirable stereoisomer from a racemic mixture of enantiomers • Hydrolysis of triglycerides • Purification of mevinolinic acid (MK-819)
Food industry	• Fruit and vegetable juice concentration, production of low alcohol wine
Fermentation product extraction	• Cytochrome-c, myoglobin, α-chymotrypsin, catalase, urease, and α-amylase • Ethanol production through extractive fermentation • Production of acetic, lactic, succinic, citric acids
Bioproduct extraction	• Production of cytochrome-c, myoglobin, α-chymotrypsin, catalase, urease, α-amylase
Gas separation	• CO_2, SO_2, NO_x, Br_2, H_2S, I_2, N_2, NH_3, $(NH_4)_2S$, acetic acid, HCl, lactic acid, ethane, ethylene

Source: Based on Gabelman et al. [1] and Prasad and Sirkar [12].

artificial kidney. For decades, a lot of research has been carried out on membrane contactors to improve its operation and to bring it in use for another certain number of purposes. Prasad and Sirkar [12], Kiani et al. [7], Frank and Sirkar [50], D'elia et al. [51], Sirkar [52], Prasad et al. [53], Kang et al. [54], and Gabelman and Hwang [55] elaborated an inspiring and thorough theoretical and experimental work on membrane contactors that has become a basic starting point in membrane contactor technology.

2.1.3 Types of Membrane Contactor Systems

2.1.3.1 Solid Porous Membrane as Medium of Contact in Membrane Contactors

Membrane contactors made of the solid porous membranes are usually fallen in the dispersion-free separation technique category. Membranes may be symmetric, asymmetric, composite, hydrophilic, or hydrophobic. Their pore sizes are usually in the range of 0.05–1.00 μm. Their pore size distribution may be uniform because pore size distribution plays an important role in achieving uniform transport along the membrane and voiding coalescence phenomena. Membranes with big pores, high

Table 2.4 Types of membrane contactor systems.

Type of membrane contactor separation	Application, example	References
Porous solid membranes		
Symmetric/asymmetric porous membranes Hydrophobic	• Extraction of aroma compounds and copper ions from aqueous streams using PP hydrophobic membrane in hollow fiber membrane • Recovery of heavy metals from aqueous streams using PP, PVDF hydrophobic membrane in membrane contactors	[17, 19, 25–27]
Hydrophilic	• Recovery of chromium(VI) from aqueous streams using hydrophilic polyether ether ketone (PEEK)-WC membrane contactors • Hydrogen mixture separation with PEEK-WC symmetric membrane contactors • Recovery of thiol sulfur using hydrophilic cellulose membrane contactor	[28–30]
Composite/mixed matrix membranes Hydrophobic	• PP/SiO_2 composite membrane contactor for CO_2 capture with amine solutions • PP/SiO_2 composite membrane contactor for water desalination • Hierarchical fibrous composite poly(vinylidene fluoride-*co*-hexafluoropropylene-polyethylene terephthalate (HFC PH-PET) membrane for water desalination in membrane distillation	[31–33]
Hydrophilic	• PSF/sulfonated polyether ether ketone (SPEEK) composite hydrophilic membrane for water desalination • Polyamide membrane with 3D hyperbranched polyglycerol	[34, 35]
Liquid membranes		
Bulk liquid membranes	Facilitated transport of copper through bulk liquid membranes containing different carriers	[36, 37]
Supported liquid membrane	Post-combustion CO_2 capture in PVDF membrane through immobilized ionic liquids	[38]
Emulsion liquid membrane	Recovery of metal ions, phenols, organic acids	[39, 40]

Table 2.5 Recognized membrane contactor processes.

Membrane contactors
Membrane extractors
Membrane scrubbers/strippers
Membrane distillation
Osmotic distillation
Membrane emulsifiers
Membrane reactors
Membrane crystallizers
Pressure-retarded osmosis

Table 2.6 Breakthrough pressure and contact angles of solid porous membranes.

Membrane material	Wetting fluid	Breakthrough pressure ΔP_c (kPa)	Contact angle	References
Polypropylene	*N*-Methyl-2-pyrrolidone (NMP)	137	—	[12, 41, 42]
Polypropylene	Water	900	109.6°	[12, 41, 42]
PVDF	Water	350	93.5–120°	[12, 41, 42]
Nylon (ENKA)	Water	172	—	[12, 41, 42]
Cellulose acetate (Millipore)	Water	110	—	[12, 41, 42]
Cellulose acetate	Water	215	—	
PP2500 (Celgard)	Water	—	101.9°	[12, 41, 42]
PTFE1301 (Donaldson)	Water	—	113.2°	
PTFE1325 (Donaldson)	Water	—	113.2°	
Biodyne A, polyamide (Pall)	Water	—	113.2°	
Supor200, polyethersulfone (PES) (Pall)	Water	—	61.6°	
TiO_2 on stainless steel support (SpinTek)	Water	—	81°	[41]
Hydrophobic TiO_2 on stainless steel support (SpinTek + Surfactis)	Water	—	116°	[41]
Polypropylene, PP303030 (Goodfellow, France)	20% v/v ethanol in water	—	83.7°	[41]
	40% v/v ethanol in water		62.2°	
	66% v/v ethanol in water		46.0°	
	80% v/v ethanol in water		38.9°	
	Lemon essential oil		100°	
Polytetrafluoroethylene, PTFE, FP303150 (Goodfellow, France)	20% v/v ethanol in water	—	94.7°	[41]
	40% v/v ethanol in water	—	72.8°	
	66% v/v ethanol in water	—	59.4°	
	80% v/v ethanol in water	—	54.0°	
	Lemon essential oil	98	45.7°	
Polyamide, PA, AM323010 (Goodfellow, France)	20% v/v ethanol in water	—	46.3°	[41]
	40% v/v ethanol in water		30.8°	
	66% v/v ethanol in water		19.5°	
	80% v/v ethanol in water		17.4°	
	Lemon essential oil		n.d.	

(Continued)

Table 2.6 (Continued)

Membrane material	Wetting fluid	Breakthrough pressure ΔP_c (kPa)	Contact angle	References
Polytetrafluoroethylene, PTFE, FP303150 (Goodfellow, France) with TiO_2 (pore size: 0.1 μm)	66% v/v ethanol in water	55		[41]
PTFE (Gore-Tex TA004), 2.0 μm	Water	—	145°	[43]
	Methyl isobutyl ketone (MIBK)	82.7	106°	
	Ethylene glycol	41.4	91°	
	17% v/v ethanol/water	42.7	94°	
	10% v/v acetic acid/water	51.0	100°	
	n-Octane	—		
	n-Butyl alcohol	—	22°	
	o-Xylene	—	26°	
	Carbon tetrachloride	—	37°	
	Methylisobutylketone	—	31°	
		—	18°	
PVDF, 0.22 μm (Millipore)	Water	217	110°	[44]
	30% v/v methanol/water		85°	
	20% v/v ethanol/water		85°	
	10% v/v isopropanol/water		94°	
PVDF, 0.45 μm (Millipore)	Water	110	112°	
	20% v/v ethanol/water		90°	
PTFE, 0.45 μm (Gore, France)	Water	232	123°	
	30% v/v methanol/water		87°	
	72% v/v ethanol/water		87°	
PP, 0.2 μm (Gore, France)	Water	415	123°	
Teflon	Hexadecane	—	72°	[45]
	Octane	—	56°	
	Hexane	—	41°	
	Pentane	—	0°	
	Butane	—	0°	
PP (0.18 μm)	Water	—	124° 145–168°	[31]
PP (0.1 μm)	Water	—	120°	[32]
PP (0.1 μm) with 1 wt% SiO_2	1 : 50 : 3.2 mole ratio of tetraethoxysilane (TEOS):ethanol:$NH_3 \cdot H_2O$		155°	

Table 2.6 (Continued)

Membrane material	Wetting fluid	Breakthrough pressure ΔP_c (kPa)	Contact angle	References
Polysulfone (PSF)	Water	—	79°	[34]
PSF blended with SPEEK (sulfonated polysulfone [SPSF])	Water		72°	
Polyamide over poly(vinylidene fluoride-*co*-hexafluoropropylene) (PVDF-HFP), 0.8 μm	Water	160	143°	[46]
Poly(vinylidene fluoride-*co*-hexafluoropropylene)–poly(ethylene terephthalate) (PVDF-HFP-PET)	Water	13–28	135–145°	[33]
Polyamide–polyglycerol (hyperbranched polyglycerol-grafted [hPG])	Water	—	16.4°	[35]
Polyethylene (PE), 0.3 μm (Solupor, Lydall Inc., USA)	Water	120	390 kPa	[47]
PES, 0.51 μm (Supor, Pall Corporation, USA)	Water	132	400 kPa	
PTFE, 0.17 μm (Tetratex, Donaldson, Belgium)	Water	138	1080 kPa	
Nylon	Water	49	—	[48]
	Glycerol	75	—	
Polyester	Water	75	—	
	Glycerol	81	—	
Polyethersulfone	Water	54	—	
	Glycerol	69	—	
Polyurethane	Water	94.4	—	
	Glycerol	100.9	—	
Polyetheretherketone	Water	69.9	—	
	Glycerol	69.5	—	
Collagen	Water	82	—	
	Glycerol	67	—	
PTFE TF200, 0.2 μm, Pall Gelman, NY, USA	Water	282	—	[49]
PTFE TF200, 0.45 μm, Pall Gelman, NY, USA	Water	138	—	
PP TF200, 0.2 μm, Pall Gelman, NY, USA	Water	276	—	

(Continued)

Table 2.6 (Continued)

Membrane material	Wetting fluid	Breakthrough pressure ΔP_c (kPa)	Contact angle	References
PTFE TF100, 1.0 μm, Pall Gelman, NY, USA	Water	48	—	
PTFE PT20, 0.2 μm, Gore Inc., USA	Water	368	—	
PTFE PT45, 0.45 μm, Gore Inc., USA	Water	288	—	
PP, 0.2 μm, Microdyn-Nadir GmbH (Wiesbaden, Germany)	Water	140	—	
PVDF, 0.22 μm, Millipore Inc., USA	Water	229	—	
PVDF, 0.45 μm, Millipore Inc., USA	Water	110	—	
PTFE, 0.2 μm, Millipore Inc., USA	Water	280	—	
PTFE, 0.5 μm, Millipore Inc., USA	Water	124	—	

porosities, and less thickness lead to better efficiencies. Separation in such membranes occurs by the equilibrium between the two phases, the source and receiving phases [56–59]. In the dispersion-free mode of membrane separation, the feed phase flows on one side of the membrane, while the receiving phase flows on the other side of the membrane without dispersing into one another. Both phases meet on the mouth of the pore, and mass transfer occurs at the interface either physically or chemically [11]. In such cases, one phase wets the membrane, and pores of the membrane fill in depending on whether the membrane is either hydrophobic or hydrophilic. In either case, a little bit more pressure is required on the non-wetting side of the membrane to let the wetting phase pass through the pores and disperse into the other side of the membrane. The wetting phase will not penetrate into the membrane's pores as long as its pressure is less than a certain amount of pressure called critical pressure. This critical pressure is known as breakthrough pressure, which is the maximum allowable pressure across the membrane pores so that the differential pressure should be less than breakthrough pressure to avoid dispersion of wetting fluid to the other side of the membrane. The breakthrough pressure is calculated by the Young–Laplace equation [47, 60, 61]

$$\Delta P = -\frac{4\sigma \cos \theta}{D} \tag{2.1}$$

where σ is termed as interfacial tension if the contacting fluids are liquid–liquid and as the liquid's surface tension if the contacting fluids are gas–liquid, D is the maximum diameter of the pore, and θ is the contact angle. Contact angle θ is defined as the angle formed between the contacting fluid and the pore of the membrane. It is measured from the pore wall to the tangent of fluid–fluid interface. It is indicated from Eq. (2.1) that large pore size and less membrane thickness lead toward lower values of breakthrough pressure. Breakthrough pressure is the minimum required pressure of the liquid to enter the membrane pores. It could be as high as possible to prevent pore wetting. High breakthrough pressure could be achieved with a high contact angle, low energy level, and small pore size with the membrane surface. However, it could be noted that small pore sizes lead to low permeability across the membrane. Therefore, the pore sizes in the range of 0.1–0.6 μm are considered in membrane contactors. Breakthrough pressure and contact angles of various membranes are listed in Table 2.6. It can be noted from Table 2.6 that the contact angle increases with increasing the polarity of contacting the fluid with the membrane. If the contacting fluid is water, the contact angle is higher than that of the organic solvent (nonpolar). Dupuy et al. measured the contact angles and breakthrough pressure for different polymeric membranes (polytetrafluoroethylene [PTFE], polypropylene [PP], and polyamide [PA]) of different morphologies and fluids in contact. Their theoretical and experimental findings found that the contact fluids with low surface tension such as lemon essential oil and aqueous ethanol solutions strongly affected the decrease of the contact angles, thus increasing the tendency to wet. Similarly, PTFE membrane with an average pore size of 0.05 μm showed the highest breakthrough pressure with 40% v/v ethanol/water solvent. However, if the PTFE membrane surface is modified with TiO_2 (average pore size of 0.1 μm) and with 66% v/v ethanol/water, the breakthrough pressure was 55 kPa. It should be noted that for a desirable mass transfer in membrane contactors, a minimum of 50 kPa breakthrough pressure is under an acceptable target [41]. Similar behavior has also been witnessed by Kim and Harriott [43]. García-Payo in their study found that the solution's concentration and operating temperature also affect the breakthrough pressure apart from wetting surface tension, pore radius, and contact angle [44]. The schematics of the solid medium as porous membranes in membrane contactors are shown in Figure 2.2.

Symmetric Porous Membranes In symmetric porous membranes, the porous layer is symmetric and fabricated of a single polymer. These membranes could be either hydrophobic or hydrophilic. A membrane surface can be made hydrophobic if either the surface free energy is reduced or the surface roughness is increased or a combination of both. If the membrane material is hydrophobic in nature, the nonpolar phases wet the membrane's pores in the case of a liquid–liquid contactor. In case the membrane contactor acts as gas–liquid or gas–gas contactor, then pores are filled with the gas phase. Hydrophobic membranes repel the polar (aqueous) phase. Likewise, if the membrane is hydrophilic, the membrane's pores are filled by the polar (phase aqueous phase), and the nonpolar (organic phase) does not penetrate into the pores of the membranes. Generally, almost all the microporous membranes are essentially symmetric. In this case, pore sizes are uniform throughout, and thus, the breakthrough pressure is uniform. However, in case the membranes are asymmetric,

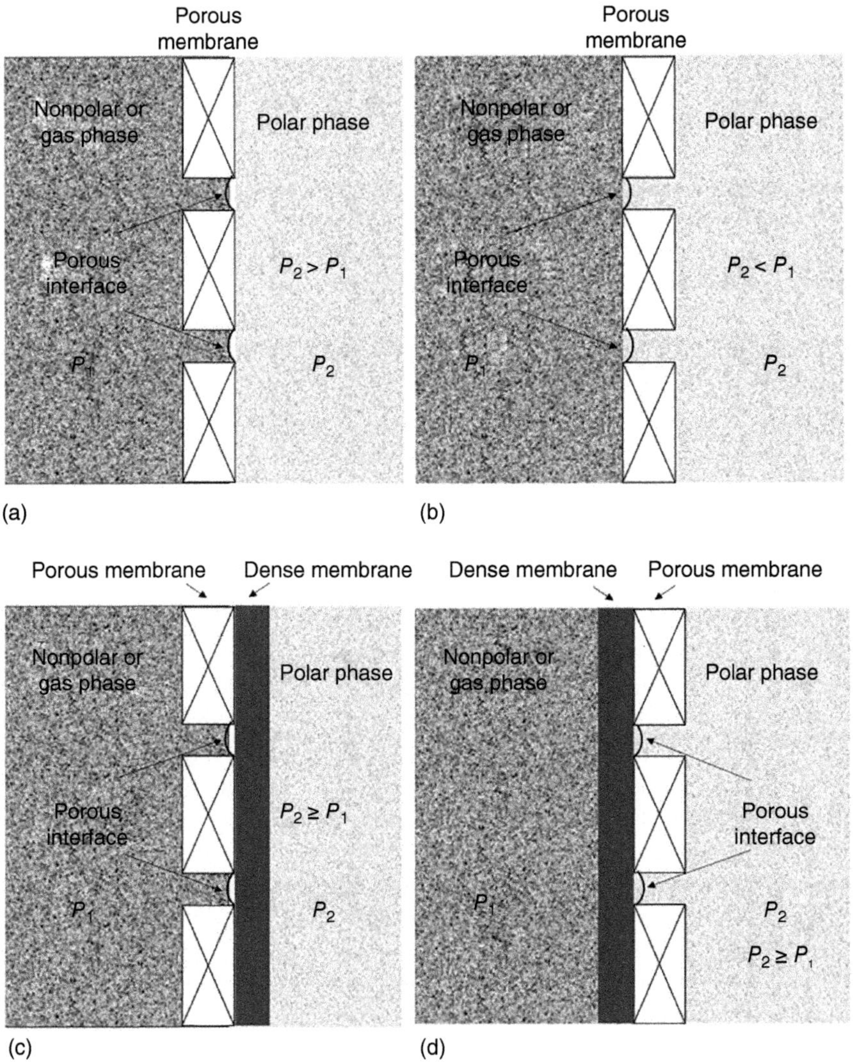

Figure 2.2 Schematics of solid porous membranes as mediums of membrane contactors. (a) Symmetric hydrophobic membrane. (b) Symmetric hydrophilic membrane. (c) Composite hydrophobic membrane. (d) Composite hydrophilic membrane.

pore sizes are not uniform. As the breakthrough pressure is inversely proportional to pore diameter, asymmetric membranes with different sizes exhibit different breakthrough pressure values. Thus, in asymmetric membranes, a single pore with a higher pore size diameter may exhibit breakthrough pressure lower than the applied pressure, which can further cause wetting of the membrane pores (Table 2.7).

Composite Membranes Polymeric membranes are very popular and dominant in the membrane contactor market due to their easy fabricating ability, low manufacturing cost, and wider applications. However, because of the inherent problems of wetting

Table 2.7 Differences between solid and liquid membrane contactors.

Solid membrane contactor	Liquid membrane contactor
• Medium of contact between source and receiving phases is solid surface	• Medium of contact for source and receiving phases is liquid
• Process is controlled by equilibrium between the phases	• Process is driven by chemical potential gradient
• Nature of membrane material is important	• Nature of membrane material is irrelevant
• Diffusion coefficient in solid surface is low	• Diffusion coefficient is an order of magnitude for liquids than in solid surfaces
• Relative low flux	• Relative high flux
• Absorption and stripping occur in different steps	• Absorption and stripping occur in one step
• Membrane itself provides the surface for contact	• Membrane is used only for support of immiscible liquid, and liquid membrane provides the surface for contact
• Extraction (absorption) and regeneration (stripping) takes place in two membrane contactor modules	• Extraction (absorption) and regeneration (stripping) takes place in a single membrane contactor module
• A good amount of receiving phase (solvent) is needed because extraction and separation is not possible in one step	• Small quantity of receiving phase (organic solvents) is needed due to their continuous regeneration associated with reverse bile reaction
• Operating cost is increased because of lack of regeneration of receiving phase on one step	• Operating cost is reduced because of the presence of regeneration of receiving phase on one step
• Investment (capital) cost is high because two separate equipment are required for extraction and separation	• Investment (capital) cost is low because one equipment is required for simultaneous extraction and separation
• There are chances of back mixing of phases and loss of receiving phase	• There are less chances of back mixing of phases and loss of receiving phase
• Separation for low solute solution is difficult or not cost and time effective	• Separation for low solute solution is possible and is even efficient

Source: Based on Coelhoso et al. [37].

and low mechanical, thermal, and chemical stability of porous polymer membranes, composite membranes have been introduced in membrane contactor devices. The inorganic fillers' addition into the polymeric membranes is considered a promising strategy to counteract the disadvantages of porous symmetric membranes due to their excellent intrinsic properties of thermal and chemical resistance. Composite membranes are expected to improve the membrane properties and facilitate long-term operation exhibiting the dual properties of organic and inorganic materials. These properties are enhanced hydrophobicity, improved chemical and thermal resistance, and excellent adaptation to the severe operating conditions. A thin layer is deposited on the porous support. These layers could have different contact angles and hence different levels of hydrophobicity or hydrophilicity. Both layers in the composite membrane may be hydrophobic or hydrophilic or both. One layer may be thicker, thinner, or dense than the other one. In several membrane processes involving the separations of organic solutes from aqueous streams such as membrane distillation, pore size is not considered important. Rather the main requirement is the prevention of membrane wetting. Highly hydrophobic membranes play a vital role in gas–liquid membrane contactors. For such separation processes, composite membranes are considered important in preventing the membrane wetting. If inorganic layers are added to hydrophobic membranes, the surface tensions are greatly increased, leading to high breakthrough pressure and contact angle.

On the other hand, the addition of organic solutes decreases the surface tension and thus causes the membrane wetting due to a decrease in contact angle. Another advantage of composite membranes is that dense layer is not prone to wetting. Because the liquid cannot pass through the dense (blocked) layers, polyvinylidene difluoride (PVDF) ultrafiltration membrane was modified by dispersing nanosized titanium(IV) oxide (TiO_2) particles in a PVDF solution. It was observed that the membrane fouling was reduced by changing the membrane surface from hydrophobic to hydrophilic after TiO_2 addition [62]. An et al. synthesized the hierarchical fibrous composite membranes and successfully tested for water desalination using membrane distillation [33]. In another study, they applied the nanocomposite membrane of polyamide over poly(vinylidene fluoride-*co*-hexafluoropropylene) (PVDF-HFP) in membrane distillation (MD)-forward osmosis (FO) for apple juice concentration. Both membranes were hydrophobic and exhibited a water contact angle of more than 140° [46]. Similarly, Liu et al. fabricated the superhydrophilic (water contact angle [WCA] < 17) polyamide membrane with 3D hyperbranched polyglycerol and tested for water filtration [35].

2.1.3.2 Liquid Membrane Contactors

When membranes are viewed as semipermeable phase separators, then the traditional concept of membranes as polymer films can be extended to include liquids consisting of a thin film of organic reagent. They are defined as liquid membranes (LMs). A liquid membrane system involves a liquid that is immiscible with the source (feed) and receiving (product) solutions that serves as a semipermeable barrier between these two liquid and gas phases. Liquid membranes provide a medium from a liquid phase as contact and serve as semipermeable barrier between the two immiscible feed and receiving phases. They are being investigated and applied in various fields such as wastewater treatment (recovery of toxic metals, organic compounds from wastewater), biomedical engineering (development of sensing devices, recovery of fermentation products), chemical engineering, and analytical chemistry (gas separation) [63–65]. Separation in liquid membranes occurs due to the chemical potential gradient. Absorption and stripping occur in one step. However, the main disadvantages of liquid membranes are the chemical instability of liquid membrane and porous support's mechanical instability. The processes where liquid membranes (LMs) are employed are called bulk liquid membranes (BLMs), supported liquid membranes (SLMs), and emulsion liquid membranes (ELMs). They have been shown schematically in Figure 2.3.

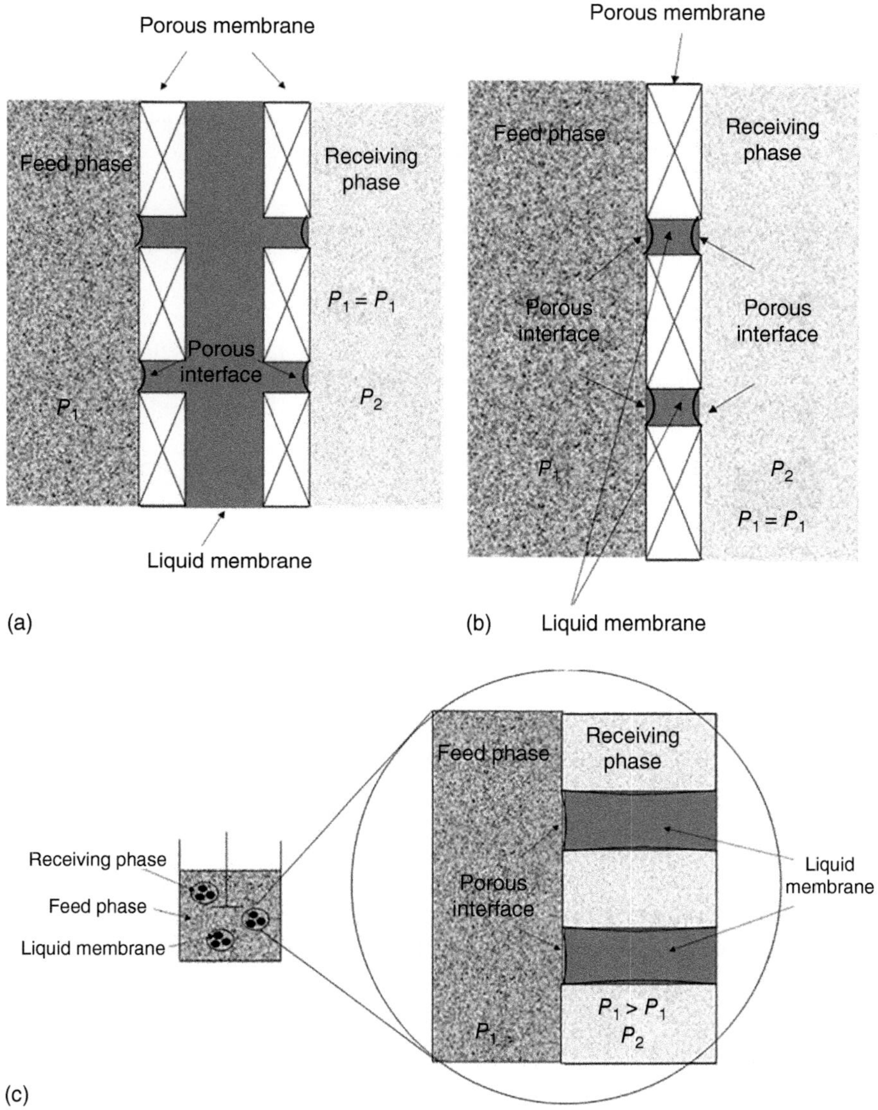

Figure 2.3 Schematics of liquid membrane as medium of contact. (a) Bulk liquid membrane (BLM). (b) Supported liquid membrane (SLM). (c) Emulsion liquid membrane (ELM).

Bulk Liquid Membranes (BLMs) In BLM contactors, the two phases are sandwiched by bulk of the liquid membrane. This liquid membrane is usually a liquid phase immiscible with water. The bulk of liquid membrane is itself surrounded by a porous solid membrane. Thus, the liquid membrane acts as a carrier, and the selected component passes through the pores of the membrane using liquid membrane as medium. BLMs are relatively thick with thickness in the range of 10^{-4}–10^{-2} m. Thus, the thickness of BLMs is more than twofold as compared with that of SLM.

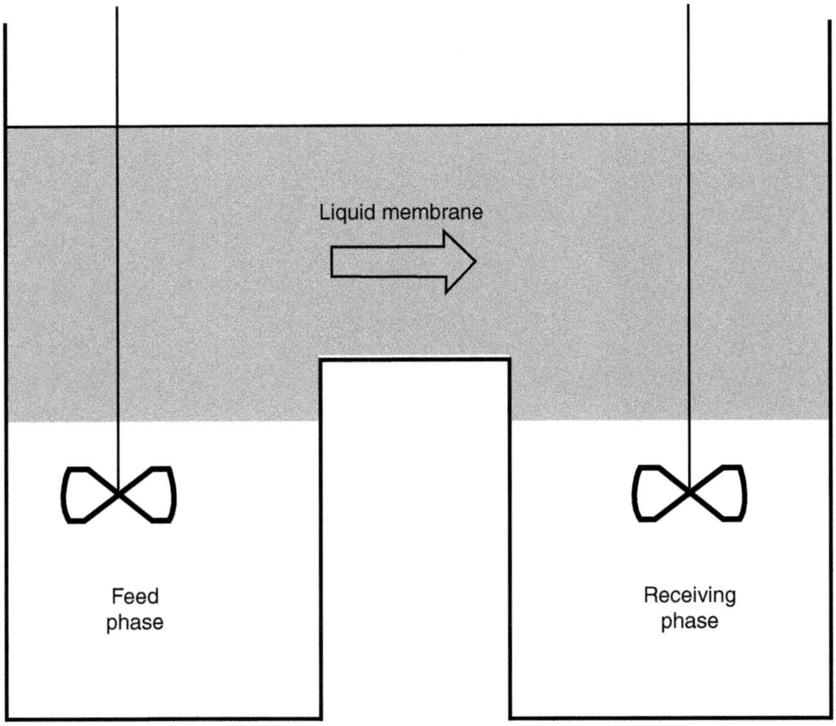

Figure 2.4 Stirred-type Lewis cell representing facilitated transport through bulk liquid membrane (BLM).

BLM is also known as carrier-facilitated membranes because other solvents or thinners can also be mixed to dilute the liquid membrane solution [36, 37]. Another schematic to represent the BLM is shown in Figure 2.4, which uses a stirred transfer Lewis cell. The liquid membrane in bulk separates the feed and receiving phases. The liquid membrane is immiscible in both fluids and facilitates the transfer of solute from the feed to the receiving phase. The BLM can be diluted in various diluents. Leon and Guzmán described the facilitated transport of copper through BLM containing various carriers. The solvent in the receiving phase diffuses from the bulk membrane to the interface of the feed phase and membrane. Copper ions are exchanged with protons of the solvent in the receiving phase at feed–membrane interface, and copper–solvent complex is formed. This complex then diffuses through the bulk membrane to the receiving phase–membrane interface where protons are exchanged for copper ions and are consequently released in the receiving phase. The solvent in the receiving phase is regenerated and the cycle is continuous [36].

Supported or Immobilized Liquid Membrane (SLM) or (ILM) Extraction Process In supported or immobilized liquid membrane (SLM or ILM) contactors, the liquid membrane is impregnated or immobilized in the porous membrane's pores. The porous membrane is filled by the liquid membrane, which separates both phases.

The ILM is usually selected as immiscible with the source and receiving phases. These membranes have also been named as gel liquid membrane (GLM), swollen polymeric membranes (SPMs), and polymeric inclusion membranes (PIMs). SLMs are preferred over BLM and ELM due to its applicability in wastewater treatment. However, concerns have been shown over the stability of the membrane layer sandwiched between the membranes. Another possible disadvantage of SLMs is the possibility of short circuiting of feed and receiving phases, leading to the loss of organic solution.

The SLM mode of extraction is used to extract and separate valuable metal ions from industrial waste solutions. It possesses important features such as high selectivity, extraction, and stripping in one single stage. Its operation is independent of conditions of equilibrium because it has nonequilibrium mass transfer characteristic. The operational limitations like phase disengagement, aqueous/organic phase ratio and emulsification, etc. can be avoided in the SLM mode of extraction [11]. SLM has applications in industrial and analytical fields for the separation, treatment of wastewater, and preconcentration [66–70]. Thus SLM technology is considered as a better alternative to the traditional technologies for the concentration of metal ions in the industrial hydrometallurgical processes [11, 69–73].

SLM is a nondispersive form of liquid membrane. The membrane phase of SLM, which is an organic reagent is immobilized in the porous polymeric membrane pores. The polymeric support is always hydrophobic in nature, and it only provides structural support to the membrane phase and does not play any role in the separation [74]. The hydrophobic nature of the membrane helps in retaining the membrane phase (organic reagent) by capillary action, or else the reagent will leak out of membrane pores [75]. The chosen polymeric material must be chemically and thermally stable on exposure to the receiving and feed phases and the organic reagent. The polymers such as PTFE, PP, and polysulfone (PSF) are used to manufacture the SLMs [11]. The organic solvent, which is present inside the supporting polymer's pores, is selected based on its affinity for the selected component in the feed solution. This organic reagent is immobilized inside the pores. The chemistry involved here for the metal ion extraction is the same as it is involved in the liquid–liquid extraction [68].

Emulsion Liquid Membrane (ELM) Extraction Process ELM is usually used in water-in-oil applications, where water and oil are the source and receiving phases, respectively. The receiving phase is emulsified in liquid membrane, which is then dispersed in the source phase. As compared with the other membrane techniques, at present times, ELM is being proved to be a very promising and attractive separation technique because of its high mass transfer rates, cheapness, and energy-saving characteristic [76]. It has its applications in various areas like separation of hydrocarbons; extraction/removal of low concentrated solutes like amines, metal ions, and organic acids from the industrial effluents; and biomedical application for artificial kidney [77–80]. ELM was used for the first time at a commercial level to remove zinc from industrial wastewater. It was observed that the separation efficiency was 99.5% [81, 82].

ELM is like a bubble within a bubble. The inside bubble contains the internal phase reagent that has the affinity for the target source element/compound, while the outer bubble acts as a membrane and contains carriers or extractants. Outside the outer bubble, everything is called source phase or continuous phase. The solute molecules' movement from the continuous/source phase to the internal phase through the ELM occurs through the concentration gradient only [83–85].

ELMs are of two types and are essentially double emulsions, i.e. oil/water/oil system and water/oil/water system. The system's selection depends upon the nature of the internal and external phases [86]. Surfactants play a significant role in the ELM, and that is why ELM is also called a liquid surfactant membrane. It has a strong influence on the transport rate of solute and emulsion stability. The emulsion's stability improves by increasing the concentration of surfactants, but the mass transfer decreases due to excess surfactant molecules at the aqueous/organic interface [87–89]. Surfactants (emulsifying agents) adsorb on the aqueous/organic interface, which causes a decrease in the interfacial surface tension, and thus, physical resistance is created, which does not allow the coalesce of the emulsion droplets [90]. ELM usually works on the water-in-oil emulsion concept. The emulsion consists of an organic extractant (liquid membrane). The surfactant is suspended in an aqueous feed containing the solute (metal) to be extracted. The typical size of a droplet and emulsion globule in ELM is 1–10 μm and 0.5–3 mm, respectively. The solute reacts with the organic extractant at the aqueous/organic phase interface, and the complexant (ligand) transfers across the organic extractant (behaving like a membrane) to the strip interface of aqueous feed by breaking up the emulsion. The solute then strips out from the extractant complexant, and the organic extractant is regenerated to restart the extraction process [91].

2.1.4 Membrane Contactor Integrated Systems

Membrane contactors are also getting popularity in integrated processes in different industrial applications due to their product quality and plant compactness. They are best suited in combining membrane operations and conversion due to their synergic effects. Membrane contactors are suitably used in integrated processes in which the feed mixture is separated into its components in a series of independent unit operations and separation processes. Membrane contactor integrated plants are advantageous in terms of size, energy consumption, use of chemicals, flexibility, and modularity. Typical examples of integrated separation systems integrating the membrane contactors with chemical or biochemical processes and separation processes in fermentation and solvent extraction processes that describes a hybrid fermentation–separation process for butyric acid production with immobilized biomass taken from [92] are shown in Figure 2.5. Another example of an integrated system is the water treatment and power production from river water or seawater using pressure-retarded osmosis, which uses a membrane contactor for the regeneration of brine solution and concentration of salt solution or production of clean water in reverse osmosis system.

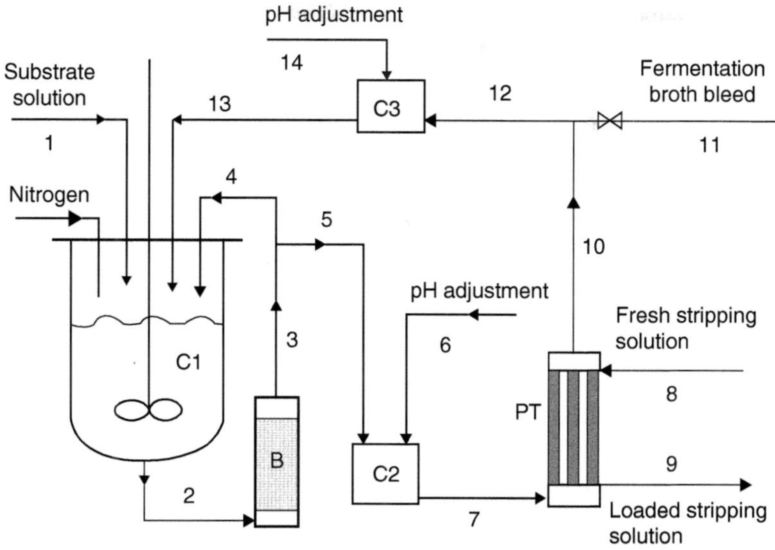

Figure 2.5 Schematic flow sheet of a hybrid fermentation–separation process for butyric acid production with immobilized biomass taken from [92]. C1, fermenter container; B, bioreactor with supported biomass; C2, container of broth for pH adjustment; PT, hollow fiber pertractor with SLM; C3, container of lean broth with pH adjustment; 1, concentrated substrate solution; 2, input into the bioreactor; 3, bioreactor output; 4, recycled broth; 5, broth entering the pertraction loop; 6, acid for pH adjustment; 7, feed into the pertractor; 8, fresh stripping solution; 9, loaded stripping solution; 10, 12, lean fermentation solution; 11, fermentation broth bleed; 13, recycled lean fermentation solution with adjusted pH; 14, ammonia solution for pH adjustment. Source: Blahušiak et al. [92].

2.1.5 Potential of Membrane Contactor in Concentration, Temperature Polarization, Wetting, and Fouling of Membranes

Pore size is the main parameter if a certain membrane is selected or synthesized for a specific separation process using membrane contactors, which could be limited to the wetting of pores with the fluid in contact with membrane pores. Unlike the conventional membrane separation processes that are based on dense membranes, membrane contactors are more prone to wetting due to their wide pores because, for wetting to occur, only active pores should be present, which should be connected through both end of the membrane. Therefore, the interaction of membrane with the fluid in contact is an important parameter in addition to pore size. Another important parameter is the nature of fluid. If organic solutes are present in water, it greatly decreases the contact angle (hydrophobicity), leading to wettings of membrane pores. However, if there are inorganic solutes in the water, then surface tension increases, causing enhanced hydrophobicity (contact angle) of membranes, thus decreasing the chances of membrane wetting. In membrane contactor, the fluid–fluid systems cannot be controlled; alternatively, membrane surfaces could be modified. The hydrophobic membrane with the impregnation of inorganic nanoparticles can increase the hydrophobicity of the membranes. Contact angle is the best measure for the membrane wetting. If the water contact angle is $<90°$

then membrane is said to be hydrophilic, and the fluid in contact is considered as wetting phase. The water contact angle >90° points toward hydrophobicity of the membrane, and the liquid is said to be non-wetting phase. Thus, the higher the value of water contact angle, the more the membrane will be hydrophobic and less prone to wetting by water. Thus, low water contact angle values indicate that the membranes are more prone to wetting membrane by water.

Concentration, temperature, and pressure polarization have been considered in membrane contactors due to their severe effects on transmembrane flux and membrane stability. The schematics of concentration, temperature, and pressure polarization variation on either side of the membrane contactor is shown in Figure 2.6. As shown, temperature, concentration, and vapor pressure gradients are created between the bulk solution and interface on non-wetting side of the membrane, causing the solute from bulk to membrane interface if the non-wetting fluid is fed and the membrane is hydrophobic. However, if the membrane is a block or fouled, or there is no more permeability across the membrane, it negatively affects the boundary layer. Concentration is accumulated in the boundary layer adjacent to the membrane, causing a decrease in driving force because of the lower

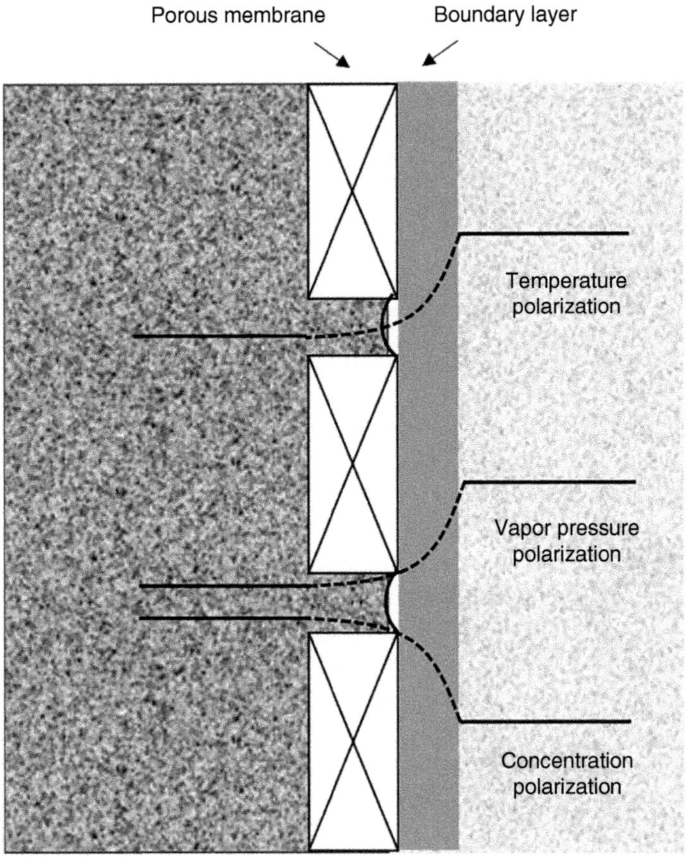

Figure 2.6 Schematic diagram of temperature, pressure, and concentration polarization in hydrophobic membrane contactor process.

flow rate of the boundary layer. However, boundary layer conditions may be used to the positive side by improving the interface diffusion, mixing, and intensifying the mass transfer driving forces due to high flow rate and temperature [32]. Similarly, aggressive solutes present in the feed phase cause aggravated scaling and deposition both on the membrane surfaces and pores called fouling, consequently causing obvious membrane damage. The causes of fouling vary depending on the nature of the solute and solute–membrane interactions. Furthermore, fouling causes an irreversible flux reduction, unlike concentration polarization, which causes a reversible flux reduction. Mostly fouling is caused by the hydrophobic substances present in the feed carrying a surface charge such as proteins in bioseparations or fruit juice concentration through membrane distillation. So, these particles can easily accumulate and be adsorbed at the hydrophobic membranes due to their hydrophobic interactions. It gives insight into looking for hydrophilic membrane materials or mixed with other materials to reduce the fouling.

2.2 Conclusion and Future Trends of Membrane Contactors

Membrane contactors have proved to be applied with L–L, L–G, and G–G interfaces in various separation, reaction, and absorption fields. The use of membrane contactors in the industry might convert the traditional industry into a compact plant in process intensification and sustainable production. The membrane contactor system presents highly efficient species mass transport achieving high separations. Membrane contactor equipment is of small size with minimum use of organic solvents and can be used in diverse applications. The state of knowledge indicates that the independent use of membrane contactors is still looking for its commercial use at an industrial scale and thus has great potential in developing various successful applications. Polymeric membrane wetting by the liquid phase is one of the main challenges in membrane contactors. New research in developing the membrane free of fouling and wetting with increased stability and lifetime by tailoring its materials' properties can bring the membrane contactors at the forefront of the modern sustainable industry. This could also be achieved by surface modification and fabrication of dual or triple selective layers. Fabricating or modifying the membranes that are superhydrophobic to avoid membrane wetting or hydrophilic, which are less prone to fouling, or a blend of hydrophobic–hydrophilic such mixed matrix membranes is the key priority in recent research in further improving the performance of membrane contactors. Other properties in material chemistry that can be fouling resistant are as follows:

- Smooth surface.
- Chemistry of polymers that resist adhesion of protein or organic macromolecules.
- Zwitterions such as poly(sulfobetaine).
- Sugar-derived molecules.
- Polyglycerol.

Moreover, there is a need to develop less toxic and environment-friendly organic solvents (receiving phase), making the membrane contactor separation processes a green process. Another advancement for transferring the membrane contactor science from laboratory scale and pilot scale to commercial reality is developing high-efficiency membrane contactors modules with large surface area economically and reliably.

Deeper understanding of mass transfer phenomena across the membrane interface in multiphase systems in diverse applications using separation science principles is required to increase the long-term operation of membrane contactor separation systems. Pilot-scale experimental studies under different operation conditions may enhance the applications of membrane contactors. A similarly deeper understanding of computational studies of membrane contactor modules can help design efficient modules with enhanced mass transfer efficiency.

References

1 Gabelman, A., Hwang, S.-T., and Krantz, W.B. (2005). Dense gas extraction using a hollow fiber membrane contactor: experimental results versus model predictions. *Journal of Membrane Science* 257 (1–2): 11–36.

2 Li, J.-L. and Chen, B.-H. (2005). Review of CO_2 absorption using chemical solvents in hollow fiber membrane contactors. *Separation and Purification Technology* 41 (2): 109–122.

3 Mansourizadeh, A. and Ismail, A.F. (2009). Hollow fiber gas–liquid membrane contactors for acid gas capture: a review. *Journal of Hazardous Materials* 171 (1–3): 38–53.

4 Noble, R.D. and Stern, S.A. (1995). *Membrane Separations Technology: Principles and Applications*. Elsevier.

5 Wang, R., Zhang, H., Feron, P. et al. (2005). Influence of membrane wetting on CO_2 capture in microporous hollow fiber membrane contactors. *Separation and Purification Technology* 46 (1–2): 33–40.

6 Sirkar, K.K. (2008). Membranes, phase interfaces, and separations: novel techniques and membranes – an overview. *Industrial and Engineering Chemistry Research* 47 (15): 5250–5266.

7 Kiani, A., Bhave, R., and Sirkar, K.K. (1984). Solvent extraction with immobilized interfaces in a microporous hydrophobic membrane. *Journal of Membrane Science* 20 (2): 125–145.

8 Pabby, A.K., Rizvi, S.S., and Requena, A.M.S. (2008). *Handbook of Membrane Separations: Chemical, Pharmaceutical, Food, and Biotechnological Applications*. CRC Press.

9 Kim, G.-B., Kim, S.-J., Kim, M.-H. et al. (2009). Development of a hollow fiber membrane module for using implantable artificial lung. *Journal of Membrane Science* 326 (1): 130–136.

10 Tsuji, T., Suma, K., Tanishita, K. et al. (1981). Development and clinical evaluation of hollow fiber membrane oxygenator. *ASAIO Journal* 27 (1): 280–284.

11 Alonso, A.I., Galán, B., González, M., and Ortiz, I. (1999). Experimental and theoretical analysis of a nondispersive solvent extraction pilot plant for the removal of Cr(VI) from a galvanic process wastewaters. *Industrial & Engineering Chemistry Research* 38 (4): 1666–1675.

12 Prasad, R. and Sirkar, K.K. (1992). Membrane-based solvent extraction. In: *Membrane Handbook* (eds. W. Ho and K. Sirkar), 727–763. Springer.

13 Sohaib, Q., Muhammad, A., Younas, M. et al. (2021). Rigorous non-isothermal modeling approach for mass and energy transport during CO_2 absorption into aqueous solution of amino acid ionic liquids in hollow fiber membrane contactors. *Separation and Purification Technology* 254: 117644.

14 Younas, M., Tahir, T., Wu, C. et al. (2020). Post-combustion CO_2 capture with sweep gas in thin film composite (TFC) hollow fiber membrane (HFM) contactor. *Journal of CO_2 Utilization* 40: 101266.

15 Ahmad, S., Marson, G.V., Zeb, W. et al. (2020). Mass transfer modelling of hollow fiber membrane contactor for apple juice concentration using osmotic membrane distillation. *Separation and Purification Technology* 250: 117209.

16 Rezakazemi, M., Darabi, M., Soroush, E., and Mesbah, M. (2019). CO_2 absorption enhancement by water-based nanofluids of CNT and SiO_2 using hollow-fiber membrane contactor. *Separation and Purification Technology* 210: 920–926.

17 Muhammad, A., Younas, M., and Rezakazemi, M. (2018). CFD simulation of copper(II) extraction with TFA in non-dispersive hollow fiber membrane contactors. *Environmental Science and Pollution Research* 25 (12): 12053–12063.

18 Hajilary, N. and Rezakazemi, M. (2018). CFD modeling of CO_2 capture by water-based nanofluids using hollow fiber membrane contactor. *International Journal of Greenhouse Gas Control* 77: 88–95.

19 Muhammad, A., Younas, M., and Rezakazemi, M. (2017). Quasi-dynamic modeling of dispersion-free extraction of aroma compounds using hollow fiber membrane contactor. *Chemical Engineering Research and Design* 127: 52–61.

20 Razavi, S.M.R., Rezakazemi, M., Albadarin, A.B. et al. (2016). Simulation of CO_2 absorption by solution of ammonium ionic liquid in hollow-fiber contactors. *Chemical Engineering and Processing: Process Intensification* 108: 27–34.

21 Rezakazemi, M., Shirazian, S., and Ashrafizadeh, S.N. (2012). Simulation of ammonia removal from industrial wastewater streams by means of a hollow-fiber membrane contactor. *Desalination* 285: 383–392.

22 Rezakazemi, M., Niazi, Z., Mirfendereski, M. et al. (2011). CFD simulation of natural gas sweetening in a gas–liquid hollow-fiber membrane contactor. *Chemical Engineering Journal* 168 (3): 1217–1226.

23 Rehman, W.U., Muhammad, A., Khan, Q.A. et al. (2019). Pomegranate juice concentration using osmotic distillation with membrane contactor. *Separation and Purification Technology* 224: 481–489.

24 Shirazian, S., Marjani, A., and Rezakazemi, M. (2012). Separation of CO_2 by single and mixed aqueous amine solvents in membrane contactors: fluid flow and mass transfer modeling. *Engineering with Computers* 28 (2): 189–198.

25 Younas, M., Bocquet, S.D., and Sanchez, J. (2008). Extraction of aroma compounds in a HFMC: dynamic modelling and simulation. *Journal of Membrane Science* 323 (2): 386–394.

26 Younas, M., Druon-Bocquet, S., and Sanchez, J. (2011). Experimental and theoretical mass transfer transient analysis of copper extraction using hollow fiber membrane contactors. *Journal of Membrane Science* 382 (1–2): 70–81.

27 Bey, S., Criscuoli, A., Figoli, A. et al. (2010). Removal of As(V) by PVDF hollow fibers membrane contactors using Aliquat-336 as extractant. *Desalination* 264 (3): 193–200.

28 Bey, S., Criscuoli, A., Simone, S. et al. (2011). Hydrophilic PEEK-WC hollow fibre membrane contactors for chromium(VI) removal. *Desalination* 283: 16–24.

29 Yang, X., Cao, Y.-M., Wang, R. et al. (2007). Study on highly hydrophilic cellulose hollow fiber membrane contactors for thiol sulfur removal. *Journal of Membrane Science* 305 (1–2): 247–256.

30 Brunetti, A., Simone, S., Scura, F. et al. (2009). Hydrogen mixture separation with PEEK-WC asymmetric membranes. *Journal of Membrane Science* 69 (2): 195–204.

31 Amirabedi, P., Akbari, A., Yegani, R.J.S. et al. (2019). Fabrication of hydrophobic PP/CH_3SiO_2 composite hollow fiber membrane for membrane contactor application. *Separation and Purification Technology* 228: 115689.

32 Shao, Y., Han, M., Wang, Y. et al. (2019). Superhydrophobic polypropylene membrane with fabricated antifouling interface for vacuum membrane distillation treating high concentration sodium/magnesium saline water. *Journal of Membrane Science* 579: 240–252.

33 An, X., Xu, G., Xie, B. et al. (2019). Structural tailoring of hierarchical fibrous composite membranes to balance mass transfer and heat transfer for state-of-the-art desalination performance in membrane distillation. *Journal of Materials Chemistry A* 7 (5): 2376–2384.

34 Chen, G., Liu, R., Shon, H.K. et al. (2017). Open porous hydrophilic supported thin-film composite forward osmosis membrane via co-casting for treatment of high-salinity wastewater. *Desalination* 405: 76–84.

35 Liu, Z., An, X., Dong, C. et al. (2017). Modification of thin film composite polyamide membranes with 3D hyperbranched polyglycerol for simultaneous improvement in their filtration performance and antifouling properties. *Journal of Materials Chemistry A* 5 (44): 23190–23197.

36 Leon, G. and Guzmán, M.A. (2008). Facilitated transport of copper through bulk liquid membranes containing different carriers: compared kinetic study. *Desalination* 223 (1–3): 330–336.

37 Coelhoso, I., Cardoso, M., Viegas, R. et al. (2000). Transport mechanisms and modelling in liquid membrane contactors. *Separation and Purification Technology* 19 (3): 183–197.

38 Ilyas, A., Muhammad, N., Gilani, M.A. et al. (2017). Supported protic ionic liquid membrane based on 3-(trimethoxysilyl)propan-1-aminium acetate for the highly selective separation of CO_2. *Journal of Membrane Science* 543: 301–309.

39 Mondal, S., Purkait, M.K., and De, S. (2018). Emulsion liquid membrane. In: *Advances in Dye Removal Technologies* (eds. M.K. Purkait and S. De), 313–323. Springer.

40 Anarakdim, K., Gutiérrez, G., Cambiella, Á. et al. (2020). The effect of emulsifiers on the emulsion stability and extraction efficiency of Cr(VI) using emulsion liquid membranes (ELMs) formulated with a green solvent. *Membranes* 10 (4): 76.

41 Dupuy, A., Athes, V., Schenk, J. et al. (2011). Experimental and theoretical considerations on breakthrough pressure in membrane-based solvent extraction: focus on citrus essential oil/hydro-alcoholic solvent systems with low interfacial tension. *Journal of Membrane Science* 378 (1–2): 203–213.

42 Wang, L., Zhang, Z., Zhao, B. et al. (2013). Effect of long-term operation on the performance of polypropylene and polyvinylidene fluoride membrane contactors for CO_2 absorption. *Separation and Purification Technology* 116: 300–306.

43 Kim, B.-S. and Harriott, P. (1987). Critical entry pressure for liquids in hydrophobic membranes. *Journal of Colloid and Interface Science* 115 (1): 1–8.

44 García-Payo, M.d.C. et al. (2000). Wetting study of hydrophobic membranes via liquid entry pressure measurements with aqueous alcohol solutions. *Journal of Colloid and Interface Science* 230 (2): 420–431.

45 Good, R.J. (1992). Contact angle, wetting, and adhesion: a critical review. *Journal of Adhesion Science and Technology* 6 (12): 1269–1302.

46 An, X., Hu, Y., Wang, N. et al. (2019). Continuous juice concentration by integrating forward osmosis with membrane distillation using potassium sorbate preservative as a draw solute. *Journal of Membrane Science* 573: 192–199.

47 Eykens, L., De Sitter, K., Dotremont, C. et al. (2017). Wetting resistance of commercial membrane distillation membranes in waste streams containing surfactants and oil. *Applied Sciences* 7 (2): 118.

48 Drioli, E., Criscuoli, A., and Curcio, E. (2011). *Membrane Contactors: Fundamentals, Applications and Potentialities*. Elsevier.

49 Rácz, G., Kerker, S., Kovács, Z. et al. (2014). Theoretical and experimental approaches of liquid entry pressure determination in membrane distillation processes. *Periodica Polytechnica, Chemical Engineering* 58 (2): 81–91.

50 Frank, G. and Sirkar, K. (1986). Integrated bioreactor-separator: in situ recovery of fermentation products by a novel membrane-based dispersion-free solvent extraction technique. In: *Biotechnology and Bioengineering Symposium*, 303–316. Wiley.

51 D'elia, N.A., Dahuron, L., and Cussler, E.L. (1986). Liquid–liquid extractions with microporous hollow fibers. *Journal of Membrane Science* 29 (3): 309–319.

52 Sirkar, K.K. (1988). Immobilized-interface solute-transfer apparatus. US Patent 4, 789, 468, issued 6 December 1988.

53 Prasad, R., Kiani, A., Bhave, R., and Sirkar, K.J. (1986). Further studies on solvent extraction with immobilized interfaces in a microporous hydrophobic membrane. *Journal of Membrane Science* 26 (1): 79–97.

54 Kang, W., Shukla, R., Frank, G., and Sirkar, K.J. (1988). Evaluation of O_2 and CO_2 transfer coefficients in a locally integrated tubular hollow fiber bioreactor. *Applied Biochemistry and Biotechnology* 18 (1): 35–51.

55 Gabelman, A. and Hwang, S.-T. (1999). Hollow fiber membrane contactors. *Journal of Membrane Science* 159 (1–2): 61–106.

56 Sohaib, Q., Muhammad, A., Younas, M. et al. (2020). Modeling pre-combustion CO_2 capture with tubular membrane contactor using ionic liquids at elevated temperatures. *Separation and Purification Technology* 241: 116677.

57 Zhang, Z., Zhao, S., Rezakazemi, M. et al. (2017). Effect of flow and module configuration on SO_2 absorption by using membrane contactors. *Global NEST Journal* 19 (4): 716–725.

58 Shirazian, S., Pishnamazi, M., Rezakazemi, M. et al. (2012). Implementation of the finite element method for simulation of mass transfer in membrane contactors. *Chemical Engineering and Technology* 35 (6): 1077–1084.

59 Shirazian, S., Rezakazemi, M., Marjani, A., and Rafivahid, M.S. (2011). Development of a mass transfer model for simulation of sulfur dioxide removal in ceramic membrane contactors. *Asia-Pacific Journal of Chemical Engineering* 7 (6): 828–834.

60 Franken, A., Nolten, J., Mulder, M. et al. (1987). Wetting criteria for the applicability of membrane distillation. *Journal of Membrane Science* 33 (3): 315–328.

61 Kumar, P., Hogendoorn, J., Feron, P., and Versteeg, G.J. (2002). New absorption liquids for the removal of CO_2 from dilute gas streams using membrane contactors. *Chemical Engineering Science* 57 (9): 1639–1651.

62 Oh, S.J., Kim, N., and Lee, Y.T. (2009). Preparation and characterization of PVDF/TiO_2 organic–inorganic composite membranes for fouling resistance improvement. *Journal of Membrane Science* 345 (1–2): 13–20.

63 Noble, R.D. and Way, J.D. (1987). *Liquid Membranes: Theory and Applications.* Washington, DC: American Chemical Society.

64 Bartsch, R.A. and Way, J.D. (1996). *Chemical Separations with Liquid Membranes: An Overview.* ACS Publications.

65 Araki, T. and Tsukube, H. (1990). *Liquid Membranes: Chemical Applications.* CRC Press.

66 Kocherginsky, N. (1996). *Facilitated Transport of Alkali Metal Cations Through Supported Liquid Membranes with Fatty Acids.* ACS Publications.

67 Ochromowicz, K. and Apostoluk, W. (2010). Modelling of carrier mediated transport of chromium(III) in the supported liquid membrane system with D2EHPA. *Separation and Purification Technology* 72 (1): 112–117.

68 Parhi, P. (2012). Supported liquid membrane principle and its practices: a short review. *Journal of Chemistry* 2013: 1–12.

69 Ren, Z., Zhang, W., Meng, H. et al. (2010). Extraction separation of Cu(II) and Co(II) from sulfuric solutions by hollow fiber renewal liquid membrane. *Journal of Membrane Science* 365 (1–2): 260–268.

70 Zhang, W., Chunhua, C., and Zisu, H. (2010). Transport study of Cu(II) through hollow fiber supported liquid membrane. *Chinese Journal of Chemical Engineering* 18 (1): 48–54.

71 Kittisupakorn, P., Weerachaipichaskul, W., and Thitiyasook, P. (2007). Modeling and simulation of copper-ion extraction with hollow fiber supported liquid membrane. *Journal of Industrial and Engineering Chemistry* 13 (6): 903–910.

72 Mondal, D.N., Sarangi, K., Pettersson, F. et al. (2011). Cu–Zn separation by supported liquid membrane analyzed through multi-objective genetic algorithms. *Hydrometallurgy* 107 (3–4): 112–123.

73 Swain, B., Sarangi, K., and Prasad Das, R. (2005). Separation of cadmium and zinc by supported liquid membrane using TOPS-99 as mobile carrier. *Separation Science and Technology* 39 (9): 2171–2188.

74 Dutta, B.K. (2009). Principles of mass transfer and separation processes. *The Canadian Journal of Chemical Engineering* 87 (5): 818–819.

75 Boyadzhiev, L. and Lazarova, Z. (1995). Liquid membranes (liquid pertraction). In: *Membrane Science and Technology* (eds. R.D. Noble and S.A. Stern), 283–352. Elsevier.

76 Kumbasar, R.A. (2010). Selective extraction of chromium(VI) from multicomponent acidic solutions by emulsion liquid membranes using tributhylphosphate as carrier. *Journal of Hazardous Materials* 178 (1–3): 875–882.

77 Abbassian, K. and Kargari, A. (2016). Modification of membrane formulation for stabilization of emulsion liquid membrane for extraction of phenol from aqueous solutions. *Journal of Environmental Chemical Engineering* 4 (4): 3926–3933.

78 Kargari, A., Kaghazchi, T., Sohrabi, M. et al. (2006). Application of experimental design to emulsion liquid membrane pertraction of gold(III) ions from aqueous solutions. *Iranian Journal of Chemical Engineering* 3: 77–91.

79 Kumar, A., Thakur, A., and Panesar, P.S. (2018). Lactic acid extraction using environmentally benign green emulsion ionic liquid membrane. *Journal of Cleaner Production* 181: 574–583.

80 Ravanchi, M.T., Kaghazchi, T., and Kargari, A. (2010). Supported liquid membrane separation of propylene–propane mixtures using a metal ion carrier. *Desalination* 250 (1): 130–135.

81 Chakraborty, M., Bhattacharya, C., and Datta, S. (2010). Emulsion liquid membranes: definitions and classification, theories, module design, applications, new directions and perspectives. In: *Liquid Membranes* (ed. V.S. Kislik), 141–199. Elsevier.

82 Rathore, N., Sastre, A.M., and Pabby, A.K. (2016). Membrane assisted liquid extraction of actinides and remediation of nuclear waste: a review. *Journal of Membrane Science and Research* 2 (1): 2–13.

83 Busca, G., Berardinelli, S., Resini, C. et al. (2008). Technologies for the removal of phenol from fluid streams: a short review of recent developments. *Journal of Hazardous Materials* 160 (2–3): 265–288.

84 Daneshfar, A. and Khezeli, T. (2009). Extraction of phenolic compounds from environmental water samples using oil-in-water emulsions. *Microchimica Acta* 167 (3–4): 211.

85 Kazemi, P., Peydayesh, M., Bandegi, A. et al. (2013). Pertraction of methylene blue using a mixture of D2EHPA/M2EHPA and sesame oil as a liquid membrane. *Chemical Papers* 67 (7): 722–729.

86 Laki, S. and Kargari, A. (2016). Extraction of silver ions from aqueous solutions by emulsion liquid membrane. *Journal of Membrane Science and Research* 2 (1): 33–40.

87 Feng-Jee, C., Bao-Long, T., Ming-Xia, X. et al. (1985). A study on a two-component liquid membrane system. *Journal of Membrane Science* 23 (2): 137–154.

88 Uddin, M.S. and Kathiresan, M. (2000). Extraction of metal ions by emulsion liquid membrane using bi-functional surfactant: equilibrium and kinetic studies. *Separation and Purification Technology* 19 (1–2): 3–9.

89 Wodzki, R., Wyszynska, A., and Narebska, A. (1990). Two-component emulsion liquid membranes with macromolecular carriers of divalent ions. *Separation Science and Technology* 25 (11–12): 1175–1187.

90 Apenten, R.K.O. and Zhu, Q.-H. (1996). Interfacial parameters for selected Spans and Tweens at the hydrocarbon–water interface. *Food Hydrocolloids* 10 (1): 27–30.

91 Aguilar, M. and Cortina, J.L. (2008). *Solvent Extraction and Liquid Membranes: Fundamentals and Applications in New Materials*. CRC Press.

92 Blahušiak, M., Schlosser, Š., and Marták, J. (2010). Simulation of a hybrid fermentation-separation process for production of butyric acid. *Chemical Papers* 64 (2): 213–222.

3

Transport Theory in Membrane Contactor: Operational Principle

Mohammad Younas[1], Waheed Ur Rehman[2], and Mashallah Rezakazemi[3]

[1] *University of Engineering and Technology, Department of Chemical Engineering, Peshawar 25120, Pakistan*
[2] *PCSIR Laboratories Complex, Jamrud Road, Peshawar 25120, Pakistan*
[3] *Shahrood University of Technology, Faculty of Chemical and Materials Engineering, Shahrood, Iran*

3.1 Diffusional Mass and Heat Transfer Modeling

Membrane transport processes are rate processes where the process performance is measured through relative transport rates of the solute and solvent. When a species is transferred from the feed phase to the receiving phase in the membrane contactor, mass transport occurs. Species are transferred through the membrane by several mechanisms. However, this transport may face several resistances by membrane pores and the membrane layers on both sides. Mass transport theory is expressed depending upon the contact system. The rate of species transfer across the membrane relies on the driving force, membrane pore characteristics, and affinity of species in the feed phase to the receiving phase at the fluid–membrane interface or in the membrane pores.

There are two leading transport mechanisms developed in earlier times of membrane development. These are the preferential sorption–capillary flow (PSCF) and solution–diffusion (SD) model. Sourirajan and coworkers first presented the PSCF model, which was relied upon the porous membrane surface. However, SD model adopts the pore-free membrane surface. Species transport through porous membranes is characterized by their pore sizes:

I. Convection flow through large pores with the size of >1000 Å separation does not occur.
II. Knudsen diffusion through the pores with the size of 100–1000 Å.
III. Molecular sieving or surface diffusion through narrow pores with the size of 5–1000 Å.

Other important mathematical relationships used in membrane transport are Fick's, Hagen–Poiseuille's, and Ohm's laws. In membranes with big pores, the species transport is due to the convection. However, the transport process switches from convection to diffusion as the pore size reduces and leads toward a nonporous skin layer [1].

Membrane Contactor Technology: Water Treatment, Food Processing, Gas Separation, and Carbon Capture, First Edition. Edited by Mohammad Younas and Mashallah Rezakazemi.

3.2 Membrane Characterization Models

Transport phenomena in membrane contactor separation processes strongly depend on the structural, surface chemistry, and morphological characteristics of the membrane like thickness, roughness, porosity, composition, surface charge, hydrophobicity or hydrophilicity, and membrane pore size distribution [2–6]. Therefore, before describing the transport phenomena of membrane contactors processes, the theory of membrane structural characteristics must be described.

3.2.1 Contact Angle and Liquid Entry Pressure

Contact angle is an important measure of hydrophobicity or hydrophilicity of a membrane. It indicates the relative affinity of the membrane surface toward the liquid spreading on its surface. Relative affinity is due to the unbalanced molecular interactions that depend upon the surface chemistry of the membrane's surface and surface tension of the spreading liquid. Contact angle is a measure of the angle among the surface of the membrane and the tangent of the liquid droplet's inner curvature. However, surface porosity, roughness, and heterogeneity may deviate from the original readings of contact angle. For contact angle values, >90° indicates the hydrophobicity while <90° leads toward hydrophilicity for liquid–solid surface contact, as shown schematically in Figure 3.1.

As shown in Figure 3.1, at point C the thermodynamic balance is defined and is shown by Young's equation [1, 7]

$$W_S = W_A - W_C \tag{3.1}$$

Here, W_S stands for the force required to spread the liquid over the surface of the membrane, where W_A and W_C show the adhesive and cohesive forces, respectively, and are correlated with the following relations:

$$W_A = \gamma_{\mathrm{LV}}(1 + \cos\,\theta) \tag{3.2}$$

$$W_C = 2\gamma_{\mathrm{LV}} \tag{3.3}$$

γ_{LV} stands for the contacting liquid's surface tension at the membrane interface.

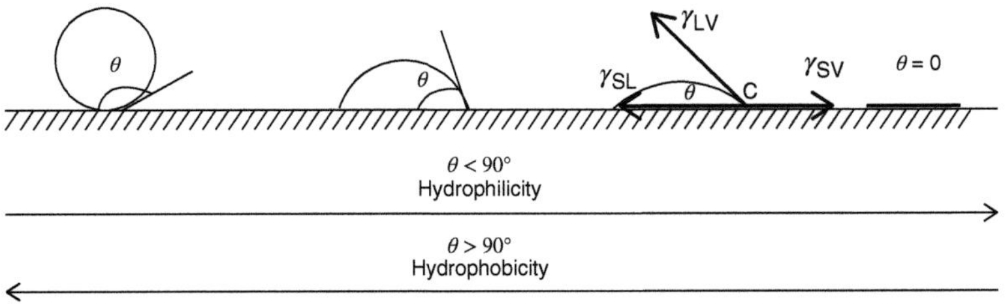

Figure 3.1 General representation of contact angle of a liquid on solid surface. Source: Based on Strathmann et al. [1] and Cuperus and Smolders [7].

Equation (3.1) thus becomes

$$W_S = \gamma_{LV}(\cos\theta - 1) \tag{3.4}$$

$$\cos\theta = \frac{W_S - \gamma_{LV}}{\gamma_{LV}} \tag{3.5}$$

Similarly, force balances at point C can be expressed as

$$\gamma_{LV}\cos\theta = \gamma_{SV} - \gamma_{SL} \tag{3.6}$$

$$\cos\theta = \frac{\gamma_{SV} - \gamma_{SL}}{\gamma_{LV}} \tag{3.7}$$

where γ_{SV} and γ_{SL} denote the membrane material's surface energy and membrane–liquid interface's surface tension, respectively. The surface tension of different liquids and surface energy of various polymer materials are given in Tables 3.1 and 3.2. Equations (3.5) and (3.6) indicate that an increase in surface tension of contacting fluid (γ_{LV}) and a decrease in surface energy of membrane material (γ_{SV}) cause an increase in contact angle and thus the hydrophobicity. These equations are only valid for a smooth membrane surface. For heterogeneous rough surfaces, a roughness parameter r is introduced, defined as the ratio of surface areas of the real to the geometrical (heterogeneous) surfaces:

$$r = \frac{A_r}{A_g} \tag{3.8}$$

where A_r and A_g show the real and the geometrical surface area of the membrane. For $r \gg 1$, Eq. (3.8) is replaced with Wenzel equation and is shown as

$$r = \frac{\cos\theta_w}{\cos\theta} \tag{3.9}$$

where θ_w denotes the Wenzel contact angle for heterogeneous membrane surface. Thus, an increase in the roughness parameter causes an increase in hydrophilicity. Equation (3.9) is valid for the condition when the surface nonhomogeneity such

Table 3.1 Surface tension of different liquids.

Liquid	**Surface tension, γ_{LV} (mN/m)**
Water	72.8
Glycerol	64
Formamide	58
Dimethyl sulfoxide	44
Chloroform	27.2
Isopropanol	21.7
Methanol	22.6

Source: Based on Cuperus and Smolders [7] and Ahmad et al. [8].

Table 3.2 Surface tension of different polymers.

Membrane	Surface tension, γ_{SV} (mN/m)	Water contact angle (°)
Nylon	75.0	—
Polyester	67.0	—
Polyurethane (PU)	56.0	—
Polyetheretherketone [9]	51.0	—
Poly(dimethylsiloxane) (PDMS)	22.0	120
Polytetrafluoroethylene (PTFE)	19.1	114
Polyvinylidenefluoride (PVDF)	30.3	85
Polyethylene (PE)	33.2	93–94
Polypropylene (PP)	33.2	95
Polyethyleneterephthalate (PET)	44.6	70
Polyvinylidene chloride (PVDC)	45.0	80
Polysulfone (PSf)	33.2	69–74
Polyethersulfone (PES)	113.7	60–70
Polyamide (PA)	35.9	53–57

Source: Based on Strathmann et al. [1] and Ahmad et al. [8].

as grooves, scratches, and cavities is completely filled with the liquid in contact. However, when air is entrapped in the roughness area beneath the contacting liquid, the contact angle θ_{CB} is given by the Cassie–Baxter equation shown as follows:

$$\cos\,\theta_{CB} = r_{f'}f'\,\cos\,\theta + f' - 1 \tag{3.10}$$

where f' refers to the fraction of the anticipated area wetted by the liquid and $r_{f'}$ denotes the roughness ratio of the wet area. If $f' = 1$, then $r_{f'} = r$, and Cassie–Baxter equation (3.9) becomes the Wenzel equation (3.10). The representation of these cases is schematically described in Figure 3.2. Keeping in view the various configuration for measuring contact angle, it has been found that the same membrane from different manufacturers achieved the different contact angle as described in Table 2.5. The occurrence of surfactants significantly decreases the surface tension in the fluid medium and therefore the contact angle.

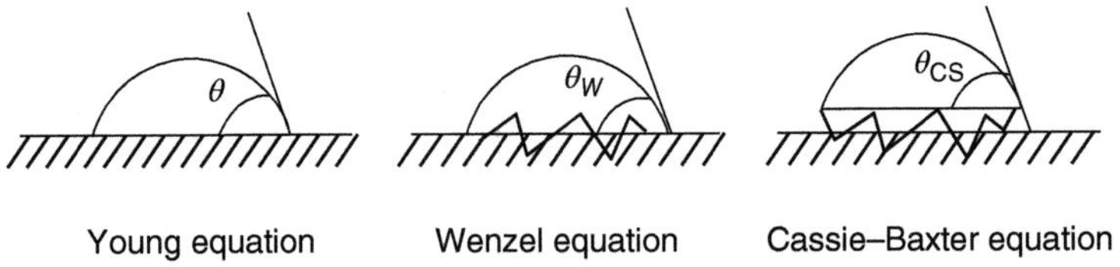

Figure 3.2 Representation of contact angle under smooth and rough surfaces.

3.2.2 Liquid Entry Pressure (LEP)

Liquid entry pressure (LEP) is an important macroscopic property of porous membranes, particularly in membrane contactor separation systems. Other terms used for this are breakthrough pressure (BTP), threshold pressure, and bubbling pressure. Bechhold first initiated it based on the capillary pressure required to make fluid–fluid interface stable and avoid dispersion [7]:

$$\Delta P_c = -\frac{2\Theta\sigma \cos\theta}{r_{\max}} \tag{3.11}$$

$$\Delta P_c = -\frac{2\Theta\gamma \cos\theta}{r_{\max}} \tag{3.12}$$

where ΔP_c denotes the liquid entry or BTP, σ is the surface tension of membrane if fluid system is gas–liquid, γ is interfacial tension for liquid–liquid system, θ is the contact angle, $r_{\max}$ is radius of the pore or capillary, and Θ is the geometric element correlated with the pore's shape, introduced by Franken et al. [10]. Equation (3.12) does not consider the tortuosity of the membrane pores. So, the estimated value of LEP measured by Eq. (3.12) may be lower than the real one. For cylindrical pores, $\Theta = 1$, and for non-cylindrical pores, $0 < \Theta < 1$. The pore size distribution is often not uniform; thus, the pore with maximum pore radius is considered. Pore size can be measured by taking air–water where contact angle is 0 and surface tension is 72 mN/m; then the pore radius in micrometers can be determined using the following equation if P is taken in bar:

$$r = \frac{1.4}{P} \tag{3.13}$$

In the case of a liquid–liquid membrane contactor system, the Cantor equation is used to estimate the pore radius by taking superhydrophilic wetting fluid ($\theta = 0$):

$$r = \frac{2\gamma}{P} \tag{3.14}$$

Kim and Harriott modified the Laplace–Young model for reticular fiber structure [11]:

$$\Delta P = -\frac{2\delta}{r} \cos\theta_{\text{ef}} \tag{3.15}$$

where r is described as half of the fibers' mean distance and θ_{ef} is the effective contact angle and is always $\theta_{\text{ef}} > \theta$.

If the pore geometry is axially odd, as defined in Figure 3.3, the BTP can be expressed by the following relation [12]:

$$\Delta P = -\frac{2\delta}{r} \frac{\cos(\theta_A - \alpha)}{1 + \frac{R}{r}(1 - \cos\alpha)} \tag{3.16}$$

where α is a structural angle in the axial direction, between the pore wall part and normal to the surface of the membrane, r is the mean pore radius, θ_A is the advancing contact angle, and R is the mean curvature radius of an element of the pore wall.

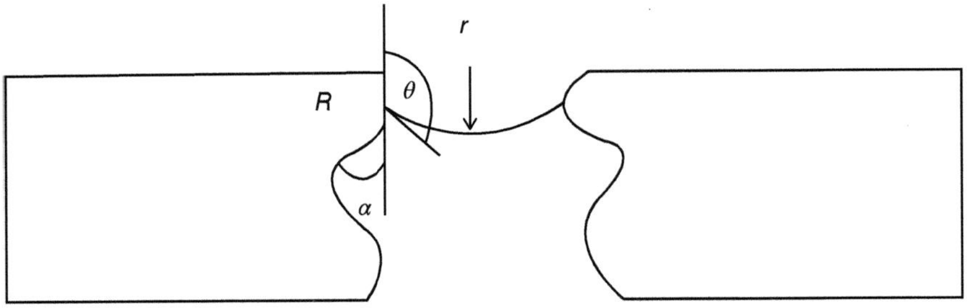

Figure 3.3 Representation of irregular pore element in the hydrophobic membrane. Source: Based on García-Payo et al. [12].

If the pores are cylindrical, then transmembrane volume flux (J_v) can be calculated from Hagen–Poiseuille equation

$$j_{vi} = \sum_{k=1}^{i} \frac{N_k \pi r_k^4 \Delta p_i}{8\mu x} \tag{3.17}$$

where μ is the dynamic viscosity of the fluid that displaces, N_k is the unit surface's number of pores, and x is the length of the pore (membrane thickness).

In the hydrophobic or hydrophilic membrane and for the liquid–liquid system, one liquid (wetting liquid) wets the membrane's pores without any hydrostatic pressure. In such a case, a certain pressure should be applied from the non-wetting fluid not to let the wetting fluid pass through the membrane from the wetting liquid region to the non-wetting fluid region. This minimum pressure is termed as BTP. Similarly, gas filled the membrane pores for the gas–liquid system, so pressure should be applied from the liquid side if the liquid does not penetrate into the pores. Conversely, pressure should be applied from the gas side. If the hydrostatic pressure exceeds the BTP to a further extent, then the liquid floods the pores of the membrane. In such a case, flooding occurs in all of the pores, and the water flux through the membrane follows Darcy's law:

$$j_i = k\Delta p \tag{3.18}$$

There are different experimental techniques to measure the BTP. The most common is introducing the bubble on non-wetting fluid side and detecting bubbles' emergence on wetting fluid across the membrane. This technique was developed by MacMullin and Muccini [13] and Thomas et al. [14]. BTP becomes more important in the membrane distillation (MD) process, where two non-wettable liquids are in contact at the hydrophobic membrane interface. Water has a surface tension of 72 mN/m, but inorganic solutes further increase the surface tension. Conversely, water's surface tension drastically decreases if the organic solute is added to water [15]. García-Payo et al. tested different pore size membranes of polyvinylidenefluoride (PVDF) and polytetrafluoroethylene (PTFE) with aqueous solutions of ethanol and isopropanol in different weight percent for BTP. They found that BTP drastically reduced with the increase of organic content in water,

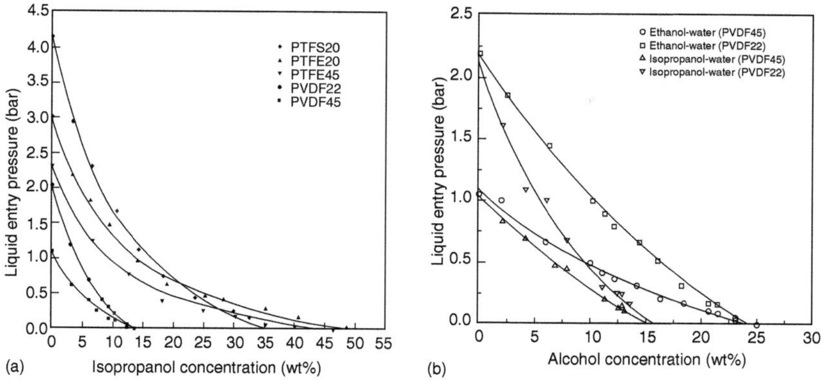

Figure 3.4 Effect of membrane type, pore size, and type of organic solutions and concentration on breakthrough pressure. (a) Effect of isopropanol concentration on liquid entry pressure, (b) Effect of alcohol concentration on liquid entry pressure.
Source: García-Payo et al. [12].

as shown in Figure 3.4. Aqueous ethanol's BTP values for solutions are larger than those for isopropanol solutions at identical aqueous concentrations and for the same membrane type. Furthermore, PTFE membrane exhibited higher BTP values than PVDF membranes. Similarly, it can also be found that BTP decreases as the pore sizes of both membrane increases [12].

Yazgan-Birgi et al. [16] showed that LEP or BTP increases with water contact angle (WCA)'s increase as shown in Figure 3.5a. Surface chemistry and surface forces like the van der Waals and electrostatic forces affect membrane wetting and fluid film stability to a significant degree. As the WCA increases, the no-slip wall condition disappears. For example, at WCA > 150°, as for superhydrophobic membrane, the surface has zero contact with the liquid, and thus no-slip condition no longer applies. Similarly, pore length also affects the LEP. Higher LEP is observed for longer pore lengths, as shown in Figure 3.5b.

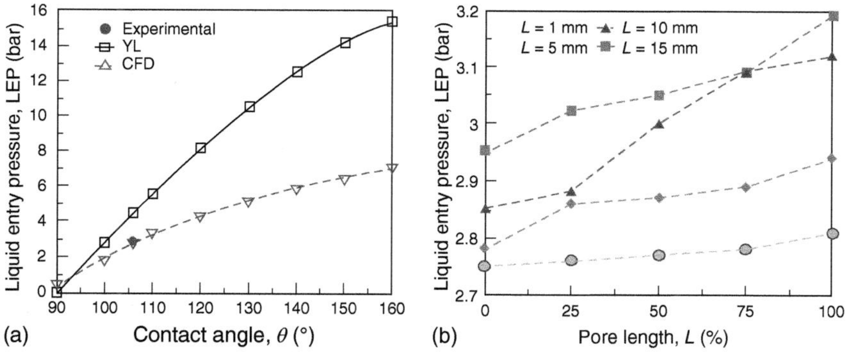

Figure 3.5 (a) Effect of contact angle on LEP. (b) Effect of pore length on LEP.
Source: Yazgan-Birgi et al. [16].

3.2.3 Permporometry (Pore Size Distribution)

Permporometry technique is applied to estimate the distribution of pore size for a gas–liquid system. Eyraud first introduced the technique, which is later modified by Katz et al. [17, 18]. It works on the vapor pressure distribution of a liquid over the surface, depending upon its curvature. The condensable (such as water or hexane) vapors are entrapped in the pores of the capillary, thus blocking the pores for the permeation of non-condensable gas (like N_2) in the feed. This is best described by the Kelvin equation [19]

$$d_k = \frac{-4v\sigma \cos \theta}{RT \ln \left(p/p_s \right)} \tag{3.19}$$

where d_k is the Kelvin pore diameter, p denotes the vapor pressure, R is the ideal gas constant, T is the temperature, v is the molar volume, θ is the contact angle, and σ is the surface tension. The pore diameter proposed by Kelvin in Eq. (3.20) is the difference of true pore diameter and thickness of the adsorbed layer:

$$d_p = d_k + t \tag{3.20}$$

The Kelvin technique excludes the closed pores, and pore size distribution of only active pores is estimated.

3.2.4 Electron Microscopy

There are two magnification techniques for visual investigation of membrane structure and porosity. Scanning electron microscopy (SEM) technique is most commonly used for polymeric and mixed matrix membranes in flat sheet and hollow fibers [20–25]. The technique is applied to determine the active and support layer thickness and visualize the pores structures on the surface of the membrane. However, SEM does not produce high-resolution images. Transmission electron microscopy [26], working on the same principle as SEM does, is used for high-resolution images, particularly in dense membranes and ultrathin layers. Both works on using high electron beam energy, which can also damage the membrane's surface [19].

3.3 Transport Models in Liquid–Liquid Contactor

The most common and widely used operation in liquid–liquid extraction that employs membrane contactors is solvent extraction. Solvent extraction using membrane contactors has been applied to various separation systems, including extraction of metals, organic acids, alcohols, proteins, pollutants, aromatics, etc. [27, 28]. Membrane-based solvent extraction is an equilibrium-based separation technique in which a solute (or solutes) in a solution, aqueous or organic, is extracted into an immiscible aqueous or organic solvent without dispersion of the phases where membrane provides the surface of contact for interfacial solute(s) transfer from one phase (feed) to the other (receiving). Dispersion-free solvent extraction in membrane contactors thus uses a microporous membrane (hydrophobic or hydrophilic) and two immiscible solutions (aqueous and organic) termed as feed and receiving phases separated by the membrane. One phase instinctively wets the membrane and transfers to the other side of the membrane through its pores. The dispersion of that phase (wetting) on the opposite membrane side can be prohibited if the other immersible liquid phase (non-wetting) is kept at a pressure equal to or greater than the wetting-phase solution. The interface of both the solutions is halted at the mouth of each pore on the non-wetting membrane side. The extent of that pressure, which does not allow the wetting phase in the membrane pores to be banished by the non-wetting phase, is called breakthrough pressure (BTP). The membrane only serves to immobilize both phases at the interface on the non-wetting-phase side of the membrane. The solute (species) interacts with the solvent at the membrane interface and transfers from one phase to the other through the membrane due to the difference in concentrations. The affinity of the solute with the solvent defines the separation factor. The greater the affinity of solute–solvent, the greater the separation. This affinity is called partition or distribution coefficient for liquid–liquid extraction and will be described in detail in Section 3.3.1.4. The concentration profiles for hydrophobic and hydrophilic membranes are shown in Figures 3.6 and 3.7, showing the transfer of a solute from the aqueous phase to the organic phase and

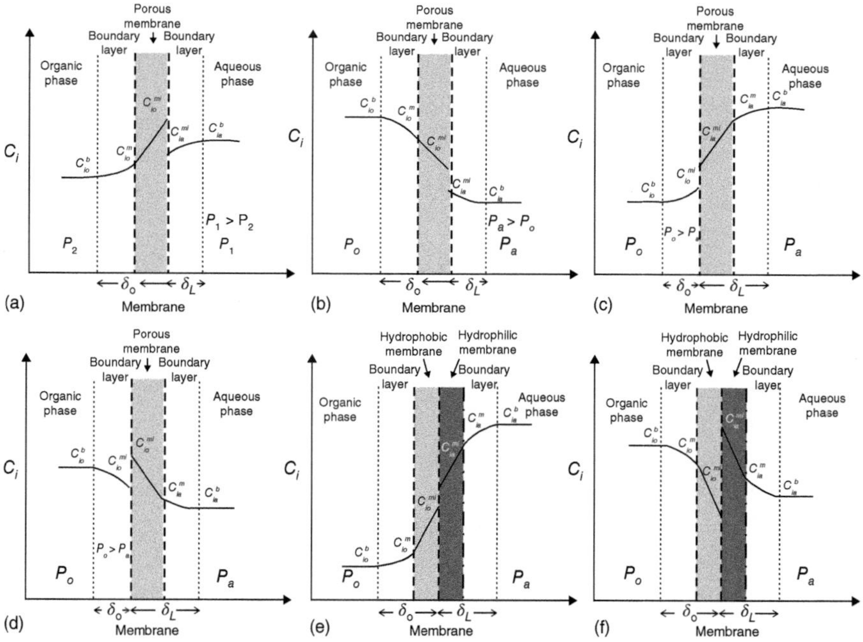

Figure 3.6 Concentration profile in liquid–liquid hydrophobic/hydrophilic symmetric membrane: (a) solute transfer from aqueous to organic phase through hydrophobic membrane, (b) solute transfer from organic to aqueous phase through hydrophobic membrane, (c) solute transfer from aqueous to organic phase through hydrophilic membrane, (d) solute transfer from organic to aqueous phase through hydrophilic membrane, (e) solute transfer from aqueous to organic phase through composite hydrophobic/hydrophilic membrane, and (f) solute transfer from organic to aqueous phase through composite hydrophobic–hydrophilic membrane.

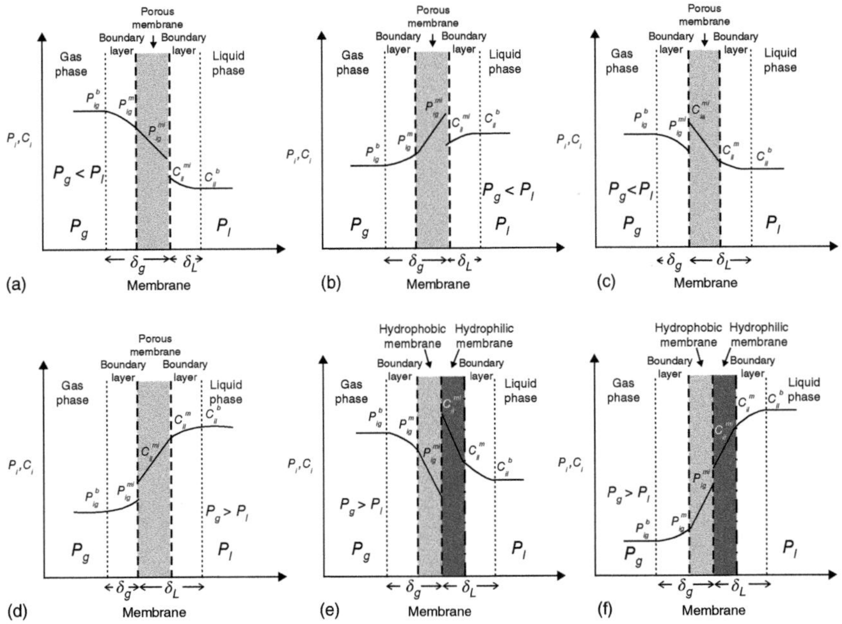

Figure 3.7 Concentration profile in gas–liquid hydrophobic/hydrophilic symmetric membrane: (a) species transfer from gas to liquid phase through the hydrophobic membrane, (b) species transfer from liquid to gas phase through the hydrophobic membrane, (c) species transfer from gas to liquid phase through the hydrophilic membrane, (d) species transfer from liquid to gas phase through the hydrophilic membrane, (e) species transfer from gas to liquid phase through the composite hydrophobic/hydrophilic membrane, and (f) species transfer from liquid to gas phase through composite hydrophobic/hydrophilic membrane.

vice versa. A generic concentration profile for a hypothetical symmetric cell in a membrane contactor or, more precisely, in the hollow fiber membrane is shown in Figure 3.8. For hydrophobic membranes, the pores are filled by the organic phase, and thus, the membrane immobilizes the phases at the membrane–aqueous phase interface. Pressure from the aqueous-phase side should be applied greater than that from the organic-phase side to avert the displacement of the organic phase into the aqueous phase. Partitioning of solute occurs at the interface of membrane and aqueous phase. The solute is then transported into the organic phase from the aqueous phase or vice versa according to the driving force because the solute will transfer from the higher to lower concentration side. In case the membrane is hydrophilic, the aqueous phase will fill the membrane pores. In this case, applied pressure on the organic side should be maintained higher than that on the aqueous-phase side to immobilize the phases at the membrane–organic phase interface. Composite membranes are also used in membrane contactor separation processes. They are composed of a minimum of two layers, hydrophobic and hydrophilic. Both layers are adjacent to each other. Pressure on both sides is maintained to keep the interfaces stable on both sides of membranes and avoid the mixing and displacement of phases. Aqueous and organic phases fill the hydrophilic and hydrophobic layers,

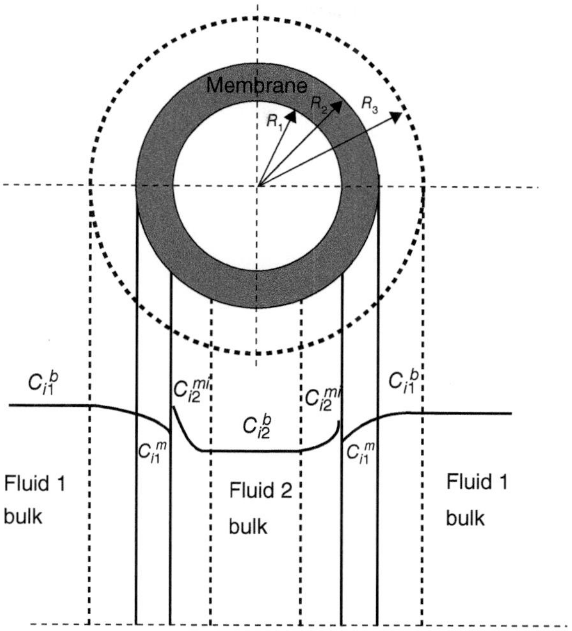

Figure 3.8 Mass transfer in hollow fiber membrane contactor.

respectively. Both phases are immobilized at the interface between both layers for the partitioning of solute from one phase with the other phase.

3.3.1 Resistance in Series Model

When aqueous and organic phases flow in the membrane contactor on the opposite sides, the mass transport of solute (species) occurs across the membrane from one phase to the other. The mass transfer follows through a series of resistances. These resistances are induced in the aqueous and organic phases and the membrane by boundary layers neighboring the porous membrane. When a solute is transported within a single phase or from one phase to the other, concentration difference is the gradient driving force that influences the overall mass transport rate. In such mass transport processes, equilibrium is of primary concern. Mass transport of the solute proceeds if the system is not at equilibrium. Furthermore, the phase equilibrium relationship quickly predicts how fast solute transport is. The following assumptions are important to consider in developing the transport models [29–34]:

I. Mass transfer occurs at steady-state conditions across the membrane.
II. Distribution of the pore size and wetting characteristics are uniform throughout the membrane.
III. Membrane wetting does not occur on any side of the membrane.
IV. Geometric representation of interface curvature does not influence mass transfer rate and species distribution coefficient.
V. Concentration profiles are linear.
VI. Thermodynamic equilibrium occurs at the fluid–fluid interfaces inside the pores of the membrane.

VII. The species partition coefficient is constant throughout the experiment.
VIII. Storage tanks for feed and receiving phases are taken as ideally mixed tanks.

3.3.1.1 Model Approach

The membrane contactor module is divided into arbitrary axial sections, and radial rings with each section in each radial ring are assumed as a perfectly ideal mixed unit. The axial stage 1 is indicated as the feed's entry, and the first radial ring corresponds to the center of the module. One radial ring is characterized by a single fiber in a hollow fiber module or a single compartment in a flat-sheet module. Each segment is then modeled based on the resistance in series (RIS) model. A schematic description of this approach is presented by Younas et al. [35, 36] and is shown in Figure 3.9 (Table 3.3).

3.3.1.2 Two Film Theory

Two film theory was presented by Whitman for interphase mass transfer problems [39]. It was based on two propositions: (i) the rate of species transfers in the boundary layers on opposite membrane sides and the membrane itself and is dominated by diffusion rate, and (ii) equilibrium occurs at the interface, and species transport is instantaneous. This theory claims that mass transfer resistance in a fluid phase resides in thin layers on both sides of the membrane, adjacent to the interface. Mass transfer resistance lies in the membrane itself.

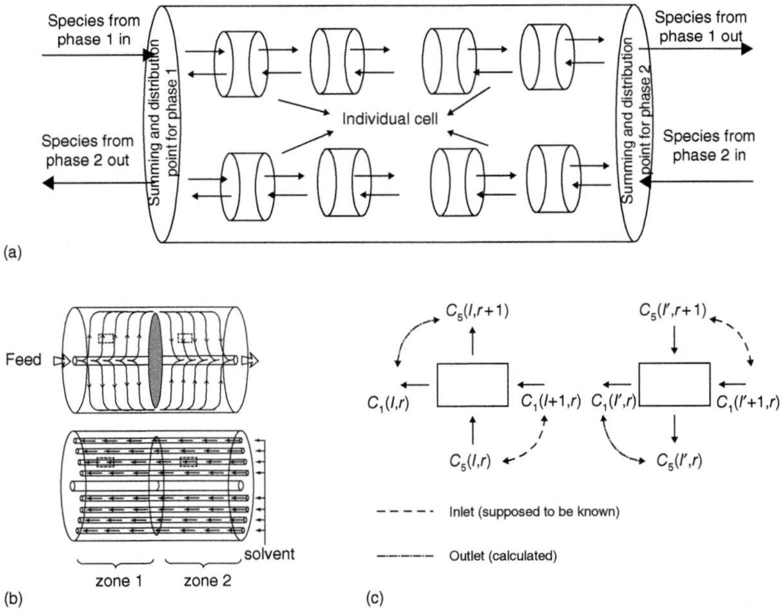

Figure 3.9 Schematic representation of steady-state model approach of membrane contactor module: (a) module axial–radial division, (b) flow pattern in module axial–radial divisions, and (c) single-cell model approach for RIS model [35–37]. Source: Bocquet et al. [37].

Table 3.3 Correlations for overall mass transfer coefficient (MTC).

Membrane/separation system	Liquid–liquid contactor: Overall MTC based on aqueous-phase side ($1/K_a$)	Liquid–liquid contactor: Overall MTC based on organic-phase side ($1/K_o$)	Gas–liquid contactor: Overall MTC based on gas-phase side ($1/K_a$)	Gas–liquid contactor: Overall MTC based on liquid-phase side ($1/K_l$)
Hydrophobic, species transport from aqueous/gas to organic/liquid phase	$\frac{A_a}{k_{ia}} + \frac{A_a}{m_i k_{im}} + \frac{A_a}{m_i k_{io}}$	$\frac{A_a}{k_{ia}} + \frac{A_a}{m_i k_{im}} + \frac{A_a}{m_i k_{io}}$	$\frac{A_g}{k_{ig}} + \frac{A_g}{k_{im}} + \frac{A_g}{H_i k_{il}}$	$\frac{A_l}{k_{ig}} + \frac{A_l}{k_{im}} + \frac{A_l}{H_i k_{il}}$
Hydrophobic, species transport from organic/liquid to aqueous/gas phase	$\frac{A_o m_i}{k_{ia}} + \frac{A_o}{k_{im}} + \frac{A_o}{k_{io}}$	$\frac{A_o m_i}{k_{ia}} + \frac{A_o}{k_{im}} + \frac{A_o}{k_{io}}$	$\frac{A_g H_i}{k_{il}} + \frac{A_g H_i}{k_{im}} + \frac{A_g}{k_{il}}$	$\frac{A_l H_i}{k_{il}} + \frac{A_l H_i}{k_{im}} + \frac{A_l}{k_{il}}$
Hydrophilic, species transport from aqueous/gas to organic/liquid phase	$\frac{A_a}{k_{ia}} + \frac{A_a}{k_{im}} + \frac{A_a}{m_i k_{io}}$	$\frac{A_o}{k_{ia}} + \frac{A_o}{k_{im}} + \frac{A_o}{m_i k_{io}}$	$\frac{A_g}{k_{ig}} + \frac{A_g}{H_i k_{im}} + \frac{A_g}{H_i k_{il}}$	$\frac{A_l}{k_{ig}} + \frac{A_l}{H_i k_{im}} + \frac{A_l}{H_i k_{il}}$
Hydrophilic, species transport from organic/liquid to aqueous/gas phase	$\frac{A_a m_i}{k_{ia}} + \frac{A_a m_i}{k_{im}} + \frac{A_a}{k_{io}}$	$\frac{A_o m_i}{k_{ia}} + \frac{A_o m_i}{k_{im}} + \frac{A_o}{k_{io}}$	$\frac{A_g H_i}{k_{ig}} + \frac{A_g}{k_{im}} + \frac{A_g}{k_{il}}$	$\frac{A_l H_i}{k_{ig}} + \frac{A_l}{k_{im}} + \frac{A_l}{k_{il}}$
Composite, species transport from aqueous/gas to organic/liquid phase	$\frac{A_a}{k_{ia}} + \frac{A_a}{k_{imph}} + \frac{A_a}{m_i k_{impl}} + \frac{A_a}{m_i k_{io}}$	$\frac{A_o}{k_{ia}} + \frac{A_o}{k_{imph}} + \frac{A_o}{m_i k_{impl}} + \frac{A_o}{m_i k_{io}}$	$\frac{A_g}{k_{ig}} + \frac{A_g}{k_{imph}} + \frac{A_g}{H_i k_{impl}} + \frac{A_g}{H_i k_{il}}$	$\frac{A_l}{k_{ig}} + \frac{A_l}{k_{imph}} + \frac{A_l}{H_i k_{impl}} + \frac{A_l}{H_i k_{il}}$
Composite, species transport from organic/liquid to aqueous/gas phase	$\frac{A_a m_i}{k_{ia}} + \frac{A_a m_i}{k_{imph}} + \frac{A_a}{k_{impl}} + \frac{A_a}{k_{io}}$	$\frac{A_o m_i}{k_{ia}} + \frac{A_o m_i}{k_{imph}} + \frac{A_o}{k_{impl}} + \frac{A_o}{k_{io}}$	$\frac{A_g H_i}{k_{ig}} + \frac{A_g H_i}{k_{imph}} + \frac{A_g}{k_{impl}} + \frac{A_g}{k_{il}}$	$\frac{A_l H_i}{k_{ig}} + \frac{A_l H_i}{k_{imph}} + \frac{A_l}{k_{impl}} + \frac{A_l}{k_{il}}$

Source: Based on Prasad and Sirkar [38].

The mass transfer rate through the membrane is correlated with the mass transfer in the respective phase boundary layer that also wet the membrane. The interface imparts no mass transfer resistance. Therefore, the concentration at the interfaces remains at equilibrium and is described by distribution law. The distance of one phase from the interface is called the "film thickness" or "boundary layer thickness" on either side of the membrane. RIS model is based on the one-dimensional species transport through the membrane. Mass transfer coefficients describe such models. Molecular diffusion ensues across the film thickness δ for either side of the membrane. For one-dimensional flow, the mass transfer rate across the particular resistance is proportional to the concentration gradient of species i:

$$j_i \propto \Delta C_i = \frac{D_{ij}}{\delta} A \Delta C_i = KA\Delta C_i \tag{3.21}$$

where D_{ij} represents the diffusion coefficient of species i in j, K is the coefficient of mass transfer, and A is the contact area of the two phases. As molecular diffusion is dominated by the flow conditions, for example, by convection, the diffusion coefficients are replaced with the mass transfer coefficient in each film on either side of the membrane. These mass transfer coefficients are related to diffusion coefficients through dimensionless correlations that will be described in Sections 3.4–3.7. The inverse of mass transfer coefficient characterizes the resistance to mass transfer. Normally the ratio $\frac{D_{ij}}{\delta}$ is unknown because the thickness of the film depends on the flow conditions. Thus, it can be substituted by a coefficient of mass transfer for each film. As a specific resistance to mass transfer represents the inverse of mass transfer coefficient for each fluid film, the method is called the resistance-in-series model.

3.3.1.3 Phase Equilibrium in Liquid–Liquid System

There are several types of equilibrium laws. The most common for the concentration-dependent operations in a single- or multiphase system is "distribution" or "partition" or "linear" law. This law states that a species' equilibrium concentrations in a single or multiphase are in linear proportion at constant temperature and pressure. The equilibrium diagrams are shown schematically in Figure 3.10 by considering a typical concentration profile of species transfer from the aqueous phase to the organic phase through the hydrophobic membrane. Mathematically it can be expressed by Eq. (3.22):

$$C_{i1} = mC_{i2} \tag{3.22}$$

where C_{i1} is the species i concentration in phase 1 and C_{i2} is the equilibrium concentration of species i in phase 2. Both phases are at equilibrium at the interface with each other. m is denoted for the distribution or partition coefficient. For liquid–liquid systems, the distribution law is referred to Raoult's law. For gas–liquid systems, Henry's law best describes the distribution law. In case the species concentration in bulk is associated with the species concentration at equilibrium, then the distribution law becomes

$$C_{i1}^{b} = mC_{i1}^{e} \tag{3.23}$$

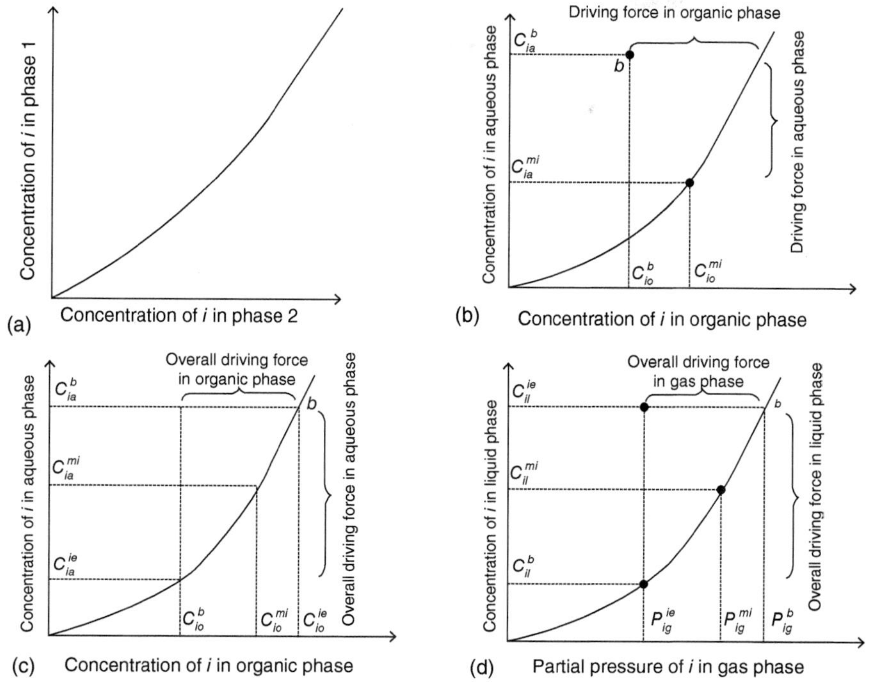

Figure 3.10 Equilibrium diagram between two phases: (a) equilibrium curve, (b) driving force description at interface, (c) equilibrium curve for liquid–liquid extraction, and (d) equilibrium curve for gas–liquid extraction.

$$C^{e}_{i2} = mC^{b}_{i2} \tag{3.24}$$

where C^{b}_{i1} and $C^{b}_{i2}$are the species concentrations in bulk of phases 1 and 2, respectively, and C^{e}_{i1} and $C^{e}_{i2}$are species concentrations of phases 1 and 2, respectively, at equilibrium with that phase. For a typical hydrophobic membrane, when species transfer from aqueous to organic phase, the relationship for the partition (distribution) coefficient becomes (Table 3.4)

$$m_i = \frac{C^{e}_{io}}{C^{b}_{ia}} = \frac{C^{mi}_{io}}{C^{mi}_{ia}} = \frac{C^{b}_{io}}{C^{e}_{ia}} \tag{3.25}$$

3.3.1.4 Overall Mass Transfer Coefficient

The main limitation of RIS model is that the interface concentrations can be measured. Therefore, the *overall mass transfer coefficients* are introduced, which account for the bulk concentrations, distribution coefficient, and individual coefficients of mass transfer:

$$j_i = K_a A_a (C^{b}_{ia} - C^{ie}_{ia}) \tag{3.26}$$

$$j_i = K_o A_o (C^{ie}_{io} - C^{b}_{io}) \tag{3.27}$$

where K_a and K_o are the overall coefficients of mass transfer for the aqueous-phase side and organic-phase side, respectively; C^{b}_{ia} and C^{b}_{io} are the species concentration in

Table 3.4 Partition coefficient values for liquid–liquid systems.

Feed phase (aqueous)	Receiving phase	Partition coefficient	References
Copper ions	LIX 84-I	4.8–1530 (varying with concentration, pH, and temperature)	[36]
Copper ions	TFA	0.5–6.9 (varying with concentration, pH, and temperature)	[40]
Dimethyl trisulfide	*n*-Hexane	463	[37]
Ethyl butyrate	*n*-Hexane	94	[37]
Benzaldehyde	*n*-Hexane	31	[37]
2-Phenylethanol	*n*-Hexane	0.5	[37]

bulk of aqueous phase and organic phase, respectively; and C_{ia}^{ie} and $C_{io}^{ie}$is the concentration of species in aqueous phase and organic phase, respectively, at equilibrium. The species concentrations in bulk and in equilibrium are shown schematically in the equilibrium distribution curve in Figure 3.1.

It has been found from Eqs. (3.7), (3.8), (3.15), (3.16), (3.23), (3.24), (3.30), (3.31), (3.36), (3.37), (3.44), (3.45), (3.53), (3.54), (3.63), (3.64), (3.72), (3.73), (3.81), (3.82), (3.89), (3.90), (3.96), and (3.97) in Appendix 3.A that partition (distribution) coefficient strongly affects the overall coefficient of mass transfer. If the transport of species is intended from aqueous to organic side, then high partition coefficient values are favored and vice versa for the transfer of species from organic to aqueous side. Equations (3.7), (3.8), (3.23), (3.24), (3.36), and (3.37) in Appendix 3.A indicate that high values of partition coefficient ($m \ggg 1$) nullify the effects of resistances due to the membrane and/or organic phase. In this case, resistance to mass transfer in aqueous phase dominates the species transport, thus controlling mass transfer. However, low values of species partition coefficient ($m \lll 1$) dominate the membrane and mass transfer resistances for overall mass transfer.

In contrast, for hydrophilic membranes as indicated by Eqs. (3.23), (3.24), (3.30), and (3.31) in Appendix 3.A, both membrane and aqueous-phase resistances control the mass transport of species if the partition coefficient values are high ($m \ggg 1$), and organic-phase resistance controls the mass transport of species for low values of the partition coefficient ($m \lll 1$). Therefore, hydrophilic membranes are recommended for the liquid–liquid systems having $m \lll 1$, and hydrophobic membranes for the liquid–liquid systems having $m \ggg 1$ [38].

3.4 Transport Model in Gas–Liquid Systems

Gas phase in gas–liquid systems typically fills the membrane pores. If the liquid phase is polar, then a hydrophobic membrane is preferred so as not wet the membrane. In case the liquid phase is nonpolar, a membrane having hydrophilic properties is employed. Thus, in each case, the interface is established at liquid-phase side of the membrane. However, in the case of using a hydrophobic membrane for nonpolar liquid phase or a hydrophilic membrane for polar liquid phase, a certain amount of pressure will be needed from gas-phase side not to let liquid phase penetrate inside the membrane pores [41–45]. The schematic is shown in Figure 3.7.

The gas molecule transfer across the nonporous polymeric membrane is founded on the model of solution diffusion. In this model, permeate diffuses through the membrane at the permeate side after dissolving in the membrane material. The selectivity of membrane in this model depends on the structure of polymer molecules. The membrane allows gas molecules to depend upon their size and kinetic energy [46–48]. This model is used for permeation in the polymer membrane. Membrane separation processes such as pervaporation, dialysis, gas separation, and reverse osmosis are founded on the model under discussion. Thomas Graham proposed the gas transport mechanism across the polymeric membrane. Various permeants are separated based on the volume of m aterial dissolved in the membrane and the diffusion rate of material across the membrane. This model has three main steps for gas transfer through the polymer: gas dissolution at the upstream face of the polymer, gas diffusion through the polymer, and gas desorption at the downstream face of the polymer. In contrast to the first and third steps, the second step is relatively slow. In mass transport, diffusion of permeate through polymer is the rate-limiting step across the membrane [49, 50]:

$$j_i = K_g A_g (p_{ig}^b - p_{ig}^e) \tag{3.28}$$

3.4.1 Phase Equilibrium for Gas–Liquid System

At the interface, for gas–liquid systems, phase equilibrium is defined by Henry's law as shown by the following mathematical relationship:

$$H_i = \frac{C_i}{p_i} \tag{3.29}$$

$$H_i = \frac{C_{il}^e}{p_{ig}^b} = \frac{C_{il}^{mi}}{p_{ig}^{mi}} = \frac{C_{il}^b}{p_{ig}^e} \tag{3.30}$$

3.4.2 Resistance in Series Model

When aqueous and organic phases flow in the membrane contactor, the mass transport of solute occurs across the membrane from one phase to the other. The mass transfer follows through a series of resistances. These resistances are caused in both the aqueous and organic phases and the membrane by boundary layers next to the porous membrane. When a solute is transported within a single phase or from one phase to the other, concentration difference is the gradient driving force that influences the overall mass transport rate. In such mass transport processes, equilibrium is of primary concern. Mass transport of the solute proceeds if the system is not at equilibrium. Furthermore, the phase equilibrium relationship quickly predicts how fast the solute transport is.

The following are the assumptions:

I. Mass transfer occurs at steady-state conditions across the membrane.
II. Distribution of pore size and wetting features are even across the whole membrane.
III. No membrane wetting occurs on either side of the membrane.
IV. Geometric representation of interface curvature does not influence rate of mass transfer and species distribution coefficient.
V. Concentration profiles are linear.
VI. Thermodynamic equilibrium occurs at the fluid–fluid boundary layers inside the pores of the membrane.
VII. The species partition coefficient is constant throughout the experiment.
VIII. Storage tanks for feed and receiving phases are taken as ideally mixed tanks.

3.5 Reactive Diffusion in Liquid-Side Boundary Layer

In membrane contactors, the rate of mass transfer is majorly affected by chemical reaction. For gas–liquid mass transfer, the liquid phase resistance controls mass transfer while gas phase resistance is insignificant in the case of no chemical reaction. In chemical reaction among solvent and absorbed gas, the gas phase resistance controls mass transfer while liquid side resistance is small. So, in the presence of chemical reactions, the mass transfer can be increased. The effect of chemical reaction is described by the enhancement factor (E) in the mass transfer rate. For identical mass transfer force, the enhancement factor is determined from the ratio of chemical absorption to physical absorption, as shown in Eq. (3.31) [51, 52]:

$$E = \frac{J_{\text{chem}}}{J_{\text{phy}}} \tag{3.31}$$

Mass transfer rate for hollow fiber membrane contactors is affected by the enhancement factor. Kumar et al. [53] have estimated enhancement factor and absorption flux for single-fiber membrane contactor. Numerical methods developed are quietly resembled DeCoursey's approximation. Enhancement factor can be calculated from a dimensionless Hatta number (Ha) and based on surface renewal theory [54, 55]:

$$E = \Gamma \frac{k_{m,n}}{K_L} D_i C^{m-1} C_j^n \tag{3.32}$$

where k stands for reaction rate constant; C is the concentration of species i, j; D_i represents the species' diffusion coefficient; m and n, with respect to "i," denote the order of partial reaction; and K_L is the mass transfer coefficient. For first reaction regime ($Ha > 2$), the enhancement factor is given by Eq. (3.33):

$$E = \frac{\sqrt{k_1 D_{il}}}{K_L} \tag{3.33}$$

where k_1 is rate constant for first-order reaction and D_{il} is the diffusion coefficient for liquid.

3.6 Mass Transfer Resistance Analysis

The resistance offered to the mass transfer is represented by taking the inverse of the coefficient of mass transfer. As described by RIS model, the resistance to mass transfer is the inverse of fluid velocity and thus can be characterized by fluid velocity. The partition coefficient can also affect the mass transfer resistances. Mass transfer resistance analysis makes it possible to estimate the efficacy of the interconnected membrane contactor extracting system. It aims to identify and thereby design an optimal method of separation. Simulation outcomes for the study of mass transfer resistance are given by Younas et al. [56] for liquid–liquid extraction in hollow fiber membrane contactor. Figure 3.8 shows the results for feed flow in the shell and inside fiber. Resistance to mass transfer was also compared at various aqueous- and organic-phase flow speeds. Flow configuration, diluent form, and hydrodynamics have been shown to have a major effect on the dominant part of mass transfer resistance for extracting about the same volume of copper(II) under identical temperature. It was noted that the partition coefficient, which varies with the diluent form, is the primary parameter for the study of mass transfer resistance. If the coefficient of partition declines, membrane resistance to mass transfer rises. In addition to that, with the growth in liquid flow speeds, mass transfer resistance in membrane pores also raises. At higher flow speeds, convective mass transfer resistance reduces. However, if the aqueous phase flows on the shell side, the aqueous-phase-side mass transfer resistance is prevailing. Likewise, organic-phase mass transfer resistance is prevalent if the organic phase flows on the shell side. It is necessary to remember that the module's shell side presents a higher surface area than the area within the fibers. Younas et al. [56] further described that the resistance of aqueous-phase-side mass transfer for decanol as diluents is lower than for methyl isobutyl ketone (MIBK) as diluents. In particular, there are distinct physicochemical properties of both diluents and thus distinct coefficients of diffusion. Organic-phase-side mass transfer resistance, especially at lower flow rates, is prevalent for organic-phase flow on the shell side.

Figure 3.7 depicts each local mass transfer resistance contribution, e.g. feed boundary layer, membrane, and receiving phase boundary layer upon total resistance for the four different aroma compounds. These species are dimethyltrisulfide (DMTS), ethyl butyrate (EB), benzaldehyde (BA), and 2-phenylethanol (PE) with partition coefficient values of 463, 94, 31, and 0.5, respectively. For species that have a high partition coefficient, it can be found that resistance to mass transfer on feed side is prevalent. The feed mass transfer resistance is greater than 90% of the overall resistance for DMTS, EB, and BA, but the primary mass transfer resistance lies inside the membrane's pores for PE. This is since the coefficient of the partition is less than 1. The same trend has been shown for different modules with different configurations, X40 and X30 modules [37] (Figures 3.11 and 3.12).

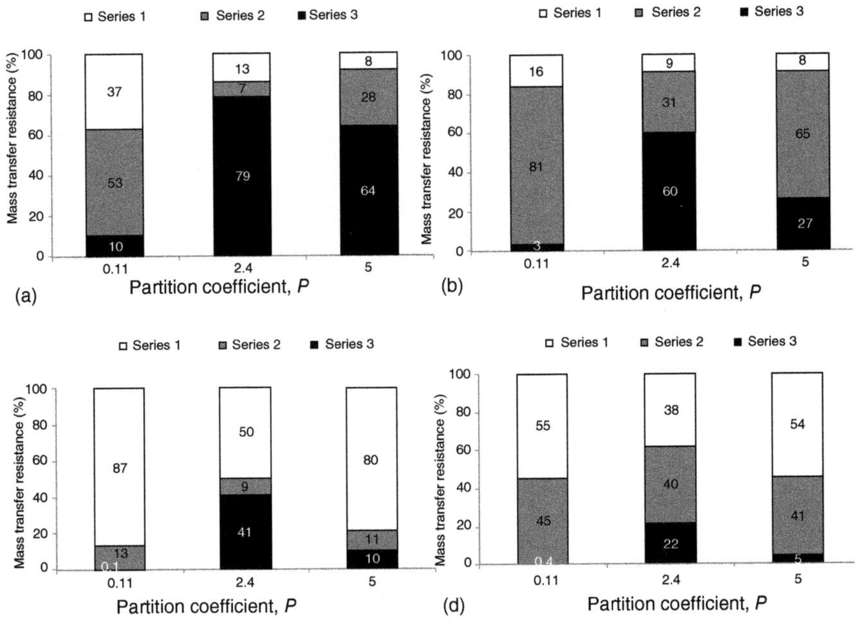

Figure 3.11 Analysis of mass transfer resistance for copper(II) extraction with trifluoroacetic acid (TFA); $V_{aq} = V_{org} = 1 \times 10^{-3}$ m^3. (a) $Q_{aq} = Q_{org} = 1.67 \times 10^{-6}$ m^3 s^{-1}, aqueous-phase flows in flow cell (shell side). (b) $Q_{aq} = Q_{org} = 16.67 \times 10^{-6}$ m^3 s^{-1}, aqueous-phase flows in flow cell (shell side). (c) $Q_{aq} = Q_{org} = 1.67 \times 10^{-6}$ m^3 s^{-1}, organic-phase flows in flow cell (shell side). (d) $Q_{aq} = Q_{org} = 16.67 \times 10^{-6}$ m^3 s^{-1}, organic-phase flows in flow cell (shell side). Source: Bocquet et al. [37].

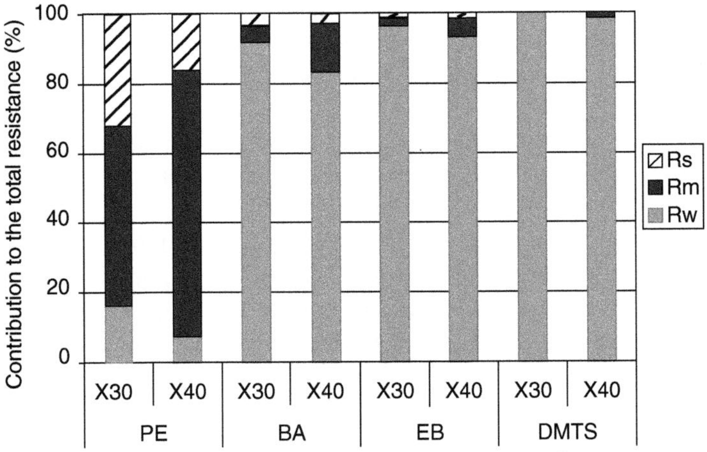

Figure 3.12 Analysis of the resistance to mass transfer for the four aroma compounds in hollow fiber membrane contactor module $Q_f = 710$ ml/min, $Q_r = 670$ ml/min. Source: Bocquet et al. [37].

3.7 Correlations for Mass Transfer Coefficients

The correlations for mass transfer coefficient are described by the following relation:

$$Sh = cRe^{\alpha}Sc^{\beta} \tag{3.34}$$

where *Re*, *Sh*, and *Sc* stand for Reynold number, Sherwood number, and Schmidt number, respectively, *c* stands for constant, and α and β denote the exponents. Reynold number, Sherwood number, and Schmidt number are the dimensionless numbers and are represented by the following expressions:

$$Re = \frac{Lv\rho}{\mu} \tag{3.35}$$

$$Sc = \frac{\mu}{\rho D_{ij}} \tag{3.36}$$

$$Re = \frac{kL}{D_{ij}} \tag{3.37}$$

L stands for the characteristic length of the system under study (it represents the dimension of a particular geometry; the detailed description will be presented in Sections 3.12 and 3.13), v is the average velocity, ρ is the fluid density, μ is the fluid viscosity, D_{ij} is the binary diffusion coefficient, and k is the mass transfer coefficient.

The mass transfer coefficient for the species transfer through the symmetric porous membrane is characterized by the unhindered diffusion of species in the pores filled by either aqueous or organic phase described as effective diffusivity (D_{ieff}):

$$k_{im} = \frac{D_{\text{ieff}}}{l} \tag{3.38}$$

where l is the thickness of the membrane. The effective coefficient of diffusion D_{ieff} is obtained by dividing the free diffusion coefficient D_i by porosity (ε) to tortuosity (τ) ratio of the membrane:

$$k_{im} = \frac{D_i \epsilon_m}{l\tau_m} \tag{3.39}$$

where D_i is the coefficient of diffusion for species in the phase that wets/fills in the pores of membrane. For example, for hydrophobic membrane, D_i will be taken as the diffusion coefficient of species in organic phase, while for hydrophilic membrane, it will be taken as the coefficient of diffusion for species in the aqueous phase. Equation (3.39) indicates that apart from species diffusion coefficient, an upsurge in porosity to tortuosity ratio and reduction in membrane thickness increases the mass transfer coefficient in the membrane's pores. However, the theoretical value of the mass transfer coefficient varies from the actual one because it also depends on the wetting fluid and BTP. As the aqueous–organic phase interface is uncertain, the actual membrane thickness filled by the wetting fluid is unpredictable due to the uneven distribution of BTP.

3.7.1 Correlation for Flow in Shell Side

Hollow fiber geometries are usually found in tube and shell configuration, having a pack of hundreds of pore fibers. These fibers are arranged axially in a cross section. These modules have significant performance. Performance depends on various factors such as fiber diameters, the impact of the module wall, fiber movement, and outlet and inlet effects. Fibers are packed uniformly or randomly in the shell side. Mass and heat transfers are explained in several studies for hollow fiber arrays. In the case of most hollow fiber modules, uniform distribution of fibers is an impractical supposition. For comparing and determining the efficiency of the module, the mass transfers through shell side are calculated for transverse and parallel flow across fibers. The coefficient of mass transfer in the shell or tube side is calculated applying the following typical correlation:

$$Sh = f(\varphi Re^{\alpha} Sc^{\beta}) \tag{3.40}$$

where α and β are the exponents and φ is packing density of the module (Table 3.5).

3.7.2 Correlation for Flow in Tube Side

Mass transfer can be accurately determined for known hydrodynamic for liquid passing in hollow fiber lumen. Liquid flows in the laminar region ($Re < 2100$) because of the small diameter of fibers, so near the interface hydrodynamic conditions are determined. Boundary conditions and differential equation set can be determined for mass balance in fibers. External resistance to the mass transfer is insignificant for equilibrium conditions at gas–liquid interface. Lévêque and Graetz suggested a method for deriving the solution for mass transfer in the hollow fiber modules [67, 68]. Lévêque found another solution by supposing that the boundary layer concentration is only related to a thin zone adjacent to the fiber wall. This assumption is only true for higher mass velocities in short fiber for laminar flow. Based on this assumption, Lévêque solution can be used for Graetz number greater than 20. Lévêque solution is given by Eq. (3.9):

$$Sh = 1.62(Gz)^{1/3}, \quad Gz > 20 \tag{3.41}$$

Mass transfer on fiber side can be estimated from Graetz–Lévêque solution. Kreulen et al. [51] have modified the Graetz–Lévêque equation that can be used for the transition region:

$$Sh = (3.673 = 1.623Gz)^{1/3}, \quad 10 < Gz > 20 \tag{3.42}$$

Flow distribution in the tube side is often not uniform. Flow distribution depends on the manifold height, fiber inner diameter, fiber packing density, manifold type, tube length, Reynolds number, and shell diameter [50]. Wickramasinghe et al. [69] obtained a power series equation for an average coefficient in the tube side:

$$Sh' = Sh\left[1 - \left(\frac{18Sh}{Gz} + 7\right)(\varepsilon_0)^2 + \dots\right] \tag{3.43}$$

where Sh' is average Sherwood number and ε_0 is standard deviation of fiber radii.

Table 3.5 Correlations of mass transfer in shell side for hollow fiber membrane modules.

Type of flow	Flow regime and limitations	References
$Sh = (0.3045\varphi^2 - 0.3421\varphi + 0.0015)Re^{0.9}Sc^{0.33}$	$32 < Re < 1287$; $0.1 < \varphi < 0.7$, parallel flow	[57, 58]
$Sh = (0.53 - 0.58\varphi)Re^{0.53}Sc^{0.33}$	$21 < Re < 324$; $0.32 < \varphi < 0.76$, parallel flow	[59]
$Sh = 5.85(1 - \varphi)(d_h/L)Re^{0.6}Sc^{0.33}$	$0 < Re < 500$; $0.04 < \varphi < 0.4$, parallel flow	[57]
$Sh = 1.25(Red_h/L^*)^{0.93}Sc^{0.33}$	$0.5 < Re < 500$; $\varphi = 0.03$, parallel flow	[60]
$Sh = 0.90Re^{0.40}Sc^{0.33}$	$1 < Re < 25$, $\varphi = 0.07$, **cross flow**	[60]
$Sh = 1.38Re^{0.34}Sc^{0.33}$	$1 < Re < 25$, $\varphi = 0.7$, **cross flow**	[60]
$Sh = 0.61Re^{0.363}Sc^{0.333}$	$0.6 < Re < 49$, $\varphi = 0.003$, **cross flow**	[61]
$Sh = 1.615(d_h/L)^{0.33}Re^{0.33}Sc^{0.33}$	Laminar flow	[62]
$Sh = 1.5(d_h/L)^{0.33}Re^{0.33}Sc^{0.33}$	Laminar flow	[63]
$Sh = 0.023Re^{0.875}Sc^{0.25}$	Turbulent flow	[64]
$Sh = 0.24$	Transverse flow over loosely packed fibers	[60]
$Sh = 0.57Re^{0.31}Sc^{0.33}$	Transverse flow over fibers present in net wrapped around central feeder tube	[65]
$Sh = 0.46Re^{0.40}Sc^{0.33}$	Baffled cylindrical module with cross flow countercurrently	[66]

3.7.3 Correlation for Mass Transfer in Membrane Pores

Diffusion through the membrane mainly controls the absorption rate in gas absorption. There are three different transport mechanisms in the microporous membrane depending upon the type of species to be transported. These are molecular (Fickian) diffusion, Knudsen diffusion, and Poiseuille (connective) transport mechanism and are shown in Figure 3.9. Molecular diffusion occurs when the membrane pores support a stationary film of species and are sufficiently large for the frequency of collisions between species and solvent. Knudsen diffusion occurs when the membrane's pores are small enough for the frequency of collisions between solvent and pore walls. Poiseuille flow occurs due to the convective transport of water vapors. Qi and Cussler [70] discussed the absorption of SO_2 and H_2S separately in aqueous NaOH and NH_3 in water. They observed that 85% of total resistance in mass transfer for NH_3–air–water system was due to the membrane. In another study, they have found that during CO_2 absorption in aqueous amine and NaOH, the main resistance lies in the liquid phase during mass transfer. The hollow fiber membrane module is more advantageous compared with the conventional packed column. Mode of operation (wetted, non-wetted, or partially wetted) is an important factor for membrane resistance. In the case of wetted operation, membrane pores are occupied with liquid, and the equation below gives the mass transfer coefficient (k_m):

$$k_m = \frac{D_i \varepsilon_m}{\tau_m l_m} \tag{3.44}$$

In the case of non-wetted operation, the gas occupies the membranes' pores, and the mass transfer coefficient (k_m) is presented by the following equation:

$$k_m = \frac{D_{\mathrm{i,eff}} \varepsilon_m}{\tau_m l_m} \tag{3.45}$$

In a non-wetted operation, the effective diffusion coefficient ($D_{i,\mathrm{eff}}$) depends on molecular interaction with wall of pores and neighbor molecules and is given by the following relation [71]:

$$D_{\mathrm{i,eff}} = \left(\frac{1}{D_{\mathrm{i},m}} + \frac{1}{D_{\mathrm{i},Kn}} \right)^{-1} \tag{3.46}$$

Kinetic gas theory estimated self-diffusion coefficient of gas ($D_{i,m}$ m^2/s) by following Eq. (3.5) [72, 73]:

$$D_{\mathrm{i},m} = 1200 \left(\frac{RT}{MP} \right) \left(\frac{\Omega_\mu}{\Omega_D} \right) \mu \tag{3.47}$$

where μ is dynamic viscosity in (Pa·s), M is the gas molecular weight, and Ω_μ and Ω_D are integrals of collision that are a dimensionless function of temperature. These can be estimated from empirical equations of Neufeld et al. [73]. Knudsen diffusion coefficient ($D_{i,Kn}$) of gas can be obtained from Eq. (3.6) [74]:

$$D_{\mathrm{i},Kn} = 0.97\, r_p \left(\frac{T}{M} \right)^{0.5} \tag{3.48}$$

The diffusion coefficient for liquid-filled pores D_{il} is determined due to the interaction of molecules that can be determined by Van Swaaij and Versteeg equation [74]:

$$D_{il} = 2.35 * 10^{-6} \exp\left(-\frac{2119}{T}\right) \tag{3.49}$$

The above equations are founded upon the supposition that properties of membrane remain constant. Knudsen diffusion coefficient dominates the membrane coefficient with decreasing membrane pore size. The membrane coefficient is affected by pore size distribution. It is also observed that the porosity of the membrane has no substantial effect on the separation process. Gas and liquid are in contact at pores entrance; the coefficient of mass transfer is determined from the total area of the membrane because of the small distance between pores rather than fiber diameter.

The Knudsen number K_n is given as the ratio of the free path λ to nominal pore diameter d_p:

$$K_n = \frac{\lambda}{d_p} \tag{3.50}$$

Normally, for $K_n < 0.01$ and > 10, molecular diffusion and Knudsen diffusion dominate the species diffusion in the porous membrane. For $0.01 < K_n < 10$, the molecular–Knudsen diffusion transition region is defined [75].

3.8 Correlations for Heat Transfer Coefficients

Huang et al. [76] performed heat transfer and fluid flow study through the elliptical hollow fiber membrane tube bank. It has been observed that the elliptical hollow fiber membranes tube bank has good mass and heat transfer as opposed to the hollow fiber membrane tube bank. Fibers are numerous and arranged randomly in elliptical hollow fiber membrane tube bank. Transport phenomena across hollow fiber membrane tubes are greatly affected by random distributions of fibers. When the distribution of fibers becomes more concentrated, the Nusselt numbers are going to become small. Comprehensive heat transfer capability is more in regular fiber distribution than random distribution [77, 78]. Nusselt number across fiber length is important for investigating the heat transfer process in the case of hollow fiber membrane contactors. Nusselt number is calculated on the basis of different empirical correlations, and only one Nusselt value is estimated under the operating condition and given membrane [79]. Nusselt number is stated as the ratio of the convective to conductive heat transfer. It is expressed by the following equation:

$$Nu = \frac{h \cdot d}{k} = \frac{q \cdot d}{k \cdot (T_b - T_m)} \tag{3.51}$$

where d, q, h, T_b, and k are hydraulic diameter, heat flux, coefficient of heat transfer, bulk temperature, and thermal conductivity of fluids, respectively.

The Nusselt number for a given system is determined using the following expression as equivalent to the Sherwood number in mass transfer:

$$Nu = \alpha Re^{\beta} Pr^{\gamma} \tag{3.52}$$

Table 3.6 Heat transfer correlations for Nusselt number.

Characteristic constants	Flow regime
$\alpha = \dfrac{1.62}{\left(\dfrac{d_h}{L}\right)^{0.33}}$ $\beta = 0.33$ $\gamma = 0.33$	Laminar flow
$\alpha = \dfrac{1.86}{\left(\dfrac{d_h}{L}\right)^{0.33}}$ $\beta = 0.33$ $\gamma = 0.33$	Laminar flow, flat-sheet modules
$\alpha = 0.298$ $\beta = 0.646$ $\gamma = 0.316$	Laminar flow $150 < Re < 3500$
$\alpha = 0.023$ $\beta = 0.8$ $\gamma = 0.4$ (fluid is being heated) $\gamma = 0.3$ (fluid is being cooled)	Turbulent flow
$\alpha = \dfrac{0.036}{\left(\dfrac{d_h}{L}\right)^{0.055}}$ $\beta = 0.8$ $\gamma = 0.33$	Turbulent flow
$\alpha = \dfrac{0.036}{\left(\dfrac{d_h}{L}\right)^{0.055}}$ $\beta = 0.6$ $\gamma = 0.33$	Turbulent flow inside fibers

Source: Johnson and Nguyen [75].

where Pr is the dimensionless Prandtl number and α, β, and γ are the characteristics constants and are varied for different module design and fluid flow regime. A few typical constant values are given in Table 3.6. Prandtl number is expressed as the ratio of the momentum diffusivity to heat diffusivity and is given by (Table 3.6)

$$Pr = \frac{C_p \mu}{k} \tag{3.53}$$

3.9 Interfacial Transfer Area

The membrane modules are either in flat sheet or in a hollow fiber shape. For flat-sheet membranes, the interfacial areas on either side are the same, but in the form of a hollow fiber membrane, the interfacial areas on both sides of the membrane are not the same. Interestingly, the interfacial area also depends on flow configuration, e.g. aqueous- or organic-phase flows inside fibers and moreover the membrane hydrophobicity. Thus, the overall mass transfer coefficient correlation differs for aqueous-phase flows inside fibers than that for organic-phase flows for the same type of membrane, e.g. hydrophobic or hydrophilic.

3.10 Axial Pressure Drop in Membrane Contactor Module

When a fluid flows across the module length, it loses pressure. Axial pressure drop is important because it indicates BTP to decrease. Pressure drop calculations are more significant in hollow fiber modules, particularly on the module's shell side. The flow inside membrane contactor (MC) modules' fibers is usually laminar flow and can be estimated using Hagen–Poiseuille's law. However, the pressure drop in the shell side of the module is a little bit intricate phenomenon. Pressure drop correlations for laminar flow in the circular tube can be used to calculate the pressure drop in membrane contactor assuming the tube like flow configuration in flat-sheet membrane contactor or in fiber side in hollow fiber membrane contactor:

$$\Delta P = \frac{32\mu L v_t}{d_t^2} \tag{3.54}$$

3.11 Dynamic Modeling

Membrane contactor systems are usually modeled in recycle-based recirculation of feed and receiving phases. A typical experimental setup is presented in Figure 3.13, where a hollow fiber module is connected with recirculating feed and receiving phases.

Unsteady-state differential equations are developed across the storage tank's feed and receiving phases. The species enters the membrane contactor module with the exit species concentration of C_5^e at time t and leaves the module with a new concentration C_5^e at time $t + \Delta t$. The species from the module's exit reenters the storage tanks and is mixed in the storage tank, giving out the new concentration to enter the module again. The unsteady-state mass balance across the storage tanks, which act as a perfectly mixed tank, is as follows [35]:

$$C_5^e(t+\Delta t) = \frac{\Delta t}{\Gamma_f} C_5^o(t) + C_5^e(t)\left[1 - \frac{\Delta t}{\Gamma_f}\right] \tag{3.55}$$

The equation measures the species concentration in the feed tank for a certain time period. Likewise, the unsteady-state equation presenting the species concentration in the receiving-phase storage tank can be obtained by taking the unsteady-state mass balance across the receiving-phase storage tank:

$$C_1^e(t+\Delta t) = \frac{\Delta t}{\Gamma_r} C_1^o(t) + C_1^e(t)\left[1 - \frac{\Delta t}{\Gamma_r}\right] \tag{3.56}$$

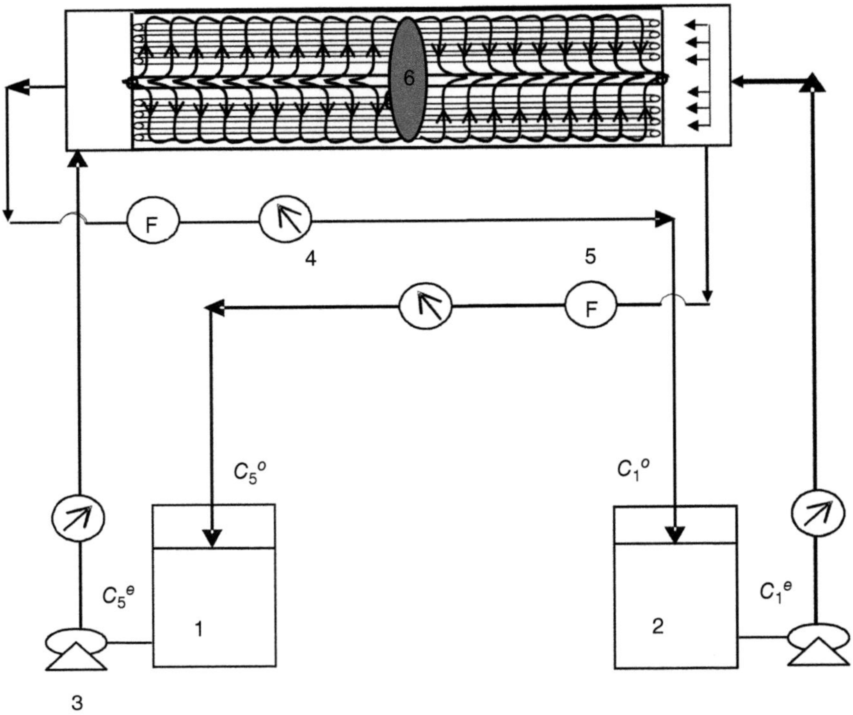

Figure 3.13 Experimental setup of recycled based continuous liquid–liquid extraction system. Source: Younas et al. [35].

The extraction efficiency of the hollow fiber module for the species is given as the ratio of the species amount transported to the initial species concentration entering the module at any time and is expressed by the following mathematical expression:

$$\emptyset = \left[1 - \frac{C_5^o(t)}{C_5^e(t)}\right] \times 100 \tag{3.57}$$

Extraction efficiency can also be expressed in terms of partition coefficient by the following relation [36]:

$$\emptyset = \frac{m}{1+m} \times 100 \tag{3.58}$$

3.12 Transfer Units and Module Design Length

The height of transfer unit (HTU) is another key parameter for evaluating the membrane contactors' performance to obtain the desired species extraction. It is given by the following mathematical equation [80, 81]:

$$\text{HTU} = \frac{Q_f}{K_f a A} \tag{3.59}$$

where a stands for the interfacial area per unit volume and A represents the module surface area. On the contrary, the number of transfer units (NTUs) measure the separation stages and are defined as a change in species concentration for the unit driving force responsible for the species transfer. Mathematically it can be expressed for inlet and exit species concentrations across the membrane contactor module by Eq. (3.60):

$$\mathrm{NTU} = \int_{C_e}^{C^o} \frac{dC}{C - C^e} \tag{3.60}$$

Thus, the theoretical length of the module is calculated by multiplying HTU and NTU. It produces the required module length for the desired species extraction under the specified operating conditions:

$$Z = \mathrm{HTU} \times \mathrm{NTU} \tag{3.61}$$

3.13 Numerical Modeling of Mass Transport in Membrane Contactor Modules

Computational fluid dynamic (CFD) tools offer a thorough examination of the fields of flow and the species diffusion on macroscale that occurs inside the membrane packs [82, 83]. The numerical (CFD) modeling of membrane contactor devices is based on differential equations. CFD investigation results in a proper design and mass transfer characterization within the system. To bring down the cost and time for computation, simple assumptions are made, and the modeling is performed based on a single fiber or a module section. Several studies have used the CFD tools to examine the consequences of operational parameters on the performance of various applications applying diverse hollow fiber module (HFM) geometries. Usually, the finite element approach and finite volume approach is used in CFD simulations [27, 84]. The laminar flow model characterizes the flow of fluid within the contactor fibers. Happel's free surface model is assumed in the shell side of the contactor and is developed on the assumption that fibers inside the hollow fiber module are distributed uniformly in the shell. It is also presumed that each fiber is encircled by a specified area. It was supposed that the shell side fluid moves parallel to the fibers, and every fiber is encircled by a specific area of the fluid on the shell side. This encircled area is known as "flow cell" (Figure 3.14). Thus, the number of flow cells equates the number of fibers. A further assumption is made that the fibers are organized in the ring, originating from the single fiber in the first ring of the module center. The number of fibers is increased in the succeeding rings till the edge of the shell. This configuration is termed as "flow cell." Thus each "flow cell" itself represents the fiber-in-shell configuration where the sum of the surface area of total flow cells is equal to the inside the shell surface area. Thus, fibers are uniformly spread in the space of the shell, which makes the results of the symmetry of a single fiber to be comprehended to the whole module. As shown in Figure 3.8b, the radial position of $r = 0$ is the center of a fiber. r_i, r_o, and r_e denotes the radial distances for inner radius, outer radius, and flow cell (Happel's free area) of the fiber, respectively [27, 28, 56, 85]. Similarly, the axial positions are denoted by z.

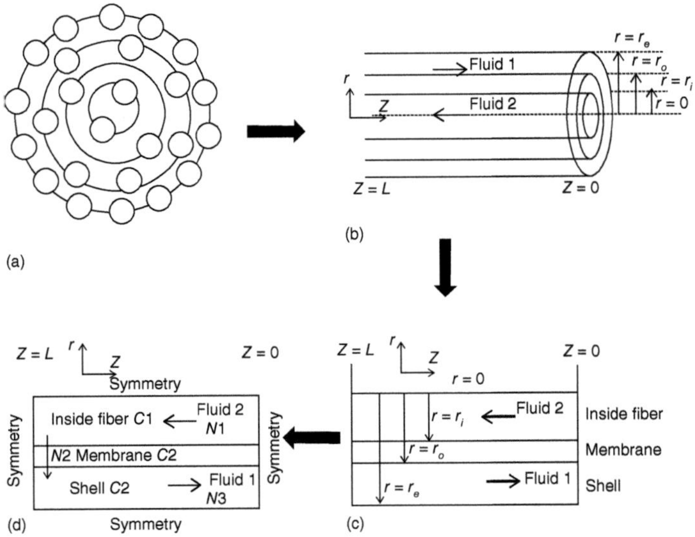

Figure 3.14 Membrane contactor module hypothetical cell formation: (a) fiber rings, (b) single hypothetical flow cell, (c) geometry of hypothetical flow cell, and (d) symmetry of flow cell. Source: Adapted from Muhammada et al. [40].

$z = 0$ represents the fiber inlet position and $z = L$ indicates the outlet position of a fiber. The diameter of the flow cell (Happel's free surface) can be calculated from the following relation:

$$d_e = \frac{d_o}{\sqrt{p_f}} \tag{3.62}$$

where d_o and p_f are the external fiber diameter and packing fraction of the module, respectively. p_f is the packing fraction of the module of diameter d_s containing n fibers and is measured by the following equation:

$$p_f = \frac{n(d_o)^2}{(d_s)^2} \tag{3.63}$$

The following assumptions are taken to illustrate the mass transport of species in the shell of the module, inside fibers, and membrane pores [86] (Figure 3.14):

I. Mass transport occurs under steady-state conditions.
II. The system is at constant temperature (isothermal conditions).
III. The physical and transport properties of the fluid are constant.
IV. Velocity profile in shell side is measured by considering only the liquid surrounding the fiber, e.g. the concentric portion of the Happel's free surface (flow cell).
V. The velocity profile inside fiber is supposed to be under laminar flow conditions, and a parabolic profile that is fully developed is assumed.
VI. Axial diffusion inside fibers is neglected.
VII. Fiber-side pressure is constant.
VIII. Only diffusion occurs in membrane thickness.

3.13.1 Mass Transfer in Shell Side

The shell-side mass transfer is characterized by taking the partial differential equation of the cylindrical coordinates mass balance via Fick's law of diffusion and is explained as [86, 87]

$$u_{z,1}\frac{\partial C_{i,1}}{\partial z} = D_{i,1}\left[\frac{\partial^2 C_{i,1}}{\partial r^2} + \frac{1}{r}\frac{\partial C_{i,1}}{\partial r} + \frac{\partial^2 C_{i,1}}{\partial z}\right] \tag{3.64}$$

$u_{z,1}$ is the fluid 1 velocity in the shell side where velocity distribution is determined through Navier–Stokes equations:

$$\rho u_{z,1} \cdot \nabla u_{z,1} - \nabla \cdot \mu(\nabla u_{z,1} + (\nabla u_{z,1})^T) + \nabla p = 0 \tag{3.65}$$

$$\nabla \cdot u_{z,1} = 0 \tag{3.66}$$

where ρ denotes density, μ is the dynamic viscosity, and p is the pressure. The density and viscosity of the fluid are considered to be constant in the Navier–Stokes equations.

It can be further stated as a function of dimensionless geometric parameters:

$$u_{z,1} = u_{\max,1}f(r) = 2u_{m,1}f(r) \tag{3.67}$$

$$f(r) = \left[1 - \left(\frac{r_o}{r_e}\right)^2\right]\left[\frac{\left(\frac{r}{r_e}\right)^2 - \left(\frac{r_o}{r_e}\right)^2 + 2\ln\left(\frac{r_o}{r}\right)}{3 + \left(\frac{r_o}{r_e}\right)^4 - 4\left(\frac{r_o}{r_e}\right)^2 + 4\ln\left(\frac{r_o}{r_e}\right)}\right] \quad \text{and } d = 2r \tag{3.68}$$

The equations for mass and momentum balance are then coupled to characterize the flow on shell side. The boundary conditions for Fickian mass transfer in the shell side flow are the following:

$$\text{At } r = r_e, \quad \frac{\partial C_{i,1}}{\partial r} = 0 \text{ (symmetry condition)} \tag{3.69}$$

$$\text{At } r = r_o, \quad D_{i,1}\frac{\partial C_{i,1}}{\partial r} = k_m s(C_{i,1} - C_i^*) \tag{3.70}$$

$$C_{i,1} = \frac{C_{i,\text{mem}}}{m} \tag{3.71}$$

(if the membrane is wetted by fluid 2 and interface is established at shell–membrane side)

$$\text{At } z = L, \quad C_{i,1} = C_{i,\text{in}} \tag{3.72}$$

$$\text{At } z = 0, \quad D_{i,1}\nabla C_{i,1} = 0 \text{ (convective flux)} \tag{3.73}$$

For momentum balance, boundary conditions for the Navier–Stokes equation are

$$\text{At } r = r_e, \quad u_{z,1} = 0 \text{ (no slip condition)} \tag{3.74}$$

$$\text{At } r = r_o, \quad u_{z,1} = 0 \text{ (no slip condition)} \tag{3.75}$$

$$\text{At } z = L, \quad u_{z,1} = u_0 \text{ (inlet boundary)} \tag{3.76}$$

$$\text{At } z = L, \quad p = p_{\text{atm}} \text{ (outlet boundary)} \tag{3.77}$$

where $D_{i,1}$ denotes the species i diffusion coefficient in the fluid 1 phase, S stands for the outer radius-based geometric factor, and C^*_{i1} is the species i concentration in the fluid 1 in equilibrium with the reception phase at the membrane–wetting phase interface. k_m is the membrane mass transfer coefficient, indicating the linear profile of species concentration through the membrane thickness.

If the dimensionless variables are introduced for radius, length, and concentration [86, 88]

$$\overline{r} = \frac{r}{r_o}, \overline{z} = \frac{z}{L}, \overline{C_i} = \frac{C_{i,f}}{C_{i,\text{in}}} \tag{3.78}$$

then Eq. (3.78) becomes

$$u_{z,g}\frac{\partial \overline{C_i}}{\partial \overline{z}} = D_{i,f}\left[\frac{1}{\overline{r}}\frac{\partial}{\partial \overline{r}}\left(\overline{r}\frac{\partial \overline{C_i}}{\partial r}\right)\right] \tag{3.79}$$

$$\frac{2u_{m,g}f(\overline{r})}{D_{i,f}}\frac{r_o^{\,2}}{L}\frac{\partial \overline{C_i}}{\partial \overline{z}} = \frac{1}{\overline{r}}\frac{\partial}{\partial \overline{r}}\left(\overline{r}\frac{\partial \overline{C_i}}{\partial \overline{r}}\right) \tag{3.80}$$

$$\frac{u_{m,g}d_o^{\,2}}{2D_{i,f}L}f(\overline{r})\frac{\partial \overline{C_i}}{\partial \overline{z}} = \frac{1}{\overline{r}}\frac{\partial}{\partial \overline{r}}\left(\overline{r}\frac{\partial \overline{C_i}}{\partial \overline{r}}\right) \tag{3.81}$$

$$\frac{Gz_s}{2}f(\overline{r})\frac{\partial \overline{C_i}}{\partial \overline{z}} = \frac{1}{\overline{r}}\frac{\partial}{\partial \overline{r}}\left(\overline{r}\frac{\partial \overline{C_i}}{\partial \overline{r}}\right) \tag{3.82}$$

where $Gz_s = \frac{u_{m,g}d_o^{\,2}}{D_{i,f}L}$ is Graetz number in shell side of the module:

$$f(\overline{r}) = \left[1 - \left(\frac{r_o}{r_e}\right)^2\right]\left[\frac{\left(\frac{\overline{r}r_o}{r_e}\right)^2 - \left(\frac{r_o}{r_e}\right)^2 + 2\ln\left(\frac{1}{\overline{r}}\right)}{3 + \left(\frac{r_o}{r_e}\right)^4 - 4\left(\frac{r_o}{r_e}\right)^2 + 4\ln\left(\frac{r_o}{r_e}\right)}\right] \tag{3.83}$$

The boundary conditions now become as follows:

$$\text{At } \overline{r} = \frac{r_e}{r_o}, \quad \frac{\partial \overline{C_i}}{\partial \overline{r}} = 0 \tag{3.84}$$

$$\text{At } \overline{r} = 1, \quad \frac{\partial \overline{C_i}}{\partial \overline{r}} = \frac{k_m s r_o}{D_{i,f}}(\overline{C_i} - \overline{C_i^*}) = Sh_m(\overline{C_i} - \overline{C_i^{\text{rec}}}) \tag{3.85}$$

Sh_m is the Sherwood number:

$$\text{At } \overline{z} = 0, \quad \overline{C_i} = 1 \tag{3.86}$$

3.13.2 Mass Transfer Inside Fibers

The mass continuity equation for species transport inside fiber is illustrated by the following relation [86, 87]:

$$u_{z,2}\frac{\partial C_{i,2}}{\partial z} = D_{i,2}\left[\frac{\partial^2 C_{i,2}}{\partial r^2} + \frac{1}{r}\frac{\partial C_{i,2}}{\partial r} + \frac{\partial^2 C_{i,2}}{\partial z}\right] \tag{3.87}$$

where $D_{i,2}$ is the species i diffusion coefficient in the fluid 2 and $C_{i,2}$ is species concentration in fluid 2. $u_{z,2}$ represents the axial flow and is supposed to be Newtonian and fully developed. The velocity profile in this subdomain is expressed by the following equation:

$$u_{z,2} = u_{\max,2}\left[1-\left(\frac{r}{r_i}\right)^2\right] = 2u_{m,2}\left[1-\left(\frac{r}{r_i}\right)^2\right] \tag{3.88}$$

$u_{\max,2}$ and $u_{m,2}$ denote the maximum and average velocity of fluid 2 flowing inside fiber.

Equation (3.87) can be rewritten as follows by inserting Eq. (3.88):

$$2u_{m,2}\left[1-\left(\frac{r}{r_i}\right)^2\right]\frac{\partial C_{i,2}}{\partial z} = D_{i,2}\left[\frac{\partial^2 C_{i,2}}{\partial r^2} + \frac{1}{r}\frac{\partial C_{i,2}}{\partial r} + \frac{\partial^2 C_{i,2}}{\partial z}\right] \tag{3.89}$$

The boundary conditions for species transport inside fibers are defined as the following:

$$\text{At } r = 0, \quad \frac{\partial C_{i,2}}{\partial r} = 0 \text{ (axial symmetry)} \tag{3.90}$$

$$\text{At } r = r_i, \quad C_{i,2} = C_{i,\text{mem}} \text{ if fluid 2 wets the membrane} \tag{3.91}$$

$$\text{At } r = r_i, \quad C_{i,2} = \frac{C_{i,\text{mem}}}{m} \tag{3.92}$$

if fluid 2 does not wet the membrane and an interface is established between the fluid 2 and membrane.

$$\text{At } z = 0, \quad C_{i,2} = C_{i,2,\text{ini}} \tag{3.93}$$

$$\text{At } z = L, \quad D_{i,2}\nabla C_{i,2} = 0 \text{ (convective flux)} \tag{3.94}$$

The bulk average or mixing cup concentration of species i in phase 2 at $z = L$ is given by the following expression:

$$C_{i,z=0} = \frac{\int_0^{r_{\text{in}}} C_{i,2}u_{z,2}2\pi r\,dr}{\int_0^{r_{\text{in}}} u_{z,2}2\pi r\,dr} = \frac{4}{r_{\text{in}}^2}\int_0^{r_{\text{in}}} C_{i,2}\left[1-\left(\frac{r}{r_{\text{in}}}\right)^2\right] r\,dr \tag{3.95}$$

If the following dimensionless variables are demarcated,

$$\bar{r} = \frac{r}{r}, \bar{z} = \frac{z}{L}, \overline{C_i} = \frac{C_i}{C_{i,\text{sat}}} \tag{3.96}$$

Then the following Eq. (3.97) can be rewritten in the dimensionless form as follows:

$$\frac{Gz_{\text{in}}}{2}\left[1-\bar{r}^2\right]\frac{\partial \bar{C}_{i,2}}{\partial \bar{z}} = \frac{1}{\bar{r}}\frac{\partial}{\partial \bar{r}}\left(\bar{r}\frac{\partial \bar{C}_{i,2}}{\partial \bar{r}}\right) \tag{3.97}$$

where $Gz_{\text{in}} = \frac{u_{m,l}d_{\text{in}}^2}{D_{i,2}L}$ is Graetz number inside fiber.

The boundary conditions with dimensionless variables are as follows:

$$\text{At } \bar{r} = 0, \quad \frac{\partial \bar{C}_{i,2}}{\partial \bar{r}} = 0 \tag{3.98}$$

$$\text{At } \bar{r} = 1, \quad \frac{\partial \bar{C}_{i,2}}{\partial \bar{r}} = \frac{\partial \bar{C}_i}{\partial \bar{r}}\frac{D_{i,f}}{D_{i,2}}m \tag{3.99}$$

$$\text{At } \bar{z} = 1, \quad \bar{C}_{i,2} = 0 \tag{3.100}$$

3.13.3 Mass Transfer in Membrane Pores

Inside the membrane pores, axial and radial diffusional mass transfers are presumed. The following mass continuity equation is applied to define the mass transfer of species i through the membrane [86, 87]:

$$D_{i,\text{mem}}\left[\frac{\partial^2 C_{i,\text{mem}}}{\partial r^2} + \frac{1}{r}\frac{\partial C_{i,\text{mem}}}{\partial r} + \frac{\partial^2 C_{i,\text{mem}}}{\partial z^2}\right] = 0 \tag{3.101}$$

Within the membrane pores, the diffusion coefficient $D_{i,\text{ mem}}$ is stated as an effective diffusion to account for the membrane's porosity (ε) and tortuosity (τ):

$$D_{i,\text{mem}} = \frac{D_{i,*}\varepsilon}{\tau} \tag{3.102}$$

where $D_{i,*}$ is the species diffusion coefficient in fluid that wets the pores of the membrane. It may be either that of fluid 1 or fluid 2.

It should be noted that the convective term in Eq. (3.87) is vanished in Eq. (3.101) since no convectional mass transfer happens through the membrane pores.

The boundary conditions are defined as follows.

$$\text{At } r = r_i, \quad C_{i,\text{mem}} = m_i C_{i,2} \tag{3.103}$$

if fluid 2 wets the membrane, and

$$C_{i,\text{mem}} = C_{i,2} \tag{3.104}$$

if fluid does not wet the membrane.

$$\text{At } r = r_0, \quad C_{i,\text{mem}} = m_i C_{i,1} \tag{3.105}$$

if fluid 1 wets the membrane, and

$$C_{i,\text{mem}} = C_{i,1} \tag{3.106}$$

if fluid does not wet the membrane.

3.13.4 Numerical Modeling Term in the Case of Membrane Wetting

Membrane wetting occurs when both phases of fluids 1 and 2 (feed and receiving) wet the portion of the membrane. In such case, the mass balance expression for the fluid 1-filled section is described considering species diffusion in the fluid 1 phase [86]:

$$D_{i,1-\text{mem}}\left[\frac{\partial^2 C_{i,1-\text{mem}}}{\partial r^2}+\frac{1}{r}\frac{\partial C_{i,1-\text{mem}}}{\partial r}+\frac{\partial^2 C_{i,1-\text{mem}}}{\partial z^2}\right]=0 \tag{3.107}$$

The boundary conditions for fluid 1-filled membrane section are as follows:

$$\text{At } r=r_w,\quad C_{i,1-\text{mem}}=\frac{C_{i,1-\text{mem}}}{m_i} \tag{3.108}$$

where r_w is the radial distance up to which fluid 1 wets the portion of membrane.

$$\text{At } r=r_e,\quad C_{i,f-\text{mem}}=C_{i,2} \tag{3.109}$$

For the fluid 2 phase filled section of the membrane, the following mass balance expression is considered:

$$D_{i,2-\text{mem}}\left[\frac{\partial^2 C_{i,2-\text{mem}}}{\partial r^2}+\frac{1}{r}\frac{\partial C_{i,2-\text{mem}}}{\partial r}+\frac{\partial^2 C_{i,2-\text{mem}}}{\partial z^2}\right]=0 \tag{3.110}$$

The boundary conditions for fluid 2-filled membrane section are as follows:

$$\text{At } r=r_i,\quad C_{i,2-\text{mem}}=C_{i,2} \tag{3.111}$$

$$\text{At } r=r_w,\quad C_{i,2-\text{mem}}=\frac{C_{i,1}}{m_i}\ \text{for absorbed species} \tag{3.112}$$

$$\frac{\partial C_{i,2-\text{mem}}}{\partial r}=0\ \text{for the non-absorbed species} \tag{3.113}$$

Figure 3.7 illustrates the use of hollow fiber membrane contactor (HFMC) for post-combustion CO_2 capture with ionic liquids and its subsequent CFD model approach. The membrane wetting was also shown in Figure 3.7b. CO_2/N_2 mixture flows inside fiber, whereas the ionic liquid solution is recirculated counter currently in the module's shell side. For gas–liquid interface, CO_2 solubility in liquid in the current case is typically described by Henry's law constant. Henry's law constant magnitude is an indication of physical or chemical absorption. A lower value, typically less than 3 MPa, indicates high solubility of gas and chemical absorption of CO_2 in ionic liquids (ILs). Henry's law constants are determined from experimental PTx data (vapor–liquid equilibria [VLE]) [89]:

$$H(T,P)=\lim_{x\to 0}\frac{f(T,P,x_i)}{x_i}=\lim_{x\to 0}\frac{\emptyset(T_{\text{eq}},P_{\text{eq}})P_{\text{eq}}}{x_i} \tag{3.114}$$

where H (Pa) is Henry's law constant, f is the gas species (CO_2) fugacity, x_i is the mole fraction of gaseous species (CO_2) dissolved in liquid, $\emptyset$ is the fugacity coefficient, T_{eq} is the equilibrium temperature, and P_{eq} is the equilibrium pressure. The dimensionless Henry's law constant is then determined according to the following relation:

$$\acute{H} = \frac{\rho_l R_g T}{M_l H} \tag{3.115}$$

where ρ_l is the density of the liquid, R_g is the gas constant, M_l is the molar weight of liquid, and T is the temperature. Henry's constant values for different solvents are listed in Table 3.7.

The overall mass transfer coefficient K is calculated by

$$J = K\Delta C_{g-lm} = \frac{Q_{g-in}C_{i-in} - Q_{g-out}C_{i-out}}{A} \tag{3.116}$$

where $Q_{g-in}C_{i-in}$ and Q_{g-out} are the gas side inlet and outlet flow rates, respectively, and C_{i-in} and C_{i-out} are the gas side inlet and outlet concentrations of species i, respectively. A denotes the active membrane area while C_{g-lm} represents the logarithmic mean of gas-phase concentration and is calculated using Eq. (3.117):

$$\Delta C_{g-lm} = \frac{(C_{i-in} - C^*_{i-in}) - (C_{i-out} - C^*_{i-out})}{\ln\left(\frac{C_{i-in} - C^*_{i-in}}{C_{i-out} - C^*_{i-out}}\right)} \tag{3.117}$$

where C^*_{i-in} and C^*_{i-out} are the concentrations of gas phase at equilibrium with corresponding concentration of gaseous species in the liquid phase. Henry's law ($C^*_i = HC^*_l$) is applied to measure the equilibrium concentration of gas phase.

C_{i-out} is determined by

$$C_{i-out} = \frac{\iint_{r=0}^{r=r_1} C_{i-2}(r) r\, dr\, d\theta}{\iint_{r=0}^{r=r_1} r\, dr\, d\theta} \tag{3.118}$$

where $Q_{g-in}C_{i-in}$ and Q_{g-out} are the gas side inlet and outlet flow rates, respectively (Figure 3.15).

Table 3.7 Henry's constant of test compounds.

	Henry's constant	
Compound	**(atm m^3 mol^{-1})**	**(Dimensionless)**
Chloroform	3.67×10^{-3}	1.50×10^{-1}
Butyl acetate	2.81×10^{-4}	1.15×10^{-2}
Butyl acetate	1.34×10^{-4}	5.48×10^{-3}
1-Butanol	8.81×10^{-6}	3.60×10^{-4}
1-Propanol	7.41×10^{-6}	3.03×10^{-4}
Phenol	3.33×10^{-7}	1.36×10^{-7}

Source: Howard and Meylan [90].

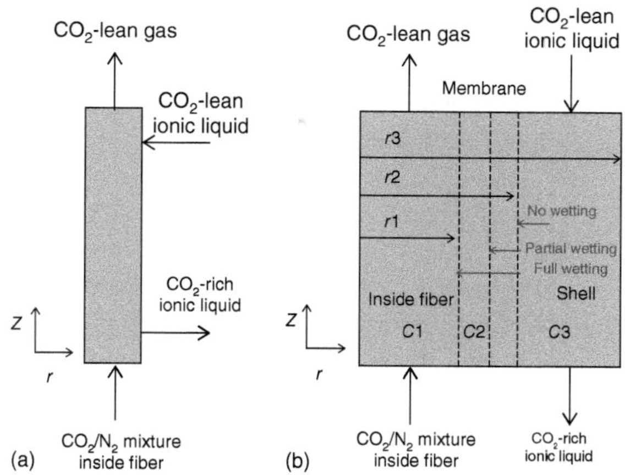

Figure 3.15 (a) Experimental set up gas–liquid separation process in hollow fiber membrane contactor and (b) CFD model approach in hollow fiber membrane contactor for gas–liquid separation. Source: Adapted from Qazi et al. [89].

3.14 Numerical Modeling of Heat Transport in Membrane Contactor Modules

MD uses membrane contactors for water desalination using temperature difference as a driving force. An inclusive two-dimensional mathematical model is applied to explore the transport phenomena of water in MD, which are mass, heat, and momentum, composing of conductive and convection terms. Three areas, i.e. hot channel, membrane, and cold channel, are used to explain the mathematical model. The saline water passes as feed in the hot channel, while the pure water comes in the cold channel. Hydrophobic membrane is used so as to prevent the water from wetting the membrane pores. Hot water evaporates and water vapors flow from hot to cold channel. Vapor pressure difference establishes at both of the membrane sides. The water vapors condense in the cold-water channel after passing through the membrane pores. The geometry of the simulation model is shown in Figure 3.16 [91].

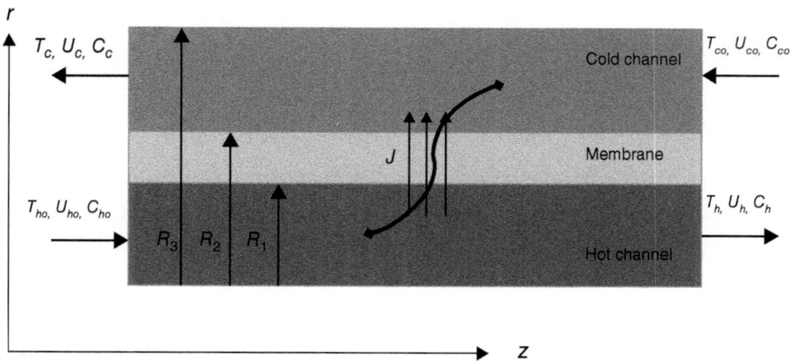

Figure 3.16 Model's geometry used in CFD simulation for heat transfer in MD. Source: Modified from Esfandiari et al. [91].

For simplifying the governing equations, some assumptions are made in the development of model:

I. Steady-state operation conditions.
II. Laminar flow streams.
III. Heat exchange occurring between the system and the environment is ignored.
IV. Water is completely impermeable in the membrane.
V. Concentrated agglomeration of salt molecules in the solution is neglected.
VI. The diffusion coefficient of water vapors is measured by taking the average of surface temperatures of hot and cold membrane surfaces.
VII. The convection term along the width of channel is ignored because the length of channels entrance is short.
VIII. Viscous flow and the mechanisms of surface diffusion in the dusty-gas model (DGM) are ignored in calculating the water vapor flux in membrane.
IX. The geometrical parameters of membrane like thickness, pore size, and porosity are assumed as constant.

3.14.1 Governing Equation in Cold and Hot Channels

The energy conservation equation is expressed by:

$$\frac{\partial^2(kT)}{\partial z^2} + \frac{\partial^2(kT)}{\partial r^2} - \frac{\partial^2(\rho C_p UT)}{\partial z} = 0 \tag{3.119}$$

where ρ is density, C_p is the heat capacity, U is the vector of flow velocity in the z-direction, k is the coefficient of thermal conductivity, and T is the temperature. Navier–Stokes equation is used to predict the velocity distribution and is given by

$$\frac{\partial(\rho V)}{\partial t} - \nabla \cdot \left(\mu(\nabla V + (\nabla V)^T\right) + V \cdot \nabla(\rho V) + \nabla p = F \tag{3.120}$$

where μ is the dynamic viscosity, V is the velocity vector in the r-direction, p is the pressure, and F is body force. The equation of continuity is given by

$$\frac{\partial(\rho U)}{\partial z} + \frac{\partial(\rho V)}{\partial r} = 0 \tag{3.121}$$

The values of density and viscosity for both fluids may take as variables for more accurate calculations or are simply supposed to be constant for the comfort of calculations.

3.14.2 Governing Equation Inside Membrane

The membrane is supposed to be a solid porous structure. Here the diffusion is the only instrument to originate heat transfer. So, the general equation for energy conservation is presented as

$$\nabla \cdot (k_m \nabla T_m) = \left(\frac{\partial^2(k_m T_m)}{\partial z^2} + \frac{\partial^2(k_m T_m)}{\partial r^2}\right) = 0 \tag{3.122}$$

where k_m is the membrane thermal conductivity and T_m is the membrane surface temperature.

The governing equations for mass transfer have already been described in Section 3.4.

Table 3.8 Boundary conditions for governing equations.

Equation/field	Hot channel	Membrane	Cold channel
Energy	At $z = 0$, $T_h = T_{ho}$ At $z = L$, $\frac{\partial T_h}{\partial z} = 0$ At $r = 0$, $\frac{\partial T_h}{\partial r} = 0$ (thermal insulation) At $r = R_1$, $-k_h \frac{\partial T_h}{\partial r} = -k_m \frac{\partial T_m}{\partial r} - JH_v$	At $z = 0$, $\frac{\partial T_m}{\partial z} = 0$ (thermal insulation) At $z = L$, $\frac{\partial T_m}{\partial z} = 0$ (thermal insulation) At $r = R_1$, $T_m = T_h$ At $r = R_2$, $-k_m \frac{\partial T_h}{\partial r} = -k_c \frac{\partial T_m}{\partial r} + JH_v$	At $z = 0$, $\frac{\partial T_c}{\partial z} = 0$ (developed flow) At $z = L$, $T_c = T_{co}$ At $r = R_2$, $T_c = T_m$ At $r = R_3$, $\frac{\partial T_c}{\partial r} = 0$ (thermal insulation)
Navier–Stokes	At $z = 0$, $U = U_o$ At $x = L$, $p = p_{\text{atm}}$ At $r = 0$, *no slip condition* At $r = R_1$, *no slip condition*	–	At $z = 0$, $p = p_{\text{atm}}$ At $x = L$, $u_x = u_o$ At $r = R_2$, *no slip condition* At $z = L$, $U = 0$

Source: Modified from Esfandiari et al. [91].

Membrane thermal conductivity k_m is determined by

$$k_m = \varepsilon k_{wp} + (1 - \varepsilon)k_{ws} \tag{3.123}$$

where the subscripts p and s denote the pore and membrane matrix, respectively, and ε and k represent the membrane porosity and the water vapor thermal conductivity, respectively.

To model the diffusion of water vapors inside the porous membrane in MD (D_{wm}) as required for mass transport, the Knudsen diffusion (D_{wk}) and Poiseuille diffusion (D_{wp}) must be combined:

$$\frac{1}{D_{wm}} = \frac{1}{D_{wp}} + \frac{1}{D_{wk}} \tag{3.124}$$

$$D_{wp} = \frac{p \times d_p}{16 \times \mu} \tag{3.125}$$

$$D_{wk} = 9.7 \times 10^3 \frac{d_p}{2} \sqrt{\frac{T}{M}} \tag{3.126}$$

$$D_{\text{eff}} = D_{wm} \left(\frac{\varepsilon}{\tau}\right) \tag{3.127}$$

where μ is the dynamic viscosity, d_p is pore diameter, and T is the temperature. D_{eff} is the effective diffusion of water transport in membrane pores. The boundary conditions are listed in Table 3.8 [91].

3.15 Model Solution Algorithm

The governing equations are solved through a fully implicit technique and then integrated through the control volume method. The solution accuracy is enhanced and calculation cost is declined using structured mesh. The upwind and central designs were applied to isolate the convective terms, according to the Peclet number. The central method is applied if the absolute power of Peclet number is in between 0 and 2; otherwise, upwind is used. For evaluating discrete diffusion term, the former design was applied. The coefficient matrices made for transfer equations of mass, energy, and momentum, respectively, are 6, 6, and 5 diagonals. Gaussian Elimination was used to solve the matrixes.

3.16 Conclusions and Perspectives

The understanding of theoretical concepts of membrane contactors is inevitable for the operation of liquid–liquid, gas–liquid, and gas–gas separations as discussed in Chapter 2. Furthermore, innovative and efficient design of membrane contactor module is getting importance in twenty-first century where advanced mathematical techniques and high-performance simulation software made was user-friendly, economical, and efficient. Micro- to macrolevel theoretical models and the understanding for the operational studies of membrane contactors have been presented in detail. It covers the analytical, algebraic, and numerical models of the studied system using membrane contactors. The mathematical concepts studied in this chapter are equally applied to flat-sheet and hollow fiber membrane modules. The algebraic, analytical, and numerical models have been derived from ab initio in different scenarios of process applications and fluid hydrodynamics. The guidelines could be obtained to model the membrane contactor module for a specific operation in liquid–liquid, gas–liquid, and gas–gas applications. Model algorithms that are best suited in MATLAB™ and Comsol Multiphysic™ have also been described. Though a good level of efforts has been given to present the theoretical aspects of membrane contactor modeling and simulation, there still remains the advanced theoretical studies. It includes the experimental design, artificial neural network, and fuzzy logic. It is pertinent that the use of artificial intelligence in theoretical modeling will further improve the theory, design, and operation of membrane contactors in different applications.

3.A Membrane Transport Theory: Operational Principle

3.A.1 Steady-State Resistance-in-Series Model Across Liquid–Liquid Contactor

3.A.1.1 Hydrophobic Membrane Based on Aqueous-Phase Side (Species Transfers from Aqueous Phase to Organic Phase)

The Fick's law equation describing the species flux with overall mass transfer coefficient and concentration gradient is

$$j_i = K_a A_a (C^b_{ia} - C^{ie}_{ia}) \tag{3.A.1}$$

$$\frac{1}{A_a K_a} = \frac{(C_{ia}^b - C_{ia}^{ie})}{j_i} \tag{3.A.2}$$

$$\frac{1}{A_a K_a} = \frac{\left(C_{ia}^b - \frac{C_{io}^b}{m_i}\right)}{j_i} \tag{3.A.3}$$

$$\frac{1}{A_a K_a} = \frac{\left(C_{ia}^b - \frac{C_{io}^b}{m_i} + C_{ia}^{mi} - C_{ia}^{mi} + \frac{C_{io}^m}{m_i} - \frac{C_{io}^m}{m_i}\right)}{j_i} \tag{3.A.4}$$

$$\frac{1}{A_a K_a} = \frac{(C_{ia}^b - C_{ia}^{mi}) + \left(C_{ia}^{mi} - \frac{C_{io}^m}{m_i}\right) + \left(\frac{C_{io}^m}{m_i} - \frac{C_{io}^b}{m_i}\right)}{j_i} \tag{3.A.5}$$

$$\frac{1}{A_a K_a} = \frac{(C_{ia}^b - C_{ia}^{mi}) + \left(\frac{C_{io}^{mi}}{m_i} - \frac{C_{io}^m}{m_i}\right) + \left(\frac{C_{io}^m}{m_i} - \frac{C_{io}^b}{m_i}\right)}{j_i} \tag{3.A.6}$$

$$\frac{1}{A_a K_a} = \frac{1}{k_{ia}} + \frac{1}{m_i k_{im}} + \frac{1}{m_i k_{io}} \tag{3.A.7}$$

3.A.1.2 Hydrophobic Membrane Based on Organic-Phase Side (Species Transfers from Aqueous Phase to Organic Phase)

$$\frac{1}{A_o K_a} = \frac{1}{k_{ia}} + \frac{1}{m_i k_{im}} + \frac{1}{m_i k_{io}} \tag{3.A.8}$$

3.A.1.3 Hydrophobic Membrane Based on Organic-Phase Side (Species Transfers from Organic Phase to Aqueous Phase)

$$j_i = K_o A_o (C_{io}^{ie} - C_{io}^b) \tag{3.A.9}$$

$$\frac{1}{A_o K_o} = \frac{(C_{io}^e - C_{io}^b)}{j_i} \tag{3.A.10}$$

$$\frac{1}{A_o K_o} = \frac{(m_i C_{ia}^b - C_{io}^b)}{j_i} \tag{3.A.11}$$

$$\frac{1}{A_o K_o} = \frac{(m_i C_{ia}^b - C_{io}^b + C_{io}^{mi} - C_{io}^{mi} + C_{io}^m - C_{io}^m)}{j_i} \tag{3.A.12}$$

$$\frac{1}{A_o K_o} = \frac{(m_i C_{ia}^b - C_{io}^{mi}) + (C_{io}^{mi} - C_{io}^m) + (C_{io}^m - C_{io}^b)}{j_i} \tag{3.A.13}$$

$$\frac{1}{A_o K_o} = \frac{(m_i C_{ia}^b - m_i C_{ia}^{mi}) + (C_{io}^{mi} - C_{io}^m) + (C_{io}^m - C_{io}^b)}{j_i} \tag{3.A.14}$$

$$\frac{1}{A_o K_o} = \frac{m_i}{k_{ia}} + \frac{1}{k_{im}} + \frac{1}{k_{io}} \tag{3.A.15}$$

3.A.1.4 Hydrophobic Membrane Based on Aqueous-Phase Side (Species Transfers from Organic Phase to Aqueous Phase)

$$\frac{1}{A_a K_a} = \frac{m_i}{k_{ia}} + \frac{1}{k_{im}} + \frac{1}{k_{io}} \tag{3.A.16}$$

3.A.1.5 Hydrophilic Membrane Based on Aqueous-Phase Side (Species Transfers from Aqueous Phase to Organic Phase)

$$j_i = K_a A_a (C_{ia}^{b} - C_{ia}^{ie}) \tag{3.A.17}$$

$$\frac{1}{A_a K_a} = \frac{(C_{ia}^{b} - C_{ia}^{ie})}{j_i} \tag{3.A.18}$$

$$\frac{1}{A_a K_a} = \frac{\left(C_{ia}^{b} - \frac{C_{io}^{b}}{m_i}\right)}{j_i} \tag{3.A.19}$$

$$\frac{1}{A_a K_a} = \frac{\left(C_{ia}^{b} - \frac{C_{io}^{b}}{m_i} + C_{ia}^{m} - C_{ia}^{m} + C_{ia}^{mi} - C_{ia}^{mi}\right)}{j_i} \tag{3.A.20}$$

$$\frac{1}{A_a K_a} = \frac{(C_{ia}^{b} - C_{ia}^{m}) + (C_{ia}^{m} - C_{ia}^{mi}) + \left(C_{ia}^{mi} - \frac{C_{io}^{b}}{m_i}\right)}{j_i} \tag{3.A.21}$$

$$\frac{1}{A_a K_a} = \frac{(C_{ia}^{b} - C_{ia}^{m}) + (C_{ia}^{m} - C_{ia}^{mi}) + \left(\frac{C_{io}^{mi}}{m_i} - \frac{C_{io}^{b}}{m_i}\right)}{j_i} \tag{3.A.22}$$

$$\frac{1}{A_a K_a} = \frac{1}{k_{ia}} + \frac{1}{k_{im}} + \frac{1}{m_i k_{io}} \tag{3.A.23}$$

3.A.1.6 Hydrophilic Membrane Based on Organic-Phase Side (Species Transfers from Aqueous Phase to Organic Phase)

$$\frac{1}{A_o K_a} = \frac{1}{k_{ia}} + \frac{1}{k_{im}} + \frac{1}{m_i k_{io}} \tag{3.A.24}$$

3.A.1.7 Hydrophilic Membrane Based on Organic-Phase Side (Species Transfers from Organic Phase to Aqueous Phase)

$$j_i = K_o A_o (C_{io}^{ie} - C_{io}^{b}) \tag{3.A.25}$$

$$\frac{1}{A_o K_o} = \frac{(C_{io}^{e} - C_{io}^{b})}{j_i} \tag{3.A.26}$$

$$\frac{1}{A_o K_o} = \frac{(m_i C_{ia}^{b} - C_{io}^{b})}{j_i} \tag{3.A.27}$$

$$\frac{1}{A_o K_o} = \frac{(m_i C_{ia}^{b} - m_i C_{ia}^{m} + m_i C_{ia}^{m} - C_{io}^{mi} - C_{io}^{mi} - C_{io}^{b})}{j_i} \tag{3.A.28}$$

$$\frac{1}{A_o K_o} = \frac{(m_i C_{ia}^{b} - m_i C_{ia}^{m}) + (m_i C_{ia}^{m} - m_i C_{ia}^{mi}) + (C_{io}^{mi} - C_{io}^{b})}{j_i} \tag{3.A.29}$$

$$\frac{1}{A_o K_o} = \frac{m_i}{k_{ia}} + \frac{m_i}{k_{im}} + \frac{1}{k_{io}} \tag{3.A.30}$$

3.A.1.8 Hydrophilic Membrane Based on Aqueous-Phase Side (Species Transfers from Organic Phase to Aqueous Phase)

$$\frac{1}{A_a K_o} = \frac{m_i}{k_{ia}} + \frac{m_i}{k_{im}} + \frac{1}{k_{io}} \tag{3.A.31}$$

3.A.1.9 Composite Membrane Based on Aqueous-Phase Side (Species Transfers from Aqueous Phase to Organic Phase)

$$j_i = K_a A_a (C_{ia}^b - C_{ia}^{ie}) \tag{3.A.32}$$

$$\frac{1}{A_a K_a} = \frac{(C_{ia}^b - C_{ia}^{ie})}{j_i} \tag{3.A.33}$$

$$\frac{1}{A_a K_a} = \frac{\left(C_{ia}^b - \frac{C_{io}^b}{m_i}\right)}{j_i} \tag{3.A.34}$$

$$\frac{1}{A_a K_a} = \frac{\left(C_{ia}^b - C_{ia}^m + C_{ia}^m - C_{ia}^{mi} + C_{ia}^{mi} - \frac{C_{io}^m}{m_i} + \frac{C_{io}^m}{m_i} - \frac{C_{io}^b}{m_i}\right)}{j_i} \tag{3.A.35}$$

$$\frac{1}{A_a K_a} = \frac{(C_{ia}^b - C_{ia}^m) + (C_{ia}^m - C_{ia}^{mi}) + \left(\frac{C_{io}^{mi}}{m_i} - \frac{C_{io}^m}{m_i}\right) + \left(\frac{C_{io}^m}{m_i} - \frac{C_{io}^b}{m_i}\right)}{j_i} \tag{3.A.36}$$

$$\frac{1}{A_a K_a} = \frac{1}{k_{ia}} + \frac{1}{k_{imph}} + \frac{1}{m_i k_{impl}} + \frac{1}{m_i k_{io}} \tag{3.A.37}$$

3.A.1.10 Composite Membrane Based on Organic-Phase Side (Species Transfers from Aqueous Phase to Organic Phase)

$$\frac{1}{A_o K_a} = \frac{1}{k_{ia}} + \frac{1}{k_{imph}} + \frac{1}{m_i k_{impl}} + \frac{1}{m_i k_{io}} \tag{3.A.38}$$

3.A.1.11 Composite Membrane Based on Organic-Phase Side (Species Transfers from Organic Phase to Aqueous Phase)

$$j_i = K_o A_o (C_{io}^{ie} - C_{io}^b) \tag{3.A.39}$$

$$\frac{1}{A_o K_o} = \frac{(C_{io}^e - C_{io}^b)}{j_i} \tag{3.A.40}$$

$$\frac{1}{A_o K_o} = \frac{(m_i C_{ia}^b - C_{io}^b)}{j_i} \tag{3.A.41}$$

$$\frac{1}{A_o K_o} = \frac{(m_i C_{ia}^b - m_i C_{ia}^m + m_i C_{ia}^m - C_{io}^{mi} + C_{io}^{mi} - C_{io}^m + C_{io}^m - C_{io}^b)}{j_i} \tag{3.A.42}$$

$$\frac{1}{A_o K_o} = \frac{(m_i C_{ia}^b - m_i C_{ia}^m + m_i C_{ia}^m - m_i C_{ia}^{mi} + C_{io}^{mi} - C_{io}^m + C_{io}^m - C_{io}^b)}{j_i} \tag{3.A.43}$$

$$\frac{1}{A_oK_o} = \frac{(m_iC_{ia}^b - m_iC_{ia}^m) + (m_iC_{ia}^m - m_iC_{ia}^{mi}) + (C_{io}^{mi} - C_{io}^m) + (C_{io}^m - C_{io}^b)}{j_i} \tag{3.A.44}$$

$$\frac{1}{A_oK_o} = \frac{m_i}{k_{ia}} + \frac{m_i}{k_{imph}} + \frac{1}{k_{impl}} + \frac{1}{k_{io}} \tag{3.A.45}$$

3.A.1.12 Composite Membrane Based on Aqueous-Phase Side (Species Transfers from Organic Phase to Aqueous Phase)

$$\frac{1}{A_aK_o} = \frac{m_i}{k_{ia}} + \frac{m_i}{k_{imph}} + \frac{1}{k_{impl}} + \frac{1}{k_{io}} \tag{3.A.46}$$

3.A.2 Steady-State Resistance-in-Series Model Across Gas–Liquid Contactor

3.A.2.1 Hydrophobic Membrane Based on Gas-Phase Side (Species Transfers from Gas Phase to Liquid Phase)

$$j_i = A_gK_g(p_{ig}^b - p_{ig}^e) \tag{3.A.47}$$

$$\frac{1}{A_gK_g} = \frac{(p_{ig}^b - p_{ig}^e)}{j_i} \tag{3.A.48}$$

$$\frac{1}{A_gK_g} = \frac{\left(p_{ig}^b - C_{il}^b/H_i\right)}{j_i} \tag{3.A.49}$$

$$\frac{1}{A_gK_g} = \frac{\left(p_{ig}^b - C_{il}^b/H_i + p_{ig}^m - p_{ig}^m + p_{ig}^{mi} - p_{ig}^{mi}\right)}{j_i} \tag{3.A.50}$$

$$\frac{1}{A_gK_g} = \frac{(p_{ig}^b - p_{ig}^m) + (p_{ig}^m - p_{ig}^{mi}) + \left(p_{ig}^{mi} - C_{il}^b/H_i\right)}{j_i} \tag{3.A.51}$$

$$\frac{1}{A_gK_g} = \frac{(p_{ig}^b - p_{ig}^m) + (p_{ig}^m - p_{ig}^{mi}) + \left(C_{il}^{mi}/H_i - C_{il}^b/H_i\right)}{j_i} \tag{3.A.52}$$

$$\frac{1}{A_gK_g} = \frac{(p_{ig}^b - p_{ig}^m)}{j_i} + \frac{(p_{ig}^m - p_{ig}^{mi})}{j_i} + \frac{1}{H_i}\frac{(C_{il}^{mi} - C_{il}^b)}{j_i} \tag{3.A.53}$$

$$\frac{1}{A_gK_g} = \frac{1}{k_{ig}} + \frac{1}{k_{im}} + \frac{1}{H_ik_{il}} \tag{3.A.54}$$

3.A.2.2 Hydrophobic Membrane Based on Liquid-Phase Side (Species Transfers from Gas Phase to Liquid Phase)

$$\frac{1}{A_lK_g} = \frac{1}{k_{ig}} + \frac{1}{k_{im}} + \frac{1}{H_ik_{il}} \tag{3.A.55}$$

3.A.2.3 Hydrophobic Membrane Based on Liquid-Phase Side (Species Transfers from Liquid Phase to Gas Phase)

$$j_i = A_l K_l (C_{il}^e - C_{il}^b) \tag{3.A.56}$$

$$\frac{1}{A_l K_l} = \frac{(C_{il}^e - C_{il}^b)}{j_i} \tag{3.A.57}$$

$$\frac{1}{A_l K_l} = \frac{(H_i p_{ig}^b - C_{il}^b)}{j_i} \tag{3.A.58}$$

$$\frac{1}{A_l K_l} = \frac{(H_i p_{ig}^b - C_{il}^b + H_i p_{ig}^m - H_i p_{ig}^m + H_i p_{ig}^{mi} - H_i p_{ig}^{mi})}{j_i} \tag{3.A.59}$$

$$\frac{1}{A_i K_i} = \frac{(H_i p_{ig}^b - H_i p_{ig}^m) + (H_i p_{ig}^m - H_i p_{ig}^{mi}) + (H_i p_{ig}^{mi} - C_{il}^b)}{j_i} \tag{3.A.60}$$

$$\frac{1}{A_i K_i} = \frac{(H_i p_{ig}^b - H_i p_{ig}^m) + (H_i p_{ig}^m - H_i p_{ig}^{mi}) + (C_{il}^{mi} - C_{il}^b)}{j_i} \tag{3.A.61}$$

$$\frac{1}{A_i K_i} = \frac{H_i (p_{ig}^b - p_{ig}^m) + H_i (p_{ig}^m - p_{ig}^{mi}) + (C_{il}^{mi} - C_{il}^b)}{j_i} \tag{3.A.62}$$

$$\frac{1}{A_i K_i} = \frac{H_i (p_{ig}^b - p_{ig}^m)}{j_i} + \frac{H_i (p_{ig}^m - p_{ig}^{mi})}{j_i} + \frac{(C_{il}^{mi} - C_{il}^b)}{j_i} \tag{3.A.63}$$

$$\frac{1}{A_l K_l} = \frac{H_i}{k_{il}} + \frac{H_i}{k_{im}} + \frac{1}{k_{il}} \tag{3.A.64}$$

3.A.2.4 Hydrophobic Membrane Based on Gas-Phase Side (Species Transfers from Liquid Phase to Gas Phase)

$$\frac{1}{A_g K_l} = \frac{H_i}{k_{il}} + \frac{H_i}{k_{im}} + \frac{1}{k_{il}} \tag{3.A.65}$$

3.A.2.5 Hydrophilic Membrane Based on Gas-Phase Side (Species Transfers from Gas Phase to Liquid Phase)

$$j_i = A_g K_g (p_{ig}^b - p_{ig}^e) \tag{3.A.66}$$

$$\frac{1}{A_g K_g} = \frac{(p_{ig}^b - p_{ig}^e)}{j_i} \tag{3.A.67}$$

$$\frac{1}{A_g K_g} = \frac{\left(p_{ig}^b - C_{il}^b/H_i\right)}{j_i} \tag{3.A.68}$$

$$\frac{1}{A_g K_g} = \frac{\left(p_{ig}^b - C_{il}^b/H_i + p_{ig}^{mi} - p_{ig}^{mi} + C_{il}^m/H_i - C_{il}^m/H_i\right)}{j_i} \tag{3.A.69}$$

$$\frac{1}{A_g K_g} = \frac{(p_{ig}^b - p_{ig}^{mi}) + \left(p_{ig}^{mi} - C_{il}^m/H_i\right) + \left(C_{il}^m/H_i - C_{il}^b/H_i\right)}{j_i} \tag{3.A.70}$$

$$\frac{1}{A_g K_g} = \frac{(p_{ig}^b - p_{ig}^m) + \left(C_{il}^{mi}/H_i - C_{il}^m/H_i\right) + \left(C_{il}^m/H_i - C_{il}^b/H_i\right)}{j_i} \tag{3.A.71}$$

$$\frac{1}{A_g K_g} = \frac{(p_{ig}^b - p_{ig}^m)}{j_i} + \frac{1}{H_i}\frac{(C_{il}^{mi} - C_{il}^m)}{j_i} + \frac{1}{H_i}\frac{(C_{il}^m - C_{il}^b)}{j_i} \tag{3.A.72}$$

$$\frac{1}{A_g K_g} = \frac{1}{k_{ig}} + \frac{1}{H_i k_{im}} + \frac{1}{H_i k_{il}} \tag{3.A.73}$$

3.A.2.6 Hydrophilic Membrane Based on Liquid-Phase Side (Species Transfers from Gas Phase to Liquid Phase)

$$\frac{1}{A_l K_g} = \frac{1}{k_{ig}} + \frac{1}{H_i k_{im}} + \frac{1}{H_i k_{il}} \tag{3.A.74}$$

3.A.2.7 Hydrophilic Membrane Based on Liquid-Phase Side (Species Transfers from Liquid Phase to Gas Phase)

$$j_i = A_l K_l (C_{il}^e - C_{il}^b) \tag{3.A.75}$$

$$\frac{1}{A_l K_l} = \frac{(C_{il}^e - C_{il}^b)}{j_i} \tag{3.A.76}$$

$$\frac{1}{A_l K_l} = \frac{(H_i p_{ig}^b - C_{il}^b)}{j_i} \tag{3.A.77}$$

$$\frac{1}{A_l K_l} = \frac{(H_i p_{ig}^b + C_{il}^{mi} - C_{il}^{mi} + C_{il}^m - C_{il}^m - C_{il}^b)}{j_i} \tag{3.A.78}$$

$$\frac{1}{A_l K_l} = \frac{(H_i p_{ig}^b - C_{il}^{mi} + C_{il}^{mi} - C_{il}^m + C_{il}^m - C_{il}^b)}{j_i} \tag{3.A.79}$$

$$\frac{1}{A_l K_l} = \frac{(H_i p_{ig}^b - H_i p_{ig}^{mi}) + (C_{il}^{mi} - C_{il}^m) + (C_{il}^m - C_{il}^b)}{j_i} \tag{3.A.80}$$

$$\frac{1}{A_l K_l} = H_i \frac{(p_{ig}^b - p_{ig}^{mi})}{j_i} + \frac{(C_{il}^{mi} - C_{il}^m)}{j_i} + \frac{(C_{il}^m - C_{il}^b)}{j_i} \tag{3.A.81}$$

$$\frac{1}{A_l K_l} = \frac{H_i}{k_{ig}} + \frac{1}{k_{im}} + \frac{1}{k_{il}} \tag{3.A.82}$$

3.A.2.8 Hydrophilic Membrane Based on Gas-Phase Side (Species Transfers from Liquid Phase to Gas Phase)

$$\frac{1}{A_g K_l} = \frac{H_i}{k_{ig}} + \frac{1}{k_{im}} + \frac{1}{k_{il}} \tag{3.A.83}$$

3.A.2.9 Composite Membrane Based on Gas-Phase Side (Species Transfers from Gas Phase to Liquid Phase)

$$j_i = K_g A_g (p_{ig}^b - p_{ig}^e) \tag{3.A.84}$$

$$\frac{1}{A_g K_g} = \frac{(p_{ig}^b - p_{ig}^e)}{j_i} \tag{3.A.85}$$

$$\frac{1}{A_g K_g} = \frac{\left(p_{ig}^b - \frac{C_{il}^b}{H_i}\right)}{j_i} \tag{3.A.86}$$

$$\frac{1}{A_g K_g} = \frac{\left(p_{ig}^b - \frac{C_{il}^b}{H_i} + p_{ig}^m - p_{ig}^m + p_{ig}^{mi} - p_{ig}^{mi} + \frac{C_{il}^m}{H_i} - \frac{C_{il}^m}{H_i}\right)}{j_i} \tag{3.A.87}$$

$$\frac{1}{A_g K_g} = \frac{(p_{ig}^b - p_{ig}^m) + (p_{ig}^m - p_{ig}^{mi}) + \left(\frac{C_{il}^{mi}}{H_i} - \frac{C_{il}^m}{H_i}\right) + \left(\frac{C_{il}^m}{H_i} - \frac{C_{il}^b}{H_i}\right)}{j_i} \tag{3.A.88}$$

$$\frac{1}{A_g K_g} = \frac{(p_{ig}^b - p_{ig}^m)}{j_i} + \frac{(p_{ig}^m - p_{ig}^{mi})}{j_i} + \frac{1}{H_i}\frac{(C_{il}^{mi} - C_{il}^m)}{j_i} + \frac{1}{H_i}\frac{(C_{il}^m - C_{il}^b)}{j_i} \tag{3.A.89}$$

$$\frac{1}{A_g K_g} = \frac{1}{k_{ig}} + \frac{1}{k_{imph}} + \frac{1}{H_i k_{impl}} + \frac{1}{H_i k_{il}} \tag{3.A.90}$$

3.A.2.10 Composite Membrane Based on Liquid-Phase Side (Species Transfers from Gas Phase to Liquid Phase)

$$\frac{1}{A_l K_g} = \frac{1}{k_{ig}} + \frac{1}{k_{imph}} + \frac{1}{H_i k_{impl}} + \frac{1}{H_i k_{il}} \tag{3.A.91}$$

3.A.2.11 Composite Membrane Based on Liquid-Phase Side (Species Transfers from Gas Phase to Liquid Phase)

$$j_i = K_l A_l (C_{il}^e - C_{il}^b) \tag{3.A.92}$$

$$\frac{1}{A_l K_l} = \frac{(C_{il}^e - C_{il}^b)}{j_i} \tag{3.A.93}$$

$$\frac{1}{A_l K_l} = \frac{(H_i p_{ig}^b - C_{il}^b)}{j_i} \tag{3.A.94}$$

$$\frac{1}{A_l K_l} = \frac{(H_i p_{ig}^b - C_{il}^b + H_i p_{ig}^m + H_i p_{ig}^m \mp C_{il}^{mi} - C_{il}^{mi} + C_{il}^m - C_{il}^m)}{j_i} \tag{3.A.95}$$

$$\frac{1}{A_l K_l} = \frac{(H_i p_{ig}^b - H_i p_{ig}^m) + (H_i p_{ig}^m - H_i p_{ig}^{mi}) + (C_{il}^{mi} - C_{il}^m) + (C_{il}^m - C_{il}^b)}{j_i} \tag{3.A.96}$$

$$\frac{1}{A_l K_l} = H_i \frac{(p_{ig}^b - p_{ig}^m)}{j_i} + H_i \frac{(p_{ig}^m - p_{ig}^{mi})}{j_i} + \frac{(C_{il}^{mi} - C_{il}^m)}{j_i} + \frac{(C_{il}^m - C_{il}^b)}{j_i} \tag{3.A.97}$$

$$\frac{1}{A_l K_l} = \frac{H_i}{k_{ig}} + \frac{H_i}{k_{imph}} + \frac{1}{k_{impl}} + \frac{1}{k_{il}} \tag{3.A.98}$$

3.A.2.12 Composite Membrane Based on Gas-Phase Side (Species Transfers from Liquid Phase to Gas Phase)

$$\frac{1}{A_g K_l} = \frac{H_i}{k_{ig}} + \frac{H_i}{k_{imph}} + \frac{1}{k_{impl}} + \frac{1}{k_{il}} \tag{3.A.99}$$

3.A.3 Dynamic Modeling Across the Storage Tank

Rate of mass flow rate entering the tank – rate of mass flow rate exiting the tank = rate of mass accumulation within the tank.

At time t,

$$Q_f(C_5^o - C_5^e) = V_f \frac{dC_5^e}{dt} \tag{3.A.100}$$

At time $t + \Delta t$,

$$Q_f(C_5^o - C_5^e) = V_f \left[\frac{C_5^e(t + \Delta t) - C_5^e(t)}{\Delta t} \right] \tag{3.A.101}$$

$$(C_5^o - C_5^e) = \frac{V_f}{Q_f} \left[\frac{C_5^e(t + \Delta t) - C_5^e(t)}{\Delta t} \right] \tag{3.A.102}$$

$$(C_5^o - C_5^e) = \frac{\Gamma_f}{\Delta t} [C_5^e(t + \Delta t) - C_5^e(t)] \tag{3.A.103}$$

$$C_5^e(t + \Delta t) = \frac{\Delta t}{\Gamma_f} C_5^o(t) + C_5^e(t) \left[1 - \frac{\Delta t}{\Gamma_f} \right] \tag{3.A.104}$$

Equation (3.A.104) measures the species concentration in feed tank over a certain period of time. In the similar way, the unsteady-state equation showing the species concentration in the receiving-phase storage tank can be obtained by taking the unsteady-state mass balance across the receiving-phase storage tank:

$$C_1^e(t + \Delta t) = \frac{\Delta t}{\Gamma_r} C_1^o(t) + C_1^e(t) \left[1 - \frac{\Delta t}{\Gamma_r} \right] \tag{3.A.105}$$

References

1 Strathmann, H., Giorno, L., and Drioli, E. (2011). *Introduction to Membrane Science and Technology*, vol. 544. Weinheim: Wiley-VCH.
2 Sohaib, Q., Muhammad, A., Younas, M. et al. (2021). Rigorous non-isothermal modeling approach for mass and energy transport during CO_2 absorption into aqueous solution of amino acid ionic liquids in hollow fiber membrane contactors. *Separation and Purification Technology* 254: 117644.
3 Younas, M., Tahir, T., Wu, C. et al. (2020). Post-combustion CO_2 capture with sweep gas in thin film composite (TFC) hollow fiber membrane (HFM) contactor. *Journal of CO_2 Utilization* 40: 101266.
4 Sohaib, Q., Muhammad, A., Younas, M. et al. (2020). Modeling pre-combustion CO_2 capture with tubular membrane contactor using ionic liquids at elevated temperatures. *Separation and Purification Technology* 241: 116677.
5 Pishnamazi, M., Nakhjiri, A.T., Taleghani, A.S. et al. (2020). Molecular investigation into the effect of carbon nanotubes interaction with CO_2 in molecular separation using microporous polymeric membranes. *Scientific Reports* 10 (1): 13285.
6 Pishnamazi, M., Nakhjiri, A.T., Marjani, A. et al. (2020). Computational study on SO_2 molecular separation applying novel EMISE ionic liquid and DMA aromatic amine solution inside microporous membranes. *Journal of Molecular Liquids* 313: 113531.
7 Cuperus, F. and Smolders, C.A. (1991). Characterization of UF membranes: membrane characteristics and characterization techniques. *Advances in Colloid and Interface Science* 34: 135–173.
8 Ahmad, N.A., Leo, C.P., Ahmad, A.L. et al. (2015). Membranes with great hydrophobicity: a review on preparation and characterization. *Separation and Purification Reviews* 44 (2): 109–134.
9 Demirci, M.A., Ipek, Y., Gul, F. et al. (2018). Extraction, isolation of heat-resistance phenolic compounds, antioxidant properties, characterization and purification of 5-hydroxymaltol from Turkish apple pulps. *Food Chemistry* 269: 111–117.
10 Franken, A., Nolten, J., Mulder, M. et al. (1987). Wetting criteria for the applicability of membrane distillation. *Journal of Membrane Science* 33 (3): 315–328.
11 Kim, B.-S. and Harriott, P. (1987). Critical entry pressure for liquids in hydrophobic membranes. *Journal of Colloid and Interface Science* 115 (1): 1–8.
12 García-Payo, M.d.C., Izquierdo-Gil, M.A., and Fernández-Pineda, C. (2000). Wetting study of hydrophobic membranes via liquid entry pressure measurements with aqueous alcohol solutions. *Journal of Colloid and Interface Science* 230 (2): 420–431.
13 MacMullin, R. and Muccini, G.A. (1956). Characteristics of porous beds and structures. *AIChE Journal* 2 (3): 393–403.
14 Thomas, L., Katz, D., and Tek, M.R. (1968). Threshold pressure phenomena in porous media. *Society of Petroleum Engineers Journal* 8 (02): 174–184.

15 Eykens, L., De Sitter, K., Dotremont, C. et al. (2017). Wetting resistance of commercial membrane distillation membranes in waste streams containing surfactants and oil. *Applied Sciences* 7 (2): 118.

16 Yazgan-Birgi, P., Hassan Ali, M.I., and Arafat, H.A. (2018). Estimation of liquid entry pressure in hydrophobic membranes using CFD tools. *Journal of Membrane Science* 552: 68–76.

17 Katz, M. and Baruch, G. (1986). New insights into the structure of microporous membranes obtained using a new pore size evaluation method. *Desalination* 58 (3): 199–211.

18 Mey-Marom, A. and Katz, M.G. (1986). Measurement of active pore size distribution of microporous membranes – a new approach. *Journal of Membrane Science* 27 (2): 119–130.

19 Tylkowski, B. and Tsibranska, I. (2015). Overview of main techniques used for membrane characterization. *Journal of Chemical Technology and Metallurgy* 50 (1): 3–12.

20 Sheikh, M., Pazirofteh, M., Dehghani, M. et al. (2020). Application of ZnO nanostructures in ceramic and polymeric membranes for water and wastewater technologies: a review. *Chemical Engineering Journal* 391: 123475.

21 Mirqasemi, M.S., Homayoonfal, M., and Rezakazemi, M. (2020). Zeolitic imidazolate framework membranes for gas and water purification. *Environmental Chemistry Letters* 18 (1): 1–52.

22 Hajilary, N., Rezakazemi, M., and Shirazian, S. (2019). Biofuel types and membrane separation. *Environmental Chemistry Letters* 17 (1): 1–18.

23 Rezakazemi, M., Sadrzadeh, M., and Matsuura, T. (2018). Thermally stable polymers for advanced high-performance gas separation membranes. *Progress in Energy and Combustion Science* 66: 1–41.

24 Rezakazemi, M., Khajeh, A., and Mesbah, M. (2018). Membrane filtration of wastewater from gas and oil production. *Environmental Chemistry Letters* 16 (2): 367–388.

25 Rezakazemi, M., Dashti, A., Riasat Harami, H. et al. (2018). Fouling-resistant membranes for water reuse. *Environmental Chemistry Letters* 16 (3): 715–763.

26 Firouzjaei, M.D., Seyedpour, S.F., Aktij, S.A. et al. (2020). Recent advances in functionalized polymer membranes for biofouling control and mitigation in forward osmosis. *Journal of Membrane Science* 596: 117604.

27 Muhammad, A., Younas, M., and Rezakazemi, M. (2018). CFD simulation of copper(II) extraction with TFA in non-dispersive hollow fiber membrane contactors. *Environmental Science and Pollution Research* 25 (12): 12053–12063.

28 Muhammad, A., Younas, M., and Rezakazemi, M. (2017). Quasi-dynamic modeling of dispersion-free extraction of aroma compounds using hollow fiber membrane contactor. *Chemical Engineering Research and Design* 127: 52–61.

29 Rezakazemi, M., Darabi, M., Soroush, E. et al. (2019). CO_2 absorption enhancement by water-based nanofluids of CNT and SiO_2 using hollow-fiber membrane contactor. *Separation and Purification Technology* 210: 920–926.

30 Rehman, W.U., Muhammad, A., Khan, Q.A. et al. (2019). Pomegranate juice concentration using osmotic distillation with membrane contactor. *Separation and Purification Technology* 224: 481–489.

31 Rezakazemi, M. (2018). CFD simulation of seawater purification using direct contact membrane desalination (DCMD) system. *Desalination* 443: 323–332.

32 Hajilary, N. and Rezakazemi, M. (2018). CFD modeling of CO_2 capture by water-based nanofluids using hollow fiber membrane contactor. *International Journal of Greenhouse Gas Control* 77: 88–95.

33 Razavi, S.M.R., Rezakazemi, M., Albadarin, A.B. et al. (2016). Simulation of CO_2 absorption by solution of ammonium ionic liquid in hollow-fiber contactors. *Chemical Engineering and Processing: Process Intensification* 108: 27–34.

34 Rezakazemi, M., Shirazian, S., and Ashrafizadeh, S.N. (2012). Simulation of ammonia removal from industrial wastewater streams by means of a hollow-fiber membrane contactor. *Desalination* 285: 383–392.

35 Younas, M., Bocquet, S.D., and Sanchez, J. (2008). Extraction of aroma compounds in a HFMC: dynamic modelling and simulation. *Journal of Membrane Science* 323 (2): 386–394.

36 Younas, M., Druon-Bocquet, S., and Sanchez, J. (2010). Kinetic and dynamic study of liquid–liquid extraction of copper in a HFMC: experimentation, modeling, and simulation. *AIChE Journal* 56 (6): 1469–1480.

37 Bocquet, S., Gascons Viladomat, F., Muvdi Nova, C. et al. (2006). Membrane-based solvent extraction of aroma compounds: choice of configurations of hollow fiber modules based on experiments and simulation. *Journal of Membrane Science* 281 (1): 358–368.

38 Prasad, R. and Sirkar, K.K. (1992). Membrane-based solvent extraction. In: *Membrane Handbook* (eds. W.S.W. Ho and K.K. Sirkar), 727–763. Springer.

39 Whitman, W.G. (1962). The two film theory of gas absorption. *International Journal of Heat and Mass Transfer* 5 (5): 429–433.

40 Muhammada, A., Younasa, M., Druon-Bocquetb, S. et al. (2017). Numerical modelling and simulation of membrane-based extraction of copper (II) using hollow fiber contactors. *Desalination and Water Treatment* 63: 113–123.

41 Rezakazemi, M. and Shirazian, S. (2018). Computational simulation of mass transfer in molecular separation using microporous polymeric membranes. *Chemical Engineering & Technology* 41 (10): 1975–1981.

42 Zhang, Z., Zhao, S., Rezakazemi, M. et al. (2017). Effect of flow and module configuration on SO_2 absorption by using membrane contactors. *Global NEST Journal* 19 (4): 716–725.

43 Shirazian, S., Pishnamazi, M., Rezakazemi, M. et al. (2012). Implementation of the finite element method for simulation of mass transfer in membrane contactors. *Chemical Engineering & Technology* 35 (6): 1077–1084.

44 Shirazian, S., Marjani, A., and Rezakazemi, M. (2012). Separation of CO_2 by single and mixed aqueous amine solvents in membrane contactors: fluid flow and mass transfer modeling. *Engineering with Computers* 28 (2): 189–198.

45 Rezakazemi, M., Niazi, Z., Mirfendereski, M. et al. (2011). CFD simulation of natural gas sweetening in a gas–liquid hollow-fiber membrane contactor. *Chemical Engineering Journal* 168 (3): 1217–1226.

46 Lin, H. and Freeman, B.D. (2004). Gas solubility, diffusivity and permeability in poly(ethylene oxide). *Journal of Membrane Science* 239 (1): 105–117.

47 Fasihi, M., Shirazian, S., Marjani, A. et al. (2012). Computational fluid dynamics simulation of transport phenomena in ceramic membranes for SO_2 separation. *Mathematical and Computer Modelling* 56 (11): 278–286.

48 Shirazian, S., Rezakazemi, M., Marjani, A. et al. (2011). Development of a mass transfer model for simulation of sulfur dioxide removal in ceramic membrane contactors. *Asia-Pacific Journal of Chemical Engineering* 7 (6): 828–834.

49 Wijmans, J.G. and Baker, R.W. (1995). The solution-diffusion model: a review. *Journal of Membrane Science* 107 (1): 1–21.

50 Pandey, P. and Chauhan, R.S. (2001). Membranes for gas separation. *Progress in Polymer Science* 26 (6): 853–893.

51 Kreulen, H., Smolders, C.A., Versteeg, G.F. et al. (1993). Microporous hollow fibre membrane modules as gas-liquid contactors. Part 2. Mass transfer with chemical reaction. *Journal of Membrane Science* 78 (3): 217–238.

52 Kim, Y.-S. and Yang, S.-M. (2000). Absorption of carbon dioxide through hollow fiber membranes using various aqueous absorbents. *Separation and Purification Technology* 21 (1): 101–109.

53 Kumar, P.S., Hogendoorn, J.A., Feron, P.H.M. et al. (2003). Approximate solution to predict the enhancement factor for the reactive absorption of a gas in a liquid flowing through a microporous membrane hollow fiber. *Journal of Membrane Science* 213 (1): 231–245.

54 DeCoursey, W.J. (1974). Absorption with chemical reaction: development of a new relation for the Danckwerts model. *Chemical Engineering Science* 29 (9): 1867–1872.

55 DeCoursey, W.J. (1982). Enhancement factors for gas absorption with reversible reaction. *Chemical Engineering Science* 37 (10): 1483–1489.

56 Younas, M., Druon-Bocquet, S., and Sanchez, J. (2011). Experimental and theoretical mass transfer transient analysis of copper extraction using hollow fiber membrane contactors. *Journal of Membrane Science* 382 (1–2): 70–81.

57 Prasad, R. and Sirkar, K.K. (1988). Dispersion-free solvent extraction with microporous hollow-fiber modules. *AIChE Journal* 34 (2): 177–188.

58 Wu, J. and Chen, V. (2000). Shell-side mass transfer performance of randomly packed hollow fiber modules. *Journal of Membrane Science* 172 (1): 59–74.

59 Costello, M.J., Fane, A.G., Hogan, P.A. et al. (1993). The effect of shell side hydrodynamics on the performance of axial flow hollow fibre modules. *Journal of Membrane Science* 80 (1): 1–11.

60 Yang, M.-C. and Cussler, E.L. (1986). Designing hollow-fiber contactors. *AIChE Journal* 32 (11): 1910–1916.

61 Côté, P., Bersillon, J.-L., and Huyard, A. (1989). Bubble-free aeration using membranes: mass transfer analysis. *Journal of Membrane Science* 47 (1): 91–106.

62 Gawroński, R. and Wrzesińska, B. (2000). Kinetics of solvent extraction in hollow-fiber contactors. *Journal of Membrane Science* 168 (1): 213–222.

63 Dahuron, L. and Cussler, E.L. (1988). Protein extractions with hollow fibers. *AIChE Journal* 34 (1): 130–136.

64 Gekas, V. and Hallström, B. (1987). Mass transfer in the membrane concentration polarization layer under turbulent cross flow: I. Critical literature review and adaptation of existing Sherwood correlations to membrane operations. *Journal of Membrane Science* 30 (2): 153–170.

65 Bhaumik, D., Majumdar, S., and Sirkar, K.K. (1998). Absorption of CO_2 in a transverse flow hollow fiber membrane module having a few wraps of the fiber mat. *Journal of Membrane Science* 138 (1): 77–82.

66 Wang, K.L. and Cussler, E.L. (1993). Baffled membrane modules made with hollow fiber fabric. *Journal of Membrane Science* 85 (3): 265–278.

67 Lévêque, A. (1928). *Les lois de la transmission de chaleur par convection.* Dunod.

68 Graetz, L. (1885). Ueber die Wärmeleitungsfähigkeit von Flüssigkeiten. *Annalen der Physik* 261 (7): 337–357.

69 Wickramasinghe, S., Semmens, M.J., and Cussler, E. (1992). Mass transfer in various hollow fiber geometries. *Journal of Membrane Science* 69 (3): 235–250.

70 Qi, Z. and Cussler, E.L. (1985). Microporous hollow fibers for gas absorption: II. Mass transfer across the membrane. *Journal of Membrane Science* 23 (3): 333–345.

71 Mulder, M. (2012). *Basic Principles of Membrane Technology.* Springer Science & Business Media.

72 Reid, R.C., Prausnitz, J.M., and Poling, B.E. (1987). *The Properties of Gases and Liquids.* U.S. Department of Energy, Office of Scientific and Technical Information.

73 Cussler, E.L. (2009). *Diffusion: Mass Transfer in Fluid Systems.* Cambridge University Press.

74 Versteeg, G.F. and Van Swaaij, W.P.M. (1988). Solubility and diffusivity of acid gases (carbon dioxide, nitrous oxide) in aqueous alkanolamine solutions. *Journal of Chemical & Engineering Data* 33 (1): 29–34.

75 Johnson, R.A. and Nguyen, M.H. (2017). *Understanding Membrane Distillation and Osmotic Distillation.* Wiley Online Library.

76 D'Elia, N.A., Dahuron, L., and Cussler, E.L. (1986). Liquid–liquid extractions with microporous hollow fibers. *Journal of Membrane Science* 29 (3): 309–319.

77 Zhang, L.-Z. (2011). Heat and mass transfer in a randomly packed hollow fiber membrane module: a fractal model approach. *International Journal of Heat and Mass Transfer* 54 (13): 2921–2931.

78 Jiang, R., Yang, M., Chen, S. et al. (2014). Fluid flow and heat transfer across an elliptical hollow fiber membrane tube bank with randomly distributed features. *International Journal of Heat and Mass Transfer* 76: 559–567.

79 Kays, W.M. (2012). *Convective Heat and Mass Transfer.* Tata McGraw-Hill Education.

80 Muhammad, A., Ali, W., Ahmad, I. et al. (2017). Performance evaluation of hollow fiber membrane contactors for dispersion-free extraction

of Cu^{2+} through modelling and simulation. *Periodica Polytechnica, Chemical Engineering* 61 (2): 133–143.

81 Roudet, M., Loubiere, K., Gourdon, C. et al. (2011). Hydrodynamic and mass transfer in inertial gas–liquid flow regimes through straight and meandering millimetric square channels. *Chemical Engineering Science* 66 (13): 2974–2990.

82 Tahmasbi, D., Hossainpour, S., Babaluo, A.A. et al. (2020). Hydrogen separation from synthesis gas using silica membrane: CFD simulation. *International Journal of Hydrogen Energy* 45 (38): 19381–19390.

83 Shokrollahi, M., Rezakazemi, M., and Younas, M. (2020). Producing water from saline streams using membrane distillation: modeling and optimization using CFD and design expert. *International Journal of Energy Research* 44 (11): 8841–8853.

84 Wang, Y., Brannock, M., Cox, S. et al. (2010). CFD simulations of membrane filtration zone in a submerged hollow fibre membrane bioreactor using a porous media approach. *Journal of Membrane Science* 363 (1): 57–66.

85 Gabelman, A. and Hwang, S.-T. (1999). Hollow fiber membrane contactors. *Journal of Membrane Science* 159 (1–2): 61–106.

86 Luis, P. (2018). Membrane contactors. In: *Fundamental Modelling of Membrane Systems*, Chapter 5 (ed. P. Luis), 153–208. Elsevier.

87 Bird, R.B. (2002). Transport phenomena. *Applied Mechanics Reviews* 55 (1): R1–R4.

88 Luis, P., Garea, A., and Irabien, A. (2010). Modelling of a hollow fibre ceramic contactor for SO_2 absorption. *Separation and Purification Technology* 72 (2): 174–179.

89 Qazi, S., Gómez-Coma, L., Albo, J. et al. (2020). Mathematical modeling of CO_2 absorption with ionic liquids in a membrane contactor, study of absorption kinetics and influence of temperature. *Journal of Chemical Technology and Biotechnology* 95 (7): 1844–1857.

90 Howard, P.H. and Meylan, W.P. (1997). *Handbook of Physical Properties of Organic Chemicals*. CRC Press.

91 Esfandiari, A., Hosseini Monjezi, A., Rezakazemi, M. et al. (2019). Computational fluid dynamic modeling of water desalination using low-energy continuous direct contact membrane distillation process. *Applied Thermal Engineering* 163: 114391.

4

Module Design and Membrane Materials

Nabilah Fazil[1], Sidra Saqib[2], Ahmad Mukhtar[3], Mohammad Younas[4], and Mashallah Rezakazemi[5]

[1] *Universiti Teknologi PETRONAS, Department of Chemical Engineering, Bandar Seri Iskandar, 32610, Perak, Malaysia*
[2] *COMSATS University Islamabad (CUI), Department of Chemical Engineering, Lahore Campus, Punjab, 54000, Pakistan*
[3] *NFC Institute of Engineering and Fertilizer Research, Department of Chemical Engineering, Faisalabad, Punjab, 38000, Pakistan*
[4] *University of Engineering and Technology, Department of Chemical Engineering, Peshawar, Pakistan*
[5] *Shahrood University of Technology, Faculty of Chemical and Materials Engineering, Shahrood, Iran*

4.1 Introduction

Over the last five decades, membrane technology has been grown revolutionary. It has been adopted in various applications, such as desalination reverse osmosis and water purification, ultrafiltration, gas separation, regulated drug release, genetic manipulation, food processing, and many more. This advancement led the researchers to develop various kinds of membrane materials and modules to achieve a specific application's target performance. Efficient, economical, compact, and easy-to-clean membrane contactor module designs are investable for the process applications in using liquid–liquid, gas–liquid, and gas–gas operations.

The conceptual understanding and technological advancement of membrane and module are becoming important in the recent research for module design. Separation processes based on membrane have various applications in the water, oil, chemical, petrochemical, and pharmaceutical sectors. This growth is mainly attributed to two major developments: firstly, the ability to manufacture high-permeable and relatively defect-free membranes and, secondly, the ability to assemble the membranes in compact, effective, economic, and high membrane surface modules [1, 2]. The correct choice of membrane material and module may also lead to increase in transmembrane flux, decrease in concentration and temperature polarization, and membrane fouling, which consequently increases the process efficiency, decreases the energy use, and increases the membrane material life. Reliable and stable transmembrane and axial flux rely on the permeation properties of the membrane and the hydrodynamics of the membrane module assembly. In recent decades, many efforts have been made to develop and

Membrane Contactor Technology: Water Treatment, Food Processing, Gas Separation, and Carbon Capture, First Edition. Edited by Mohammad Younas and Mashallah Rezakazemi.

manufacture the innovative membrane modules with designed geometries and/or shear-induced components to increase the species permeation and to avoid the undesirable polarization, fluid leakage, and deterioration of the membrane module [3–8]. The passive strategy in improving the membrane performance includes changing the membrane structure or improving/modifying the module geometry by incorporating distances or blankets into the membrane modules to influence the geometry of flow by forming secondary flows or edges adjacent to the membrane or/and major flow instabilities. The induced high shear rate can also render the mixing smoother and reduce the borderline surface thickness.

Well-engineered and assembled membrane modules increase the fluid's hydrodynamic conditions and greatly enhance the overall device operation. However, the production and manufacture of membrane modules have gained less recognition, despite their significance. Thus, proper selection of membrane contactor module, operating mode, and membrane materials is very crucial in making the membrane technology viable and efficient at a large scale.

This chapter reports the membrane module design and flow configurations and the membrane materials and their synthesis techniques.

4.2 Membrane Module Design Configuration

The module design is crucial for membrane contactor operation, which must operate with a large membrane surface area to achieve the desired separation. Module design varies according to the application, manufacturer, membrane material, and scale of operation. Though the specific classification of modules design is difficult, generally, two types of modules exist: flat sheet and tubular. Membrane modules are classified based on membrane material and packing and module geometry. These could be spiral wound, hollow fiber, plate and frame, tubular, and capillary. The membrane geometry is planar for the plate-and-frame and spiral wound membrane modules, while it is cylindrical for the tubular and hollow membrane module. The respective module types have distinct responsibilities. Further discussion on these four membrane module designs will be elaborated in Sections 4.2.1–4.2.4. The modules' design requirements and choice also consider the ease of cleaning, high packing density, cost-effectiveness, and replacement of membrane. The membrane's design configuration relates to the membrane geometry and its location in the space relative to fluid and permeate flow. Therefore, the membrane module design configuration defines how the membrane is being assembled in the module as most market membrane installations are flexible. The ideal characteristics of a membrane module design configuration are presented in Table 4.1.

4.2.1 Plate-and-Frame Modules

Plate-and-frame modules offer a very robust and compact design. These are one of the first and simplest modules and closely resemble the flat-sheet membranes that have been used on a laboratory scale. The key components of this module design

Table 4.1 Membrane module design characteristics and description.

Membrane module design characteristics	Description
High compactness	The capacity to fit as many membrane surfaces as practicable in a small-capacity module
Limited tangential flow resistance	Less interference lowers energy usage and significantly reduces pressure in the retained flow channel
Strong levels of turbulence	This is basically on the residual part to reduce fouling and facilitate mass transfers
Zero "dead" areas	Uniform distribution of velocity
Highly manageable	Quick to clean, store, and sustain
Low cost	Comparatively small expense per area of the membrane unit

Source: Based on Berk [2].

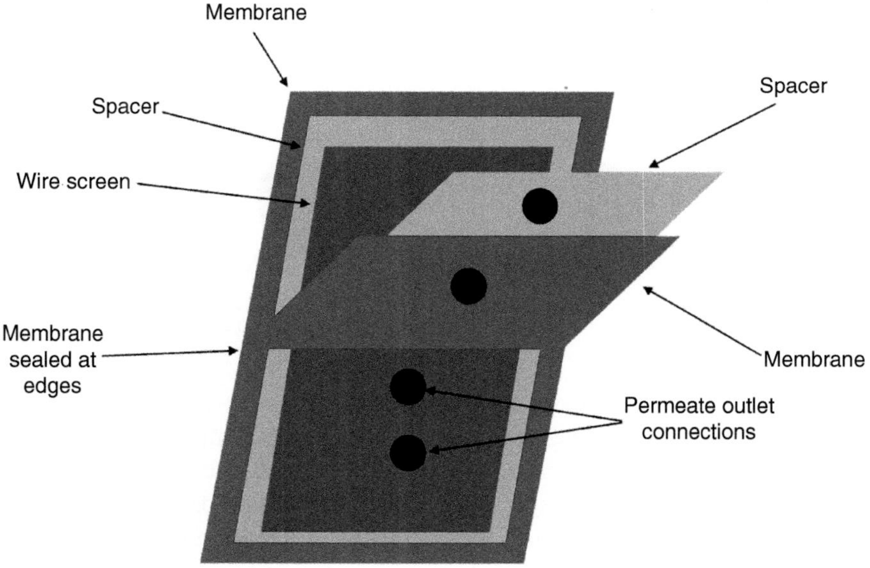

Figure 4.1 An illustration of a plate and frame module. Source: Redrawn from McKeen [5].

are flat membrane, membrane-supporting plate/spacer, and feed distribution plate [1, 3]. The earliest applications of plate-and-frame modules have been found in removing helium from natural gas using fluorinated ethylene propylene (FEP) membrane [4]. The membrane used in the plate-and-frame module was a large mesh screen on the two sides of a plate, as illustrated in Figure 4.1. Several plates are lined but separated so that they are not disrupted except at the permeated outlets. The stack is placed within a chamber that has natural gas flowing under high pressure (1000 psi). The helium is the permeate that has been separated from

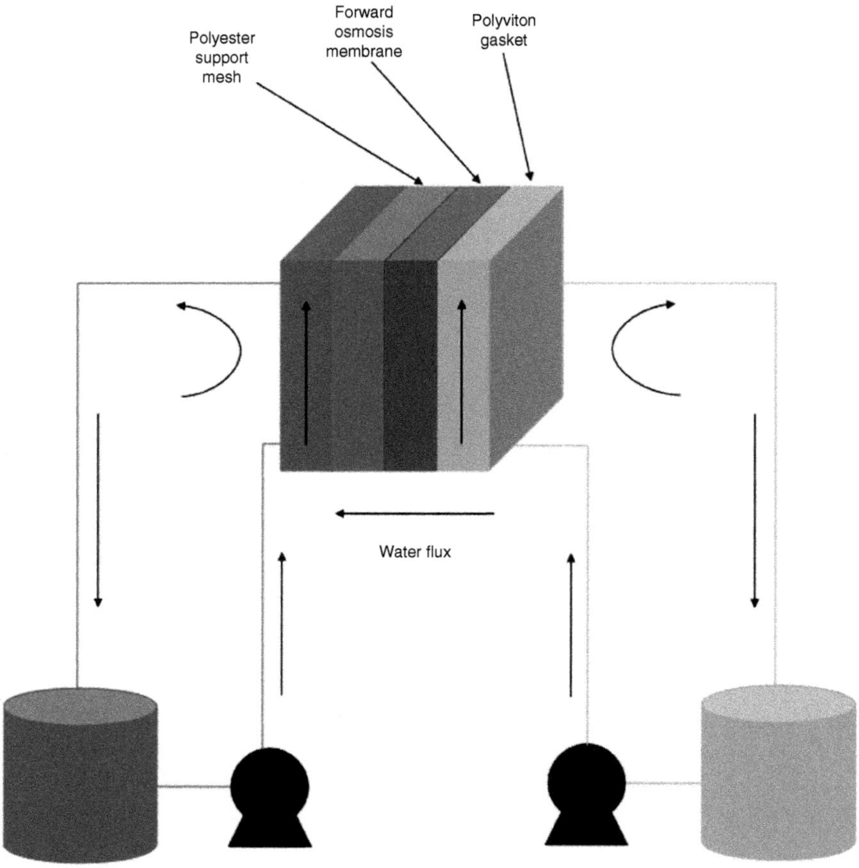

Figure 4.2 Schematic diagram of recycled-based flat-sheet membrane contactor.

the natural gas. Baffles are added to minimize the channeling (possibility to bypass sections of the membrane) and create a consistent flow distribution [1].

Plate-and-frame modules may also be engineered, created, and constructed in various orientations, sizes, and shapes based on the laboratory capacity with a single, small-sized membrane to systems comprising a set of membranes in this module. An example of a plate-and-frame application is shown in Figure 4.2. This module makes the cleaning and replacement of the membrane simple. It also allows good flow controls on both sides of the membrane [7]. Besides, this module also utilizes flat membranes without applying any adhesive. However, the plate-and-frame module has a possible leakage issue: the gaskets needed for each plate also increase due to the easily elevated fouling, only if they are washed regularly, and these also lead the interruption in the operations. Due to the numerous spacer plate and seals in this module, it is also costly relative to the other alternatives (Figure 4.3).

In general, modules are used for pervaporation, electrodialysis, ultrafiltration, and reverse osmosis [8–10]. In pervaporation applications, this module is surprisingly effective but far more popular than those of spiral and hollow fiber for gas separation

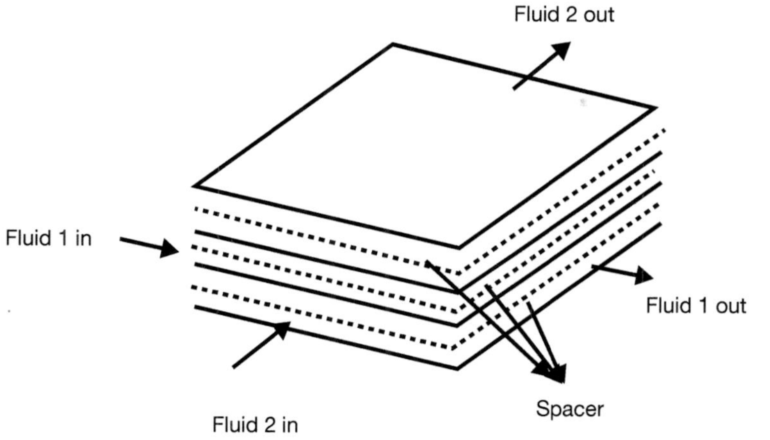

Figure 4.3 Schematic diagram of plate and frame module.

purposes because the lower membrane region is supplied per unit volume. This module's first industrial usage in gas separation is an air-to-air oxygen enrichment for small-scale medical and organic vapor recovery as invented by Borsig, with a specific prototype design (envelope type) for gas separation module [6, 11].

4.2.2 Spiral Wound Modules

The predominant membrane configuration is the spiral wound module. These were originally prepared to desalinate water but also for other uses, such as the food, pulp and paper, and gas-separating industries; the extremely compact size and low price rendered them appealing. These modules modify the general principle of plate-and-frame assembly, which selectively permeates the membrane after it allows the feed gas mixture that is pressurized through the module and forced to flow through the membrane. Spiral wound module are predominantly flat sheets and are organized in combination with fluid flow through a small break. The spiral wound module consists of a membrane envelope with spacers and a membrane wrapped around a porous tube. These modules are compact and inexpensive but are prone to scaling and fouling, and thus, pre-filtration is recommended. Typically, tubular pressure vessel and module units are found in a standard industrial spiral wound module. The spacers are used to maintain the gap between the membranes and increase the mass transfer at a low pressure drop. The significant features by the type of spacer that influence the efficiency of the module performance are the filament diameter, voidage, hydrodynamic angle, position to flow (transverse), pressure drop within the spacers, and mesh size [12].

The feed moves axially and permeates a section through the membrane envelope, spiraling through the middle and exiting the collecting channel [13]. In a laboratory scale, the membrane area of this module is commonly 0.2–1.0 m^2, while in the industrial-scale module, it is up to 20–40 m^2. This is also used to consider the pressure drops and increase the separation efficiency when the countercurrent flow is

Table 4.2 Types and variables of spiral wound elements.

Element	Description
Dimension	Element outer diameter Element length Center tube internal diameter
Outer wrap	Fiberglass Tape many designs of sanitary
End of element	Connection of female Connection of male
Center tube	Polysulphone (PSO) is the standard in industrial applications for many forms of polymers used
Feed spacer	30 mil nominal 0.75 mm 47 mil nominal 1.20 mm 90 mil nominal 2.30 mm Even higher spacers are available (1 mil = 1/1000 in.)
Spacer type	Diamond type (standard) Several free channel designs are on the market
Membrane support	Polyester (standard) Polypropylene
Brine seal	U-cup seal Lip seal No seal
Anti-telescoping devices (ATD)	Star type Hole plate type
Glue	Several types of glue used

not necessary. Large pressure systems requiring large pressure vessels and tubing raise the benefit of hollow fiber modules since it cuts device costs of parts by a factor of 10 in certain situations [14].

The positive effects of the spiral wound modules are that the construction is easy and economically viable. This module has a large volume-to-volume packing area over density, which is up to 1000 m^2/m^3. Besides, the feed spacer in this module also contributes to efficient mass transfer [15]. However, the drawbacks of this module are long permeated path and hard to clean. In general, up to 20% of industrial gas separation membranes use spiral wound modules [16–18]. On the market, there is an unbelievable variety of spiral wound elements. In Table 4.2, there are several variables mentioned.

4.2.3 Tubular Modules

Tubular modules are simple in design and are easy to study and analyze mass and heat transport phenomena in different applications. Another significant benefit of these modules is that they can tolerate very large volumes of suspended solids and mainly fibers. There are, however, some drawbacks and the usage of tubular modules:

- They need a huge amount of space.
- Membrane change can be very time consuming and difficult.
- The high internal diameter (ID) (1 in.) tubular systems consume a lot of energy.
- Large internal volumes make the use of chemicals and water costly and clean in place (CIP).
- The manufacturer can change the tubular design expensively and difficultly.

Instead, the advantages of the tubular modules sometime outweigh the disadvantages, and although rather small, the tubular membrane design is on the market.

Usually, the tubular module is a straight tube with a porous supporting sheet and tube. Tubular modules are generally made of ceramic materials and are used in very hygienic conditions. These are easily cleanable due to its larger diameter. Turbulent flow is created due to high cross-flow velocities, leading toward high pressure loss (up to 4 bar). However, due to the much lower specific membrane area as compared with fiber and spiral wound modules, the productivity of such modules is an order of magnitude lower. These modules are widely adopted in microfiltration (MF) and ultrafiltration (UF) applications. These consist of a series of membrane tubes installed inside a shell and tube arrangement structure as shown in Figure 4.4. The membrane tube comprises porous fabrics or synthetic materials with inside specific membranes. Tubes normally have inner diameters of between 5 and 25 mm, whereas tube lengths range from 0.6 to 6 m. There can be 4–18 tubes assembled in the module, though not restricted to this [20]. Tubular modules have some additional features as follows:

- Tubular modules can cope with feed streams comprising reasonably large particles because of their broad internal diameters.
- Besides, mechanical or chemical cleaning methods may be used for easy cleaning.
- This module needs a high pumping capacity since the amount of Reynolds is normally greater than 10 000 in turbulent flow conditions.

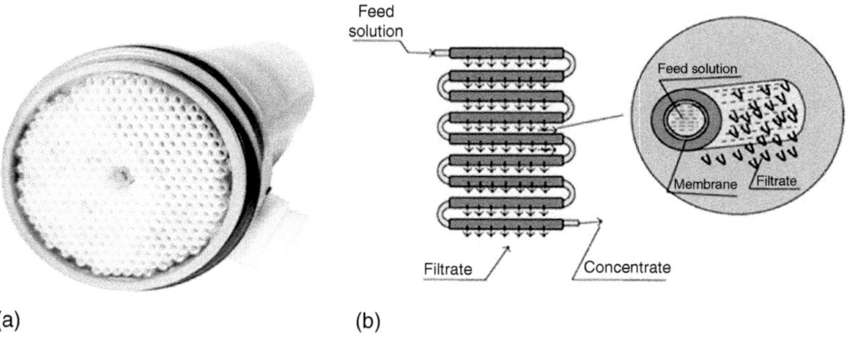

Figure 4.4 Schematics of tubular module. (a) Front view. Source: Al-Anezi et al.[19]. (b) Internal view. Source: Al-Anezi et al. [19]. Licensed under CC BY 3.0.

- The surface area-to-volume ratio of this module is the lowest of the four membrane configurations. There are also significant holdup levels of pipelines requiring massive floor space to function.

Tubular membranes are not self-supporting in comparison with capillaries and hollow fibers. The feed fluid often passes through in the middle of the tubing while the permeate moves into the module housing through the porous supporting tube. Ceramic membranes are predominantly used in the configurations of such tubular modules. However, the tubular module is very poor in packing density, which is less than 300 m^2/m^3. Tubular modules are normally confined to ultrafilters. This usage in the ultrafiltration reduced membrane resistance due to the strong fluid hydrodynamics overweighing the high costs [21].

4.2.4 Hollow Fiber Modules

Hollow fiber modules are widely used in large-scale membrane applications. DuPont pioneered the hollow fiber module for seawater desalination. It consists of an incredibly fine package of membrane fibers bundled into a cylindrical case or shell. The two ends of the fibers are put in the shell with a potting material. The fluid enters a hub, which then equally distributes in each fiber. Shell also has two ports, one for the inlet of fluid and the other for the exit of that fluid. The packing density of hollow fiber modules can be as high as 10 000 m^2/m^3. The external diameter of a single fiber is in the range of 0–200 μm. The thickness varies depending upon the porous substrate (support layer) and can be as little as 20 μm. Typical schematics of the hollow fiber module is provided in Figures 4.5–4.9. The hollow fiber properties are presented in Table 4.3.

Many efforts have been put into the research and development of new types of membrane contactors that enhance the mass transfer coefficient and decrease the capital and operating cost of contactor devices. Among them, hollow fiber membrane (HFM) contactor modules are found to be efficient and economical. Because of their low cost but, overall, higher packing density, hollow fiber membrane contactors (HFMCs) are practically more common in applications. An HFMC contains many fibers, usually in the form of a bundle joined together with a dense polymer (epoxy) at the ends of the module shell. Such modules are similar to shell and tube heat exchangers, where a bundle of tubes is placed inside the shell. The membrane modules of hollow fibers shown in Figure 4.8 are made up of two simple geometries: shell-side feed design and bore-side feed design.

Shell-side module typically operates with high-pressure applications with a maximum capacity of 1000 psig. A loop or closed bundle of fibers in a pressure vessel is enclosed in this module. This module is pressurized from the side of the shell; it permeates the fiber wall and flows through the fiber ends. The fibers generally have narrow diameters, and high-density walls, normally of 50 μm inner diameter and 100–200 μm exterior diameter, since the fiber's surface has to withstand significant hydrostatic pressure. The fibers are opened at both ends in the bore-side feed system, and the feed fluid circulates in the fibers' bores. The diameters are usually greater

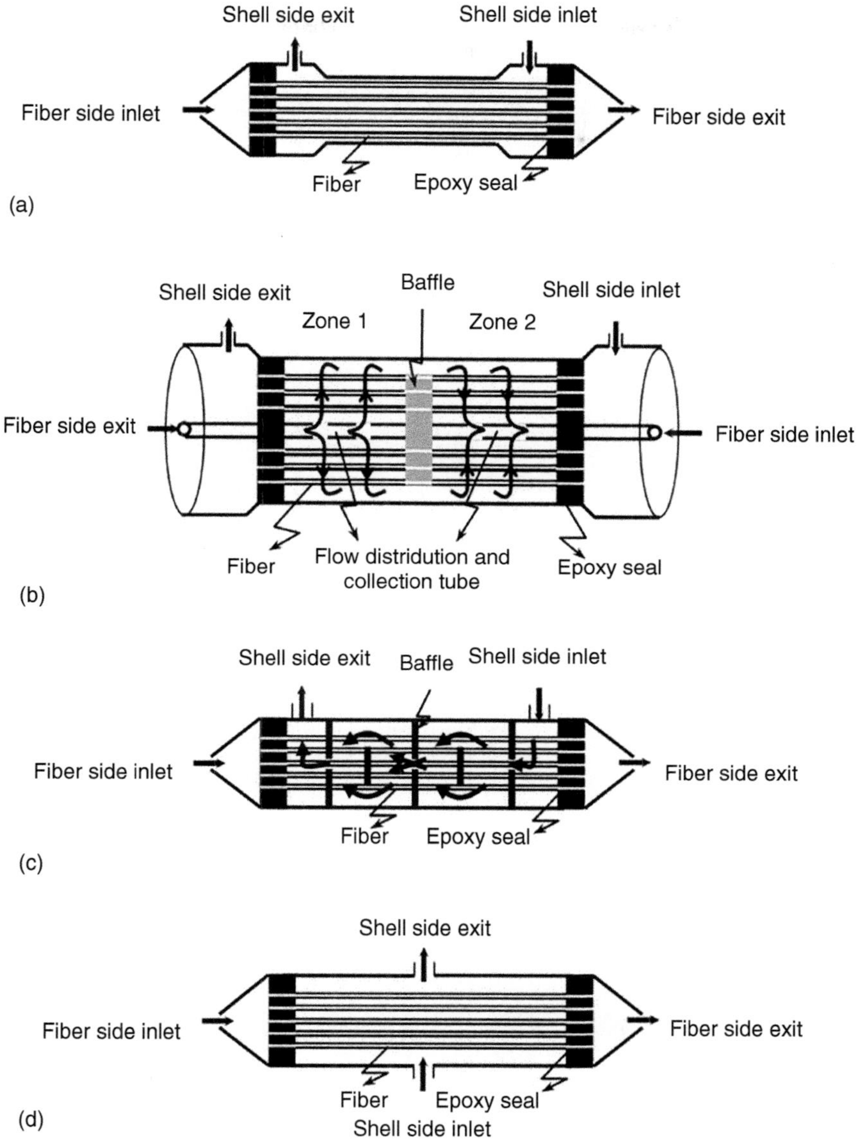

Figure 4.5 Schematic drawings of cylindrical modules: (a) parallel flow, (b) single baffled cross flow, (c) fully baffled cross flow, and (d) baffle less fully radial flow. Source: Redrawn from Wickramasinghe et al. [22].

than those of the fine fibers used in the shell-side feed method to reduce pressure drop in the fibers, which are typically rendered with a spinning solution. In this form of a module, the feed pressure is normally restricted to less than 150 psig.

A wide surface membrane bundled in a single module is one of the hollow fiber module's significant benefits. The challenges of module on the shell side are that the membrane feed often creates fouling, which requires the feed source to be pretreated. For medium-pressure feed streams of up to 150 psig, the bore-side

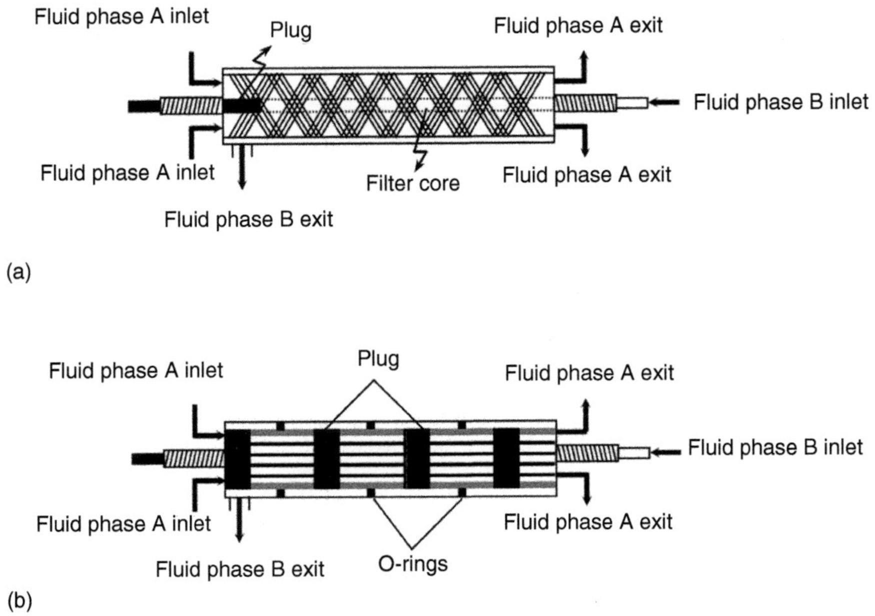

Figure 4.6 Schematic diagram of helically wound HFMC module. (a) Hollow fibers wound helically around a central core. The plug forces liquid entering the core radially outward. (b) Woven hollow fibers wound helically around a central core. The plug and O-rings provide multiple shell passes. Source: Redrawn from Wickramasinghe et al. [22].

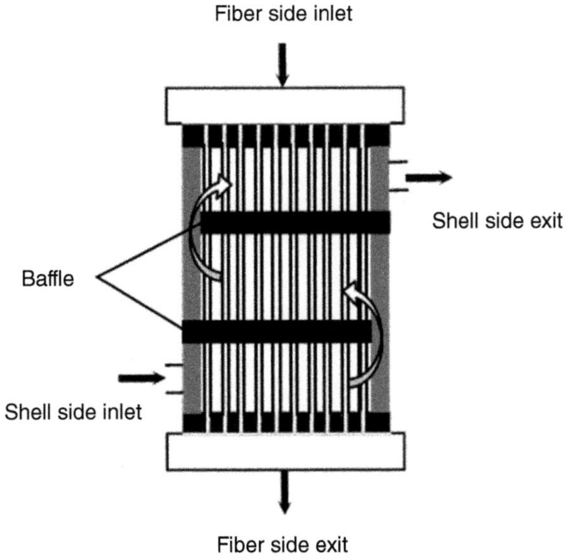

Figure 4.7 Schematic diagram of a baffled rectangular module. Source: Redrawn from Wickramasinghe et al. [22].

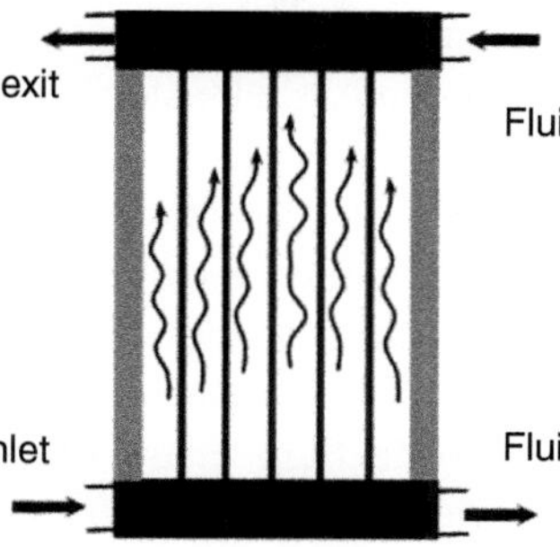

Figure 4.8 Schematic diagram of a crimped flat membrane module. Source: Redrawn from Wickramasinghe et al. [22].

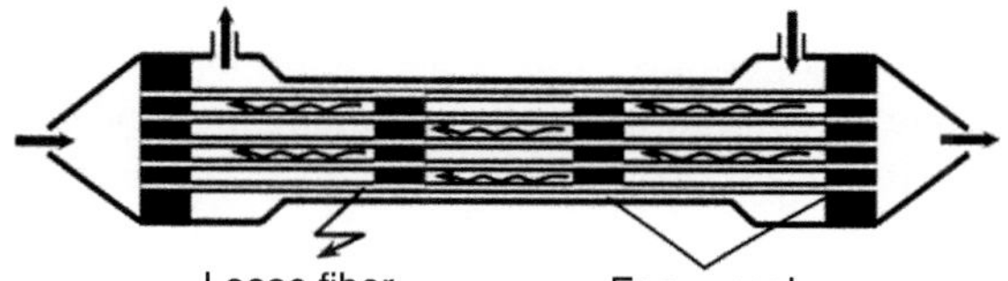

Figure 4.9 Schematic diagram of proposed parallel flow loose hollow fibers in shell with epoxy sealing at different lengths. Source: Redrawn from Wickramasinghe et al. [22].

Table 4.3 Hollow fiber properties.

Characteristics	Description
Structure	Self-supporting
Active layer	Inner–outer diameter
Feed	Lumen/shell side
Inner diameter	40–250 μm
Outer diameter	80–400 μm
Packing density	<10 000 m^2/m^3
Allowed pressure	100 bar shell side; 15 bar lumen side
Applications	Gas purification, reverse osmosis, membrane distillation, ultrafiltration, pervaporation, osmotic evaporation, pressure-retarded osmosis

Source: Iaquaniello et al. [23].

feed module is widely used to eliminate fouling and concentration on the membrane's feed side. Ultrafiltration, pervaporation, and certain low-pressure to medium-pressure gas applications are used in bore-side feed types.

HFMC are commercially found in different geometries. Among them, the usual membrane contactor geometries investigated in the literature are based on the design of a cylindrical tube bundle, helically wound fibers bundle, rectangular bed of fibers, and crimped flat membranes. The module's design varies with the inlet and exit port of fluid phases, size of the module, shell-side flow configuration, packing density, and casing of the module.

HFMC based on cylindrical tubes designs are the parallel flow fibers-in-shell baffle-less modules, cross-flow fibers-in-shell baffled modules, and radial flow fibers-in-shell baffle-less modules. The schematics of these module designs are shown in Figures 4.5 and 4.6. Figure 4.5a,d shows the conventional "fibers-in-cylindrical shell module" wherein shell-side fluid phase flows parallel and perpendicular to fibers, respectively. Cylindrically mounted fibers in shells are also made of placing baffles in the shell side to create the shell-side flow perpendicular to fibers (Figure 4.5b,c).

Helically wound fiber bundle module is an annular bed of hollow fibers wound helically around a central core as shown in Figure 4.6. Fluid phase A enters the central core and is distributed forcibly by a plug outward perpendicular to the fibers (part (a)). The second configuration of the helically wound fiber bundle module is similar to the first one except that the shell contained plugs and O-rings to provide multiple shell-side passes (part (b)).

The rectangular bed of fiber module configuration used the woven fabric fibers instead of individual fibers assembled in a rectangular shell rather than in a cylindrical shell (Figure 4.7). The baffle makes the shell side flow perpendicular to fibers rather than parallel to fibers.

The crimped flat plate membrane modules (Figure 4.8) use a flat crimped membrane channeled like the shape of isosceles triangles.

These modules' performance was analyzed to determine the better values of extraction efficiency, mass transfer coefficient, lower pressure drops, and larger surface area per unit volume. The module geometries when compared for different applications indicate different performances [7]. For example, Wickramasinghe et al. [22] stripped oxygen from water into nitrogen using the aforementioned four-module configurations. The crimped flat plate membrane module showed 72% oxygen removal efficiency for equal flow per unit of membrane area as compared with 82%, 86%, and 98% for the cylindrical bundle, helical bundle, and rectangular bundle configurations, respectively. Yeh and Chen [24] studied cross-flow rectangular hollow fiber modules for the extraction of acetic acid by methyl isobutyl ketone. They presented a theoretical analysis of mass transfer analogous to cross-flow heat exchangers and verified with experimental results. According to Jons et al. [25], high mass transfer and low pressure drop are achieved in rectangular modules with cross flow. Another study by Wickramasinghe et al. [22] argued that baffles do not affect the mass transfer coefficient. However, higher values of the mass transfer coefficient are obtained for transverse flow modules in comparison with parallel flow modules [6, 11]. Notwithstanding the proficient performance of these module configurations' special geometry, the complexity of these modules in their manufacturing and mass transfer analysis overcomes their performance advantage. In contrast, the cylindrical bundle membrane contactor modules have the advantage of low manufacturing cost. Besides, design-related specifications of such modules as similar to heat exchange units available in the literature encouraged the researchers

Table 4.4 General comparison of membrane modules.

Modules/properties	Tubular	Flat plate	Spiral wound	Hollow fiber
Packing density (m^2/m^3)	60	300	600	1200
Energy cost (pumping)	High	Moderate	Low	Low
Channel spacing (cm)	1–2.5	0.03–0.25	N/A	0.02–0.25
Particulate plugging	Low	Moderate	Very high	High
Ease of cleaning	Excellent	Good	Moderate	Fair
Holdup volume	High	Low	Low	Low
Potential for fouling	Low	Moderate	High	Very high
Relative manufacturing cost	High	High	Moderate	Low

Source: Based on Bhattacharjee et al. [26], Kucera [27], and Obotey Ezugbe and Rathilal [28].

for their vast use in membrane-based solvent extraction processes [12]. Another design of the hollow fiber module is a crimped cylindrical module where the fibers are supposed to be sealed with a dense polymer (epoxy) at different lengths within the shell. The fibers are assumed to be mounted loosely between the epoxy seal and thus are supposed to create cross flow without using a baffle. The advantage of this configuration is the lower pressure drop and higher mass transfer coefficient prediction. Moreover, efficiency can be increased while using viscous fluids. The schematic of such configuration is shown in Figure 4.9.

The general comparison, manufacturers, and applications of all the modules are tabulated in Tables 4.4–4.6.

4.3 Membrane Contactor Module Housing

For the different modules, the market has various designs of housing. These can be divided into materials (stainless steel and polymer) and functions (boxes with side port input and output/output and output via end caps). Made of glass fiber-reinforced polymer (FRP) nearly exclusively, the architecture is over 30 years old and performs well but without any difficulties. It works well with groundwater and seawater desalination systems. Its use is problematic for all other products. Most of the polymer cases are secured with a tension ring. The boxes were made solely for milk purposes. They are now becoming more commonly used and are equipped with regular side ports. Some stainless steel housings are built in the same way as reinforced glass fiber housings, making the removal of caps extremely difficult. Indoors, electric polishing of inox steel boxes should be applied. If not, elements cannot be pushed in and out of them. The glass fiber-enhanced housings have just three regular diameters: 2.5, 4, and 8 in. In a broader range of standard and nonstandard diameters, stainless steel housings are available. Unfortunately, for Europeans, the most commonly used measurement in stainless steel housings is US inch. The standard milk sizes are not too easy to obtain, but in Europe, diameters of 2.5, 4, and 8 in. are rare.

Table 4.5 Manufacturers of membrane modules.

Modules/ manufacturers	Tubular	Flat plate	Spiral wound	Hollow fiber
Company's name	Rich Kirin Environment	Chengdu Rosun Disinfection Pharmaceutical Co., Ltd.	Shandong Jinhuimo Technology Co. Ltd.	3M Science. Applied to Life
Module commercial name	YLUF-1060W	RF-0.25 Membranes	UF-001	Liqui-Cel™
Specification	1715 × 250 D/mm	465 × 365 × 7.5 mm	Dimensions (1812–8040)	Diameter (64–102 mm) Length (76–152 mm)
Conditions and limitations	3 yr lifetime	4–6 yr lifetime	N/A	N/A
Applications	0.02–0.2 MPa external pressure-based operating mode with yield capacity of 3000 litre per hour (LPH)	Sewage treatment equipment	Sugar decolorizing, fruit juice clarification and dye purification/concentration, etc.	Ideal for use with compatible solutions containing organic solvents and surfactants, such as inks, coatings, photoresists, emulsions, and plating solutions
Schematic	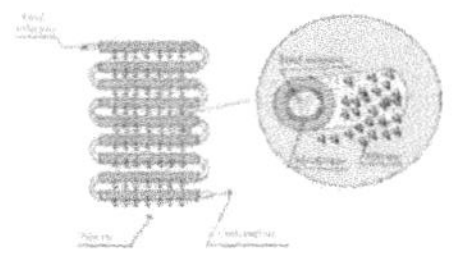	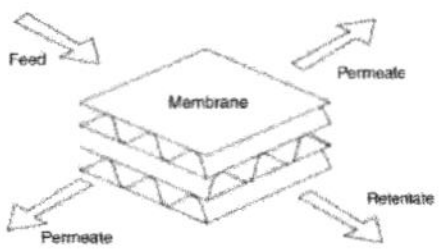	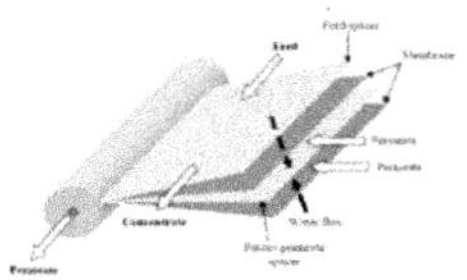	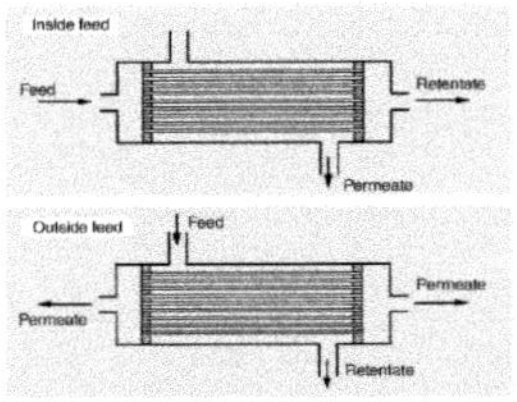

Table 4.6 Applications of membrane modules.

Module		References
Hollow fiber	**Liquid–liquid**	
	Carbon dioxide removal from water	[29]
	Copper and zinc extraction from aqueous solution	[30]
	Extraction of liquid–liquid-phase actinides	[31]
	Liquid–liquid organic solvent extraction	[32]
	Liquid–liquid extraction of copper, cobalt, cadmium, and zinc	[33] [34] [35]
	Production of porous hydrogel and hydrogel microspheres	
	Ammonia removal from water	
	Gas–liquid	
	SO_2 absorption from flue gas	[36]
	CO_2 removal using gas–liquid contactor	[37]
	Modeling of CO_2 removal using a liquid–gas contactor	[38]
	Natural gas sweetening using a liquid–gas contactor	[39] [40] [41]
	Acid gas capture	
	Olefin/paraffin separation	
	Gas–gas	
	SO_2 removal from simulated flue gas	[42]
	CO_2 removal from pressurized CO_2–CH_4 mixture	[43]
	Natural gas purification	[44]
	Post-combustion CO_2 capture	[45]
Tubular	**Liquid–liquid**	
	Hydrodynamics study of water–oil mixture separation	[46]
	Application for oil-in-water stable emulsion	[47]
	Gas–liquid	
	Mass transfer study of dissolved gases	[48]
	CO_2 recovery from liquids	[49]
	Gas–liquid-based microfiltration	[50]
	Mass transfer investigation in gas-sparged ultrafiltration	[51]
	Gas–gas	
	Water–gas shift reaction	[52]
	Selective CO_2 separation	[53]
	Partial oxidation of CH_4 in coke oven gas to syngas	[54]

(Continued)

Table 4.6 (Continued)

Module		References
Spiral wound	**Liquid–liquid**	
	Liquid membrane for the recovery of chromium and zinc	[55]
	Separation of styrene/ethylbenzene mixtures	[56]
	Liquid–liquid extraction of calcium	[57]
	Design of a 4-end spiral-wound L/L extraction membrane module	[58]
	Gas–liquid	
	Biofouling removal using two-phase flow cleaning	[59]
	Post-combustion carbon dioxide capture by gas–liquid absorption processes	[60]
	Cleaning with gas–liquid two-phase flow process	[61, 62]
	Amine-based solvents regeneration in gas–liquid membrane	[63]
	Separation of fluids and gases	[13]
	Separation of volatile materials by gas membranes	[64]
	Gas–gas	
	Natural gas processing	[65]
	CO_2/CH_4 separations	[66]
	Gas separations	[67]
	CO_2 capture from flue gas	[68]
Plate and frame	**Liquid–liquid**	
	Lactic acid recovery	[69]
	Liquid-phase microextraction	[70, 71]
	Water applications	[72]
	Gas–liquid	
	CO_2 removal using gas–liquid contactor	[37]
	Gas–liquid flow in ultrafiltration and microfiltration	[73]
	Post-combustion carbon dioxide capture by gas–liquid absorption processes	[60]
	Gas separation using a gas–liquid contactor	[74]
	Gas–gas	
	High-capacity gas separation applications	[75]
	Ultrafiltration by gas sparging	[76]

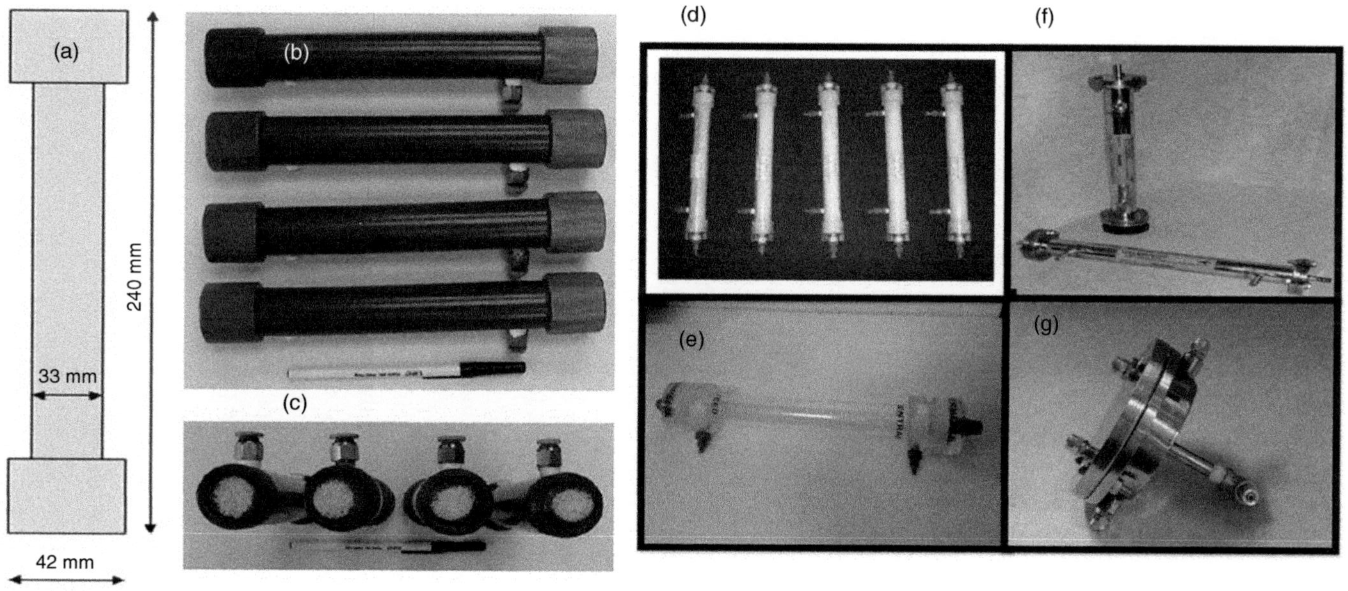

Figure 4.10 Membrane contactor module housing. (a) Font view schematic, (b) close housing, (c) in housed fibers, (d,e) acrylic type visible housing, (f) Steel housing, (g) flat sheet housing.

The 4-port style, which uses the housing as a building block without the external product manifolds, and the other type, which uses an arrangement using external manifolds, are available in two varieties. Which type is better can be argued, but most systems are built with external multiples. It allows high flow in sanitary systems, which is the major advantage of side entry. This design is seldom used in water desalination. The dimensions of the spiral wound are a jungle. The outer diameter, the length of the element, and the inner diameter of the tube are not standardized. This makes it quite difficult and creates a lot of problems for housing producers to switch from one element to another. Typical different shapes of membrane contactor housings are shown in Figure 4.10.

4.4 Membrane Module Flow Configuration

Flow configuration or mode of fluid operation is another important factor after the proper module has been selected. Module operation mode is described by the fact that the feed is allowed to flow in the shell side or inside the fiber in case an HFMC is used. Then, recirculation or continuous mode of operation is also a choice based on the degree of separation of a particular species. Cocurrent or countercurrent mode is also selected as it can affect the mass transfer coefficient. In general, when permeated and resented in the countercurrent stream or when feed is radial, cross-flow module output is better or multidirectional with the help of a distributor installed in the shell's housing. The membrane module may be regulated by either dead-end or cross-flow filtrations. Each flow configurations will be addressed in more depth.

In the radial gas flow, gas flows radially through the fiber to the residue collector's middle. The fiber penetration flows to the permeate collector axially. In the permeator, many conditions vary. In fact, for example, fibers are pressurized. Membrane permeability for the different components is a function of their structure, pressure, temperature, and temperature variations inside the membrane package. However, in many cases, these factors are constant and are ignored by their modeling variation. In the countercurrent flow, the feed is inserted into the pack from a distributor at the same end. The feed fluctuates axially through the fiber before the residue outlet is hit (ideally retaining a steady composition). The fiber penetrating is aimed into the feed oppositely. The top of the surface is at the edge of the product, and the floor is at the top. In the cocurrent flow, except that the feed or permeating flow is in the same direction, the cocurrent membrane permeator is closed to the opposite permeator. A schematic of the flow pattern in the HFMC module is shown in Figure 4.11.

4.5 Membrane Materials

The membranes are classified into anisotropic (asymmetric) and isotropic (symmetric) membranes in a transection morphology, as shown in Figure 4.12. Anisotropic membranes are composed of microporous membranes, nonporous dense membranes, and electrically charged membranes. These can also comprise the composite

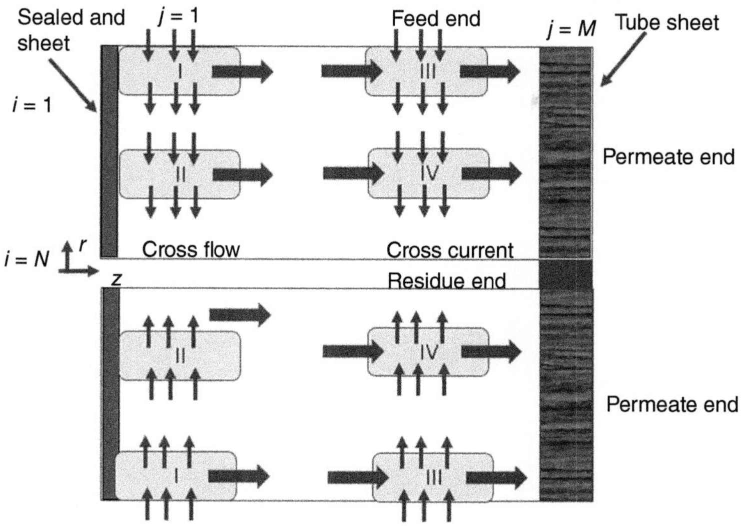

(a) Each element illustrated represent a ring radius '*R*' and thickness 'Δ *r*' and 'Δ *z*'

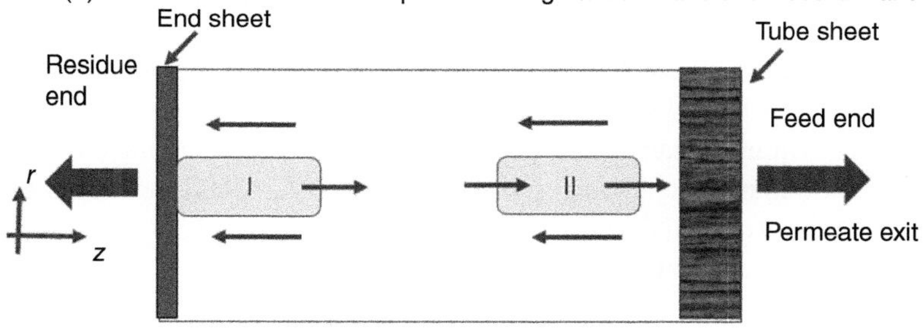

(b) Each element illustrated represent a ring radius '*R*' and thickness 'Δ *r*' and 'Δ *z*'

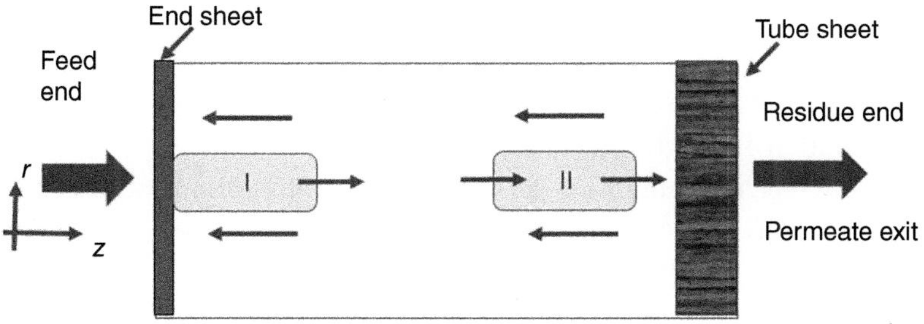

(c) Each element illustrated represent a ring radius '*R*' and thickness 'Δ *r*' and 'Δ *z*'

Figure 4.11 Flow patterns in membrane contactors: (a) radial cross flow, (b) countercurrent flow, and (c) cocurrent flow. Source: Redrawn from Dindore et al. [77].

membrane, incorporated asymmetric membranes, and assisted liquid membranes. The right membrane and membrane material can be difficult to choose for a given process. Any general process environment details must be accessible for proper collection. Initially, the process of choice (UF, reverse osmosis [RO], nanofiltration [NF], or MF) and the availability of membrane substances are to be determined.

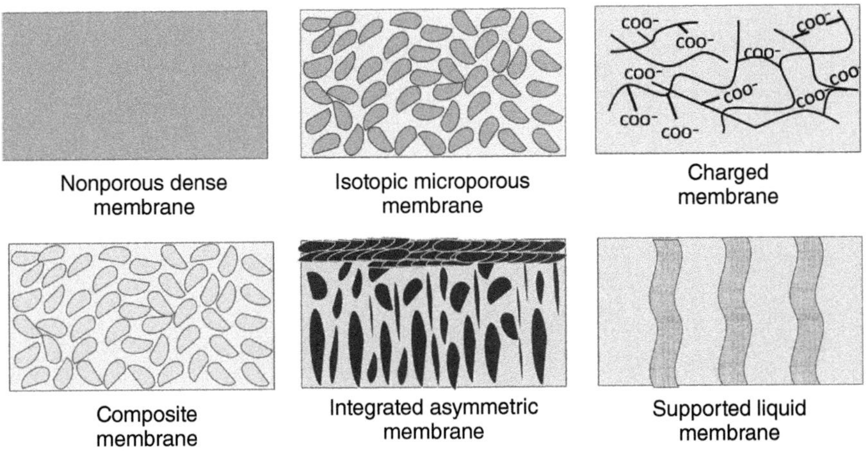

Figure 4.12 Membrane surface morphology. Source: Baker [78].

With the exception of applications, it could be difficult to choose membrane materials, and more than one membrane is often questioned. A thorough consideration must be done in selecting the best materials for membrane synthesis. It is vital to produce a membrane with the desired separation performance and good properties. The effectiveness in membrane performance is influenced by the interaction of the gas penetrant and materials making up the membrane [79]. Each selected material must have an affinity toward the targeted penetrant. Listed below are the key factors for the guidelines of materials selection (Figure 4.13; Tables 4.7 and 4.8) [101]:

i. High efficiency of separation with reasonably high fluxes for selected species as it minimizes the membrane area and capital cost.
ii. High chemical resistivity toward reactive organic vapors, aromatics, solvents, lubricating oil, and other causes that may lead to membrane swelling, to prolong the membrane's lifespan.
iii. Good mechanical stability, to maintain the membrane's performance throughout its operation due to the pressure difference.
iv. High thermal stability; ability to withstand high operating temperature.
v. Inexpensive, to produce a cost-effective membrane while possessing the intrinsic properties of high performance and durability.

4.5.1 Polymer Materials

Polymer-based membranes are typically categorized into two types: glassy and rubbery. Glassy polymers can sometimes be wholly amorphous and noncrystalline that are kept below their glass transition temperature (T_g), whereas rubbery polymers are more flexible and soft and operate above their T_gs [102]. The glass transition is a physical transition state that occurred in the polymer (i.e. hard brittle state to a rubbery state), and it possesses a reflective influence upon the gas transport properties [103]. In glassy polymers, the steric hindrance along the polymer backbone

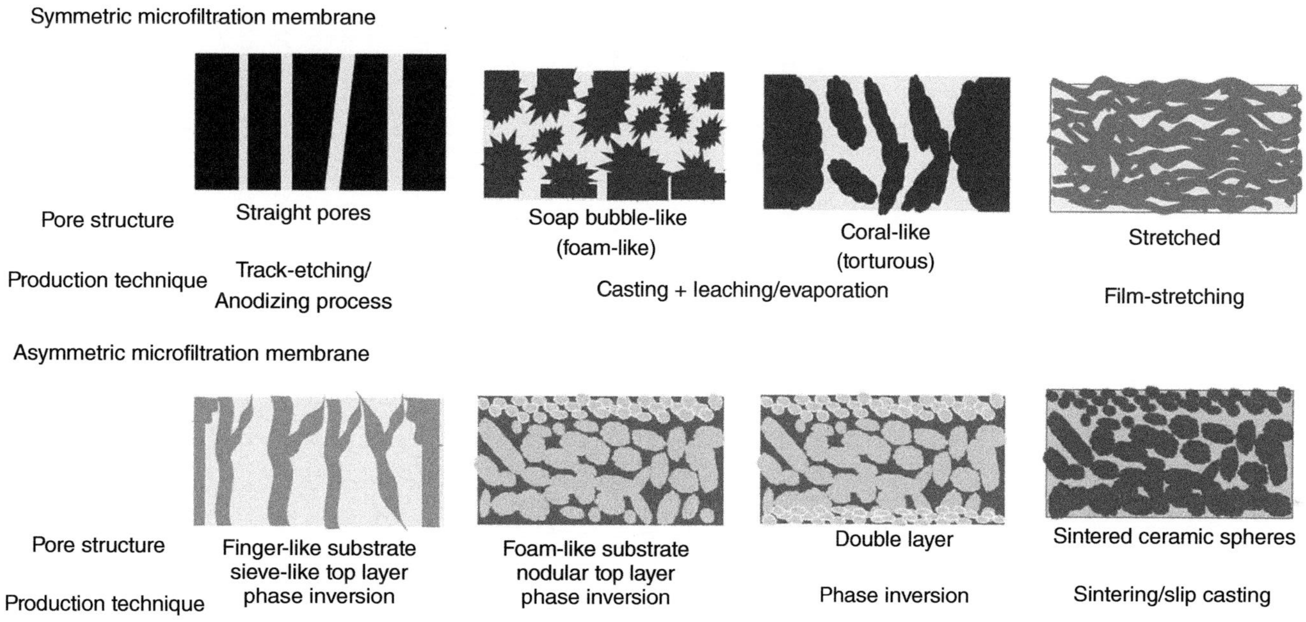

Figure 4.13 Membrane morphology. Source: Redrawn from Scott [80].

Table 4.7 Texture properties of membrane contactor materials.

Membrane material	Porosity (%)	Pore size (μm)	Thickness (μm)	References
Polyvinylidene fluoride	60–75	0.20–0.45	110–140	[81]
Polytetrafluoroethylene	Wide range	Wide range	Wide range	[81]
Polypropylene	70	0.2	45.0	[81]
Polypropylene	35–38	0.02–0.03	25	[81]
Polypropylene	N/A	0.04	40	[82]
Polyvinylidene fluoride	0.85	0.16	150	[83]
Polytetrafluoroethylene	0.64–0.75	0.20–0.45	65–175	[84, 85]
Polytetrafluoroethylene	N/A	0.45	50	[86]
Polytetrafluoroethylene	0.60	1.00	20–100	[87]
Polytetrafluoroethylene	0.75	0.45	60	[88]

has limited the mobility of polymer chains. Due to that, glassy polymers always appear as rigid, hard, and brittle. Since the thermal motion is restricted in glassy polymer, thus the permeate diffusion coefficient is low. When a glassy polymer is heated above its T_g, the rise in thermal energy will surpass the restriction of the steric hindrance rotation of polymer backbone segments and later changes its state from glassy to rubbery [104]. In recent years, glassy polymers dominate industrial membrane applications for their high selectivity and prominent mechanical properties. Table 4.9 listed a few examples of common glassy polymers that are being utilized for membrane synthesis.

Rubbery polymers that are softer and more flexible due to the rubbery polymer chains exhibit a vice versa behavior compared with the glassy polymers: higher permeability but lower selectivity. The rubbery polymer chains can rotate along their axis, thus providing more shape and size freedom to increase permeability yet reducing its selectivity. In a specific gas mixture, the permeation may vary inversely to its separation factor where a more compact molecular spacing causes an increase in the polymer's separation characteristic. The successful demonstration of rubbery polymer membranes led to the commercialization of polymeric membranes for gas separation [106]. The common rubbery polymers currently being utilized in membrane gas separation are presented in Table 4.10.

Usually, the researchers who develop gas separation membranes refer to Robeson's upper bound correlation to qualitatively conclude the process changes in permeability from solution diffusion to Knudsen diffusion. The upper bound plot was first founded by Robeson [108–113]. Later, it was modified in 2008 where the permeability–selectivity correlation follows the relationship of $P_i = k\alpha_{ij}{}^n$, where the P_i is the fast gas permeability, α_{ij} (P_i/P_j) is the separation factor, k is the "front factor," and n is the slope of the log–log plot between the relationship [114]. Most glassy polymer membranes are located close to the upper bound of Robeson's trade-off plot, indicating the higher selectivity than the rubbery polymer membranes. This

Table 4.8 Operating factors in membrane materials.

Property/factor	Description	Examples	References
Distribution	Distribution of ingredients insolvent, homogenous, heterogenous	In a comprehensive review, the pattern in solvent treatment dependent on polarity was aimed at harnessing certain results to intensify membrane processes	[89]
Shape	The contour of the material whether near spherical or rod shape	With a growing particle size, the permeability of silicalite–PDMS mixed membranes is expected to improve. At the higher zeolite rate, the differences in the permeability values are considerably more prominent with shifts in particle size	[90]
Composition of material	The weight percent of ingredients in material synthesis	Membrane pore size and porosity reduce with an increase in polymer concentration	[91]
Zeta potential	Electric charge potential across the interface between membrane materials	Determining variables for membrane fouling are their zeta potential and the subsequent electrostatic interactions	[92]
Pore size	It is identified by the particles' diameter that it can retain on the membrane and others pass through	In the clarification of grape marc extract by microfiltration (MF), the effect of membrane pores on penetrating discharge, flooding, and rejection of soluble and suspended solids, phenolic, and anthocyanin products was studied	[93]
Morphology	It indicates the pore structure such as uniform, foam or bubble-like, symmetric, asymmetric, with skin layer, etc.	As a consequence of less cake forming, the Isopore membrane results in strong filtration efficiency. Durapore membrane filtration efficiency is smaller than the Isopore membrane filtration efficiency. It is because more particles build a filter layer on the surface of the Durapore membrane. Owing to the most cake forming and extreme pore blocking, the MF-Millipore membrane results in poor filtration efficiency	[94]

(Continued)

Table 4.8 (Continued)

Property/factor	Description	Examples	References
Thermal stability	The ability of membrane material to maintain its integrity and functionality at an elevated temperature and pressure	Cellusolse acetate (CA) as a material has an inherent temperature limitation, and CA membranes are limited to an upper operating temperature of approximately 35 °C. PSO, PVDF, and polyacrylonitrile (PAN) can tolerate higher temperatures. PSO and PVDF membranes are known to operate at 95 °C without problems. PSO membranes will operate at temperatures up to 120 °C. Composite membranes can, as a general rule, operate at least up to 80 °C, and they will withstand higher temperatures at low pressure during, for instance, thermal disinfection. The temperature capability of a membrane system is, in most cases, not dictated by the temperature limitation of the membrane, but mainly through membrane configurations and other membrane system components	
Chemical resistance	The ability of membrane material to withstand upon exposure to a chemical agent. It is varied with time, temperature, pressure, and concentration	In contrast with the conventional chemistry absorbent process, there is cost rivalry from a two-stage membrane device made of the membrane	[95]
Mechanical strength	The ability of membrane material to maintain its integrity and functionality under mechanical stress	Every membrane is pressure sensitive. The term "compaction" is also used to define an irreversible "pull" of a membrane. In addition to the robustness of the membranes, it is important that the membrane is supported adequately to keep it from squeezing under strain. It is essential to read and closely observe the manufacturer's requirements	
Hydrophilic	The ability of membrane material to be soaked or wetted in water or have an affinity for water	Often hydrophilic polymers such as polyvinylpyrrolidone (PVP) and polyethylene glycol (PEG) are used for the enhancement of ultrafiltration membrane (UFM) hydrophilicity and efficiency to clean	[96]

Hydrophobic	The ability of membrane material not to be soaked or wetted in water or repel the water	Due to the surfactant–PVDF interaction, the bulk of the hydrophobic surfactants behaved poorly	[97]
Antifouling	The ability of membrane material to retain the solutes or contaminants for a longer duration without damaging the membrane material	The flow of water permeated to the surface of the membrane and the feed spacer differ during operations. Fouling reduces permeation flow and may also raise the feed pump pressure such that the same flow rate of permeate water continues to occur. Therefore, it is important to recognize temporal variables such as fouling or aging to estimate real energy use reliably over time	[98]
pH	Most membranes are extremely immune to intense pH values, with CA membranes the extraordinary exception. The biggest restriction is the usage of multi-membrane polyester support that essentially restricts the upper practical limit to pH 11.5. Often membranes modify their work with very large pH values, but they are still useful and efficient. The majority of membranes at low pH values are safe	The rejection profiles tend to be strongly dependent on pH for polymeric membranes, including polycarbonate and sulfonated membranes. These membranes can have positively and negatively charged groups, whose charges depend on the pH	[99]
Viscosity	The viscosity of the feed is not a concern in itself, but high viscosity results in a higher pressure drop for a specific flow. Unless the pressure drop is appropriate and the flow is satisfactory and steady, there is no operational issue with the viscosity. High-viscosity feeds are more an engineering challenge than a membrane issue to manage in a membrane device	Extra resistance on the membrane–liquid interface can be provided by viscous solutions	[100]

Source: Modified from Scott [80].

Table 4.9 Typical glassy polymers used in membrane development.

Polymer	T_g (°C)	Density (g/cm³)	Remark	Typical applications
Polyimide	285	1.39	High gas separation performance, thermally and mechanically stable	Separation of nitrogen/volatile organic carbon (VOC) mixtures CO_2 separation CO_2 hydrogenation Hydrogen separation
Polyethersulfone	225	1.35–1.38	High plasticization pressure, thermally stable	Ultrafiltration Production of pentyl valerate
Polysulfone	220	1.22–1.24	High gas affinity due to its polar characteristics	Food and beverages, pharmaceuticals, semiconductor water
Cellulose acetate	135	1.23–1.32	Good swelling properties, naturally hydrophilic	Air pollution, foods, and pharmaceuticals
PTFE			Hydrophobic, superior chemical and temperature resistance	Air and gases, pharmaceuticals, aggressive chemicals
Polypropylene			Hydrophobic, chemically inert, broad pH stability, high-temperature resistance	Chemicals, microelectronics, pharmaceuticals

Source: Based on Prabhakar [105].

Table 4.10 Typical rubbery polymers used in membrane development.

Polymer	T_g (°C)	Density (g/cm³)	Remark
Polyvinyl acetate	30	1.19	Has C=O polar bond to promote better affinity toward CO_2
Polydimethylsiloxane	−123.5	1.20	High gas permeability due to the availability of large free volume
Polyurethane	−73.6	1.27	High permeability but low selectivity
Polyisobutene	−73	1.52	Good for H_2/CO_2 separation

Source: Based on Ghosal and Freeman [107].

phenomenon obeys the solution–diffusion mechanism, particularly with the gases' molecular sizes. The larger gas molecular size restricts the interaction with the polymer chains, causing the diffusion coefficient to decrease. For example, a gas molecule such as CO_2 with a small kinetic diameter (33 Å) is more favorable to pass through over larger molecules like CH_4 (38 Å). Additionally, gas condensability also influences the energy required by the gas molecules (sorption coefficient) to be absorbed in the polymer. Generally, the sorption coefficient will increase as the gas condensability increases. Therefore, the sorption selectivity favors easily condensable gas like CO_2 more than CH_4 [104].

Nonetheless, the diffusion selectivity highly relies on the polymers that made up the membrane, especially their glass transition temperature. Since glassy polymers that operate below their T_g are tougher, more rigid, and deficient in chain mobility, the molecular size difference much likely affects relative mobility. On the other hand, rubbery polymers that operate above their T_g possess larger intersegmental polymer chain motion. In theory, glass polymer membrane separates gases principally by size molecule, whereas gas condensability is used for rubbery polymer membranes. Owing to the distinctive characteristic of glassy polymers such as chain mobility (to afford more space and shape-selective environment) and their good mechanical and thermal properties, they are always favorable to be applied in many industrial gas separation processes [104]. These are the key factors to be understood and adapted in obtaining the desired separation performance from different polymeric materials.

4.5.2 Inorganic Fillers

Inorganic materials are getting worldwide attention in membrane fabrication due to their exceptional separation performance over organic/polymeric materials. There are several types of inorganic materials used as fillers, especially in the development of mixed matrix membranes like carbon molecular sieves (CMSs), zeolites [115–117], silica, carbon nanotubes (CNT), zeolitic imidazolate frameworks (ZIFs) [118], metal–organic frameworks (MOFs) [119, 120], and metal oxides. These materials are well known as research incentive for their capability to surpass the upper bound separation limit of polymeric membranes.

Table 4.11 Different type of inorganic materials and their respective properties.

Inorganic material	Properties
Metal oxide (e.g. titanium, aluminum)	Nanoscale, high specific area of particle distribution, nonselective void formation [126–128]
Carbon molecular sieve (CMS)	Strong adsorptive, average mean pore size of 3–6 Å [129, 130]
Zeolite	High diffusivity, difficult to fabricate, quite expensive, brittle [131–133]
Silica	Better particle distribution, good adhesion, needing modification to enhance chemical stability [134]
Metal–organic framework (MOF)	Sufficient porosity having good affinity to certain gases, rigid, and finite pore size [135, 136]
Carbon nanotubes (CNT)	Quite hazardous, hard to attain uniform dispersion, weak interaction with polymer matrix [137–139]
Zeolitic imidazolate frame (ZIF)	Similar properties to zeolite, stable thermally and chemically, hydrophobic [140]
Nanoclay	Ability to shrink the size of phase dispersion of other domains, good mechanical property [141, 142]

The incorporation of inorganic materials within the membrane structure will produce a synergistic effect [121]. Among the advantages of having such inorganic materials as fillers are their resistivity against corrosion and high temperatures, characteristic stability when operated, ability to regenerate or multiply, and bacterial attack resistance [122–124]. Nevertheless, the challenges in developing such membranes, specifically for large-scale separation processes, are their cost of production, membranes' fragility, and other difficulties during production reproducibility [125].

Such materials as the dispersed phase onto the properties of the composite membranes relate to its types, chemical structure, and surface chemistry. These materials can be categorized into two types: porous and nonporous [9, 122, 125]. Porous filler plays its role as a molecular sieving agent within the polymer matrix. It separates the gases based on their sizes and shapes to boost the permeability and selectivity due to its concise aperture. On the other hand, the dense or nonporous filler can alter the polymer chain packing (especially toward glassy polymers) to produce a larger free volume, which results in a considerable enhancement in permeability. The different types of inorganic materials and their individual properties and applicability for membrane production are shown in Table 4.11.

4.6 Membrane and Membrane Module for Membrane Distillation (MD) and Osmotic Membrane Distillation (OMD)

Suitable pore size and their distribution in osmotic membrane distillation (OMD) applications can be generated from low surface energy hydrophilic polymers. They contain polyolefins like polyethylene (PE) and polypropylene (PP) along with perfluorocarbons like polytetrafluoroethylene (PTFE) and polyvinylidene fluoride (PVDF). The pore diameter for the membranes with given components ranges from 0.1 to 1.0 μm. This will allow maintenance of the gas phase in the membrane that is the fundamental condition in OMD operations [143]. Modification of these hydrophilic membranes can be done by two methods: (i) grafting on surface molecule containing hydrophobic fluorocarbons chains like fluoroalkyl silane and (ii) coating with thin lipid film on the surface of the alumina membrane. Both of these modified membranes have successful applications in OMD operation [144, 145].

HFMs with thin walls have great application in OMD operations because of their high surface-to-volume ratio and the absence of support or spacers. Pore size, porosity, conductivity, and thickness are few parameters to be considered during the selection of membrane. Larger pore-sized membrane can experience high evaporation flux (flux α radius) and high retention for volatile organic flavor/fragrance components [146] as compared with the membrane with smaller pore size. The bases of these results are (i) differences in feed–membrane and stripper–membrane boundary layer resistances to organic volatile transport and (ii) interfering by the liquids in membrane pores. A high porous membrane is preferred as porosity is directly proportional to evaporation flux. Moreover, the minimum heat gradient can be attributed to the high membrane conductivity.

Membrane thickness should be low as evaporation flux is indirectly proportional to the pore length. Due to the limited mechanics strength of thin membranes, they are supported by net, e.g. Gelman PTFE membranes. The overall thickness of the membrane depends upon the absence or presence of support, which ranges from 80 to 250 μm. The main disadvantages of hydrophobic membranes in OMD operations are (i) wetting of membrane, (ii) low performance, and (iii) reduced evaporation flux.

The decrease in thermal resistance consequently reduces temperature gradient (ΔT), and an increase in partial pressure gradient (ΔP) across the membrane is attributed to reducing membrane thickness. Moreover, thin hydrophilic membranes have more chances of becoming wet due to a gradual reduction in evaporation flux and separation performance. The wetting can cause the intrusion of liquid through pores in vapor flux and diffusion of nonvolatile solute from one compartment to another in OMD operations.

Peel oils and other hydrophobic compounds are present in fruit juices, e.g. citrus juice, and surface-active agents in cleaning solutions enhance the chances of wetting in OMD membranes. Prevention of liquid intrusion in membranes can be done by using hydrophilic polymer films with high intrinsic water permeability. These are impermeable to macrosolutes and colloids along with lamination on membrane surface [147]. The hydrophilic polymers include esters, ether of cellulose, cross-linked gelatin, chitin, agar, alginic acid, cross-linked polyacrylamide, cross-linked polyvinyl alcohol (PVA), and polyhydroxy 2-ethyl-methacrylate (PHEMA). Commercially available cellophane membrane forms a laminated hydrogel film at the side with a concentrated solution [148]. By the coating, one side of the commercial membrane of osmotic distillation (OD) with PVA for oily feed has been developed for OD operations. In this, the uncoated layer is continuously in contact with an oily feed even for low oil concentration in water and vice versa. The PVA-coated membranes are sustainable for 24 hours for concentration of up to 1 wt% oil solution [149].

Pall-Gelman (East Hills, NY, USA) manufactures the commercial membranes for OMD operations (TF 200, TF 450, and TF 1000) characterized by the PTFE layer supported by the PP net. The top layer of these asymmetric membranes provides resistance to gas transfer, and membrane support provides extra resistance to water transfer in the liquid form. Forty to seventy percent of the total resistance is mass transfer resistance. The sensitivity of the OMD system to concentration polarization phenomena is determined by the percentage resistance, out of which 30% resistance is provided by brine captured in PP support and 60% resistance is caused by dilute brine boundary layer. Liqui-Cel® Extra-Flow Membrane Contactor (Membrana-Charlotte, North Carolina, USA) is one of the best known module designs for OMD. It contains approximately 300 μm external diameter microporous PP HFMs with a wall thickness of 40 μm. They have a mean pore diameter of about 30 nm and porosity of 40%. The fibers are enclosed into a polyethylene tube sheet and shell casing of PP, PVDF, or 316L stainless steel [147, 150, 151]. The smallest modules are having 1.4 m^2 surface area and 2.5 in. diameter, whereas the largest module has a 130 m^2 surface area with 10 in. diameter.

4.7 Solvents Used in Membrane Synthesis

Several factors need to be considered to choose the most suitable solvent for membrane fabrication, especially for the dry phase inversion method [152]. The most important one is polymers' solubility within the solvent to achieve a homogenous dope solution [153]. Moreover, the solvent must also be less hazardous to ensure simplicity during handling and synthesizing of membranes. There are quite several solvents that can be used in membrane synthesis. Listed in Table 4.12 are the examples of solvents along with their respective characteristics and properties.

4.8 Membrane Synthesis Techniques

Permeation properties of a membrane depend on its structural morphologies and physical characteristics [158]. It is always challenging to fabricate membranes that have economically high separation performance and high durability to meet the standard industrial application requirement. Membrane fabrication processes are vital to design the materials with the appropriate methods to obtain a membrane with suitable morphologies with optimum separation. There are various methods available for membrane synthesis. Considering its applications and the type of the desired membrane (either dense or porous), the methods to be used must be selected wisely. Commonly, the phase inversion method is the most popular

Table 4.12 Solvent types and their characteristics.

Solvent	Molecular weight (g/mol)	Density (g/cm^3)	Boiling point (°C)	Solubility parameter ($J^{1/2}/cm^{3/2}$)	Remarks
N,*N*-Dimethylformamide (DMF)	73.09	0.949	153	24.9	Polar, suitable for a compound with high molecular weight, high dielectric constant, soluble with water [154]
N,*N*-Dimethylacetamide (DMAc)	87.12	0.937	166	22.8	Polar and aprotic type, suitable for polymers with high molecular weight and synthetic resins [155]
Dichloromethane (DCM)	84.94	1.336	40	19.9	Short evaporation rate, high volatility; due to its wavy structure, rapid evaporation can be avoided [156]
N-Methyl-2-pyrrolidone (NMP)	99.13	1.028	202	22.9	Very stable and high polarity, low toxicity [157]

Table 4.13 Techniques on synthesizing polymeric membranes.

Method/technique	Description	Morphological structure
Phase inversion Solvent evaporation	After casting the membrane, the solvent will naturally evaporate in an inert atmosphere	Dense and symmetric
Vapor-phase precipitation	The membrane is placed in a nonsolvent vapor atmosphere upon casting. The membrane will be formed due to the diffusion of the nonsolvent into the film	Porous and symmetric
Controlled evaporation	Dissolution of polymers into solvent and nonsolvent mixture. Solvent composition shifts into a higher nonsolvent and polymer content	Asymmetric
Thermal precipitation	After membrane casting, it is left to be cooled for phase separation to occur	Asymmetric
Immersion precipitation	The membrane film is cast on a support layer and then soaked into a nonsolvent coagulation bath	Asymmetric
Track etching	Membrane film is immersed in an acid/base bath and exposed to high-energy particle radiation exposure to create a track. The polymer will etch along the track	Porous and symmetric
Stretching	The polymer material is pulled/stretched against the direction of extrusion	Porous
Sintering	The membrane is pressed with powder particles of a certain size and then sintered at an elevating temperature	Porous

Source: Based on Ulbricht [159] and Merkel et al. [160].

technique adopted in the fabrication of polymeric membranes [157]. Table 4.13 emphasizes the common methods for membrane synthesis.

4.9 Conclusions

The membrane technology has been innovative in the last five decades and has been used in various applications, such as reverse osmosis and purification from desalination, ultrafiltration, gas isolation, medication release monitoring, genetic modification, nutritional care, etc. This development also inspired scientists to establish various membrane classifications to advanced standards of efficiency. Their classifications include the composition of the membrane module based on its geometry and flow, membrane materials, and membrane synthesis methods. Recent membrane science has considerable responsibility for the scientific awareness and technological advances of membranes.

4.10 Future Perspective

The current membrane contactor technology is focused on special features from membrane technology and traditional phase contact processes. The selectivities from traditional separation processes such as extraction or absorption are correlated with properties like the membrane-based high-specific-surface flexible nature. This provides an effective, lightweight, and scalable split technology in the new age. Gas–liquid and liquid–liquid processes are the methods for membrane contactors. For example, unique compounds from gas streams (e.g. NH_3, CO_2, SO_2, H_2S, and H_2O) are extracted, and limited hydrocarbon and heavy metal recoveries from water streams are retrieved. These are used as an end of tube technology and commodity recovery and as a process solution combined. There is now a good installation in the industry with the first full-scale membrane contactor system for gas and oil water treatment. Application efficiency depends on their specific advantages (flexible power, no phase separation, low volume, low weight, drastic phase ratios, etc.). The long-term reliability of membrane contactors under industrial conditions can be demonstrated with these plants. It will offer more trust and contribute to a whole new field of membrane contactor applications of membrane contactor technology for other end users.

References

1 Mulder, M. (2012). *Basic Principles of Membrane Technology*. Springer Science & Business Media.

2 Berk, Z. (2018). Membrane processes. In: *Food Process Engineering and Technology*, Chapter 10, 3e (ed. Z. Berk), 261–287. Academic Press.

3 Blackmer, R.H. and Hedman, J.W. (1979). Membrane oxygen enricher apparatus. *Google Patents*.

4 Stern, S., Sinclair, T.F., Gareis, P.J. et al. (1965). Helium recovery by permeation. *Industrial & Engineering Chemistry* 57 (2): 49–60.

5 McKeen, L.W. (2017). Production of films, containers, and membranes. In: *Permeability Properties of Plastics and Elastomers*, 4e (ed. L.W. McKeen), 61–82. William Andrew Publishing.

6 Scholz, M., Wessling, M., and Balster, J. (2011). Design of membrane modules for gas separations. *Membrane Engineering for the Treatment of Gases* 1: 125–149.

7 Pal, P. (2015). Arsenic removal by membrane filtration. In: *Groundwater Arsenic Remediation*, Chapter 4 (ed. P. Pal), 105–177. Butterworth-Heinemann.
8 Im, S.-J. and Jang, A. (2020). Long-term performance and initial fouling evaluation of an open-loop plate and frame forward osmosis element using wastewater treatment plant secondary effluent as a feed solution. *Journal of Water Process Engineering* 33: 101077.
9 Campione, A., Cipollina, A., Toet, E. et al. (2020). Water desalination by capacitive electrodialysis: experiments and modelling. *Desalination* 473: 114150.
10 Babaie, O. and Esfahany, M.N. (2020). Optimization and heat integration of hybrid R-HIDiC and pervaporation by combining GA and PSO algorithm in TAME synthesis. *Separation and Purification Technology* 236: 116288.
11 Melin, T. and Rautenbach, R. (2007). *Membranverfahren: Grundlagen der Modul-und Anlagenauslegung*. Springer-Verlag.
12 Baker, R.W., Wijmans, J.G., Merkel, T.C. et al. (2011). Gas separation process using membranes with permeate sweep to remove CO_2 from combustion gases. *Google Patents.*
13 Offeman, R.D. and Robertson, G.H. (2008). Spiral-wound liquid membrane module for separation of fluids and gases. *Google Patents.*
14 Koros, W.J. and Fleming, G. (1993). Membrane-based gas separation. *Journal of Membrane Science* 83 (1): 1–80.
15 Caro, J., Caspary, K.J., Hamel, C. et al. (2007). Catalytic membrane reactors for partial oxidation using perovskite hollow fiber membranes and for partial hydrogenation using a catalytic membrane contactor. *Industrial & Engineering Chemistry Research* 46 (8): 2286–2294.
16 DeJaco, R.F., Loprete, K., Pennisi, K. et al. (2020). Modeling and simulation of gas separations with spiral-wound membranes. *AIChE Journal* 66 (8): e16274.
17 Tasaki, T., Inamoto, K., Liang, C. et al. (2020). Spiral-wound gas separation membrane element, gas separation membrane module, and gas separation apparatus. *Google Patents.*
18 Chen, K.K., Salim, W., Han, Y. et al. (2020). Fabrication and scale-up of multi-leaf spiral-wound membrane modules for CO_2 capture from flue gas. *Journal of Membrane Science* 595: 117504.
19 Al-Anezi, A.A.-H., Sharif, A.O., Sanduk, M. et al. (2012). Experimental investigation of heat and mass transfer in tubular membrane distillation module for desalination. *ISRN Chemical Engineering* 2012: 1–7.
20 Fried, J.R. (1996). *Basic Principles of Membrane Technology by Marcel Mulder (University of Twente, The Netherlands)*, 564 pp. Dordrecht: Kluwer Academic. $255.00, ACS Publications (1997). ISBN: 0-7823-4247-X.
21 Baker, R. (2004). Other membrane process-charge mosaic membranes and piezodialysis. In: *Membrane Technology and Applications*, 2e, 491–522. Chichester: Wiley.
22 Wickramasinghe, S., Semmens, M.J., and Cussler, E. (1992). Mass transfer in various hollow fiber geometries. *Journal of Membrane Science* 69 (3): 235–250.

23 Iaquaniello, G., Palo, E., and Salladini, A. (2018). Membrane-assisted syngas production for gas-to-liquid processes. In: *Membrane Engineering for the Treatment of Gases: Gas-Separation Issues Combined with Membrane Reactors (2)*, Chapter 8, (eds. E. Drioli, G. Barbieri, and A. Brunetti), vol. 2, 247–272. The Royal Society of Chemistry.

24 Yeh, H. and Chen, Y. (2000). Membrane extraction through cross-flow rectangular modules. *Journal of Membrane Science* 170 (2): 235–242.

25 Jons, S.D., Johnson, J.E., and Fialkowski, M.A. (2010). Method for testing separation modules. *Google Patents*.

26 Bhattacharjee, C., Saxena, V., and Dutta, S. (2017). Fruit juice processing using membrane technology: a review. *Innovative Food Science & Emerging Technologies* 43: 136–153.

27 Kucera, J. (2015). *Reverse Osmosis: Industrial Processes and Applications*. Wiley.

28 Obotey Ezugbe, E. and Rathilal, S. (2020). Membrane technologies in wastewater treatment: a review. *Membranes* 10 (5): 89.

29 Agrahari, G.K., Verma, N., and Bhattacharya, P.K. (2011). Application of hollow fiber membrane contactor for the removal of carbon dioxide from water under liquid–liquid extraction mode. *Journal of Membrane Science* 375 (1–2): 323–333.

30 Ilias, S., Schimmel, K.A., and Yezek, P. (1999). Nondispersive liquid-liquid extraction of copper and zinc from an aqueous solution by DEHPA and LIX 984 in a hollow fiber membrane module. *Separation Science and Technology* 34 (6–7): 1007–1019.

31 Kathios, D.J. et al. (1994). A preliminary evaluation of microporous hollow fiber membrane modules for the liquid-liquid extraction of actinides. *Journal of Membrane Science* 97: 251–261.

32 He, T., Versteeg, L., Mulder, M. et al. (2004). Composite hollow fiber membranes for organic solvent-based liquid–liquid extraction. *Journal of Membrane Science* 234 (1–2): 1–10.

33 Schimmel, K.A., Ilias, S., and Akella, S. (2001). Nondispersive liquid-liquid extraction of Zn (II), Cu (II), Co (II), and Cd (II) from dilute solution with DEHPA in a hollow-fiber membrane module. *Separation Science and Technology* 36 (5–6): 805–821.

34 Elbert, D.L. (2011). Liquid–liquid two-phase systems for the production of porous hydrogels and hydrogel microspheres for biomedical applications: a tutorial review. *Acta Biomaterialia* 7 (1): 31–56.

35 Mandowara, A. and Bhattacharya, P.K. (2011). Simulation studies of ammonia removal from water in a membrane contactor under liquid–liquid extraction mode. *Journal of Environmental Management* 92 (1): 121–130.

36 Park, H.H., Deshwal, B.R., Kim, I.W. et al. (2008). Absorption of SO_2 from flue gas using PVDF hollow fiber membranes in a gas–liquid contactor. *Journal of Membrane Science* 319 (1–2): 29–37.

37 Dindore, V., Brilman, D.W.F., Geuzebroek, F. et al. (2004). Membrane–solvent selection for CO_2 removal using membrane gas–liquid contactors. *Separation and Purification Technology* 40 (2): 133–145.

38 Dindore, V. and Versteeg, G. (2005). Gas–liquid mass transfer in a cross-flow hollow fiber module: analytical model and experimental validation. *International Journal of Heat and Mass Transfer* 48 (16): 3352–3362.

39 Rezakazemi, M., Niazi, Z., Mirfendereski, M. et al. (2011). CFD simulation of natural gas sweetening in a gas–liquid hollow-fiber membrane contactor. *Chemical Engineering Journal* 168 (3): 1217–1226.

40 Mansourizadeh, A. and Ismail, A. (2009). Hollow fiber gas–liquid membrane contactors for acid gas capture: a review. *Journal of Hazardous Materials* 171 (1–3): 38–53.

41 Nymeijer, D., Visser, T., Assen, R. et al. (2004). Composite hollow fiber gas–liquid membrane contactors for olefin/paraffin separation. *Separation and Purification Technology* 37 (3): 209–220.

42 Sun, X., Meng, F., and Yang, F. (2008). Application of seawater to enhance SO_2 removal from simulated flue gas through hollow fiber membrane contactor. *Journal of Membrane Science* 312 (1–2): 6–14.

43 Marzouk, S.A., Al-Marzouqi, M.H., El-Naas, M.H. et al. (2010). Removal of carbon dioxide from pressurized CO_2–CH_4 gas mixture using hollow fiber membrane contactors. *Journal of Membrane Science* 351 (1–2): 21–27.

44 Dehkordi, J.A., Hosseini, S.S., Kundu, P.K. et al. (2016). Mathematical modeling of natural gas separation using hollow fiber membrane modules by application of finite element method through statistical analysis. *Chemical Product and Process Modeling* 11 (1): 11–15.

45 Li, S., Rocha, D.J., Zhou, S.J. et al. (2013). Post-combustion CO_2 capture using super-hydrophobic, polyether ether ketone, hollow fiber membrane contactors. *Journal of Membrane Science* 430: 79–86.

46 Motin, A., Tarabara, V.V., and Bénard, A. (2015). Numerical investigation of the performance and hydrodynamics of a rotating tubular membrane used for liquid–liquid separation. *Journal of Membrane Science* 473: 245–255.

47 Latif, M., Kasi, A.K., Kasi, J.K. et al. (2019). Strengthening of alumina tubular membrane by Al support and its application for oil-in-water stable emulsion. *Microelectronic Engineering* 218: 111134.

48 Tang, T.E. and Hwang, S.T. (1976). Mass transfer of dissolved gases through tubular membrane. *AIChE Journal* 22 (6): 1000–1006.

49 Lin, Y.-F., Chang, J.-M., Ye, Q. et al. (2015). Hydrophobic fluorocarbon-modified silica aerogel tubular membranes with excellent CO_2 recovery ability in membrane contactors. *Applied Energy* 154: 21–25.

50 Mikulášek, P., Pospíšil, P., Doleček, P. et al. (2002). Gas—liquid two-phase flow in microfiltration mineral tubular membranes: relationship between flux enhancement and hydrodynamic parameters. *Desalination* 146 (1–3): 103–109.

51 Ghosh, R. and Cui, Z. (1999). Mass transfer in gas-sparged ultrafiltration: upward slug flow in tubular membranes. *Journal of Membrane Science* 162 (1–2): 91–102.

52 Basile, A., Drioli, E., Santell, F. et al. (1996). A study on catalytic membrane reactors for water gas shift reaction. *Gas Separation & Purification* 10 (1): 53–61.

53 Aguado, S., Nicolas, C.-H., Moizan-Baslé, V. et al. (2011). Facile synthesis of an ultramicroporous MOF tubular membrane with selectivity towards CO_2. *New Journal of Chemistry* 35 (1): 41–44.

54 Zhang, Y., Su, K., Zeng, F. et al. (2013). A novel tubular oxygen-permeable membrane reactor for partial oxidation of CH_4 in coke oven gas to syngas. *International Journal of Hydrogen Energy* 38 (21): 8783–8789.

55 Teramoto, M., Tohno, N., Ohnishi, N. (1989). Development of a spiral-type flowing liquid membrane module with high stability and its application to the recovery of chromium and zinc. *Separation Science and Technology* 24 (12–13): 981–999.

56 Cao, B. and Henson, M.A. (2002). Modeling of spiral wound pervaporation modules with application to the separation of styrene/ethylbenzene mixtures. *Journal of Membrane Science* 197 (1–2): 117–146.

57 Marsousi, S., Karimi-Sabet, J., Moosavian, M.A. (2019). Liquid–liquid extraction of calcium using ionic liquids in spiral microfluidics. *Chemical Engineering Journal* 356: 492–505.

58 Bayer, C., Follmann, M., Breisig, H. et al. (2013). On the design of a 4-end spiral-wound L/L extraction membrane module. *Industrial & Engineering Chemistry Research* 52 (3): 1004–1014.

59 Wibisono, Y., El Obied, K., Cornelissen, E. et al. (2015). Biofouling removal in spiral-wound nanofiltration elements using two-phase flow cleaning. *Journal of Membrane Science* 475: 131–146.

60 Favre, E. and Svendsen, H. (2012). Membrane contactors for intensified post-combustion carbon dioxide capture by gas–liquid absorption processes. *Journal of Membrane Science* 407: 1–7.

61 Liu, Q., Wang, S., Guo, Z. et al. (2012). Enhancement of spiral nanofiltration membranes using gas sparging: application to MgSO_4 solution. *CIESC Journal*: 31.

62 Wang, S., Guo, Z., Meng, T. et al. (2011). Cleaning characteristics of spiral nanofiltration membrane with gas-liquid two-phase flow process. *CIESC Journal* 6.

63 Volkov, A., Tsarkov, S., Goetheer, E. et al. (2015). Amine-based solvents regeneration in gas-liquid membrane contactor based on asymmetric PVTMS. *Petroleum Chemistry* 55 (9): 716–723.

64 Imai, M., Furusaki, S., and Miyauchi, T. (1982). Separation of volatile materials by gas membranes. *Industrial & Engineering Chemistry Process Design and Development* 21 (3): 421–426.

65 Scholes, C.A., Stevens, G.W., and Kentish, S.E. (2012). Membrane gas separation applications in natural gas processing. *Fuel* 96: 15–28.

66 Qi, R. and Henson, M. (1998). Optimization-based design of spiral-wound membrane systems for CO_2/CH_4 separations. *Separation and Purification Technology* 13 (3): 209–225.

67 Qi, R. and Henson, M.A. (1998). Optimal design of spiral-wound membrane networks for gas separations. *Journal of Membrane Science* 148 (1): 71–89.

68 Salim, W., Vakharia, V., Chen, Y. et al. (2018). Fabrication and field testing of spiral-wound membrane modules for CO_2 capture from flue gas. *Journal of Membrane Science* 556: 126–137.

69 Matsumoto, M., Hasegawa, W., Kondo, K. et al. (2010). Application of supported ionic liquid membranes using a flat sheet and hollow fibers to lactic acid recovery. *Desalination and Water Treatment* 14 (1–3): 37–46.

70 Psillakis, E. and Kalogerakis, N. (2003). Developments in liquid-phase microextraction. *TrAC, Trends in Analytical Chemistry* 22 (9): 565–574.

71 Chimuka, L., Cukrowska, E., Michel, M. et al. (2011). Advances in sample preparation using membrane-based liquid-phase microextraction techniques. *TrAC, Trends in Analytical Chemistry* 30 (11): 1781–1792.

72 Hernández, S., Lei, S., Rong, W. et al. (2016). Functionalization of flat sheet and hollow fiber microfiltration membranes for water applications. *ACS Sustainable Chemistry & Engineering* 4 (3): 907–918.

73 Mercier-Bonin, M., Lagane, C., and Fonade, C. (2000). Influence of a gas/liquid two-phase flow on the ultrafiltration and microfiltration performances: case of a ceramic flat sheet membrane. *Journal of Membrane Science* 180 (1): 93–102.

74 Bessarabov, D., Jacobs, E., Sanderson, R. et al. (1996). Use of nonporous polymeric flat-sheet gas-separation membranes in a membrane-liquid contactor: experimental studies. *Journal of Membrane Science* 113 (2): 275–284.

75 Brinkmann, T., Pohlmann, J., Withalm, U. et al. (2013). Theoretical and experimental investigations of flat sheet membrane module types for high capacity gas separation applications. *Chemie Ingenieur Technik* 85 (8): 1210–1220.

76 Li, Q., Ghosh, R., Bellara, S. et al. (1998). Enhancement of ultrafiltration by gas sparging with flat sheet membrane modules. *Separation and Purification Technology* 14 (1–3): 79–83.

77 Dindore, V., Brilman, D.W.F., and Versteeg, G. (2005). Modelling of cross-flow membrane contactors: physical mass transfer processes. *Journal of Membrane Science* 251 (1–2): 209–222.

78 Baker, R.W. (2004). Membranes and modules. In: *Membrane Technology and Applications*, 89–160. Newark, CA, USA: Membrane Technology and Research, Inc. (MTR).

79 Aoki, T. (1999). Macromolecular design of permselective membranes. *Progress in Polymer Science* 24 (7): 951–993.

80 Scott, K. (1995). *Handbook of Industrial Membranes*. Elsevier.

81 Pal, P. (2020). *Membrane-Based Technologies for Environmental Pollution Control*. Butterworth-Heinemann.

82 Asfand, F. and Bourouis, M. (2015). A review of membrane contactors applied in absorption refrigeration systems. *Renewable and Sustainable Energy Reviews* 45: 173–191.

83 Wang, Z., Gu, Z., Feng, S. et al. (2009). Application of vacuum membrane distillation to lithium bromide absorption refrigeration system. *International Journal of Refrigeration* 32 (7): 1587–1596.

84 Ali, A.H.H. and Schwerdt, P. (2009). Characteristics of the membrane utilized in a compact absorber for lithium bromide–water absorption chillers. *International Journal of Refrigeration* 32 (8): 1886–1896.

85 Ali, A.H.H. and Schwerdt, P. (2010). For designing a compact absorber with membrane contactor at liquid-vapor Interface--influence of membrane properties on water vapor transfer. *ASHRAE Transactions* 116 (1).

86 Isfahani, R.N., Fazeli, A., Bigham, S. et al. (2014). Physics of lithium bromide (LiBr) solution dewatering through vapor venting membranes. *International Journal of Multiphase Flow* 58: 27–38.

87 Bigham, S., Yu, D., Chugh, D. et al. (2014). Moving beyond the limits of mass transport in liquid absorbent microfilms through the implementation of surface-induced vortices. *Energy* 65: 621–630.

88 Ali, A.H.H. (2010). Design of a compact absorber with a hydrophobic membrane contactor at the liquid–vapor interface for lithium bromide–water absorption chillers. *Applied Energy* 87 (4): 1112–1121.

89 Razali, M., Didaskalou, C., Kim, J.F. et al. (2017). Exploring and exploiting the effect of solvent treatment in membrane separations. *ACS Applied Materials & Interfaces* 9 (12): 11279–11289.

90 Tantekin-Ersolmaz, Ş.B., Atalay-Oral, Ç., Tatlıer, M. et al. (2000). Effect of zeolite particle size on the performance of polymer–zeolite mixed matrix membranes. *Journal of Membrane Science* 175 (2): 285–288.

91 Ghasem, N., Al-Marzouqi, M., and Duidar, A. (2012). Effect of PVDF concentration on the morphology and performance of hollow fiber membrane employed as gas–liquid membrane contactor for CO_2 absorption. *Separation and Purification Technology* 98: 174–185.

92 Breite, D., Went, M., Prager, A. et al. (2016). The critical zeta potential of polymer membranes: how electrolytes impact membrane fouling. *RSC Advances* 6 (100): 98180–98189.

93 Mora, F., Pérez, K., Quezada, C. et al. (2019). Impact of membrane pore size on the clarification performance of grape Marc extract by microfiltration. *Membranes* 9 (11): 146.

94 Hwang, K.-J. and Lin, T.-T. (2002). Effect of morphology of polymeric membrane on the performance of cross-flow microfiltration. *Journal of Membrane Science* 199 (1–2): 41–52.

95 Wang, M., Wang, Z., Li, S. et al. (2013). A high performance antioxidative and acid resistant membrane prepared by interfacial polymerization for CO_2 separation from flue gas. *Energy & Environmental Science* 6 (2): 539–551.

96 Ahmad, T., Guria, C., and Mandal, A. (2020). A review of oily wastewater treatment using ultrafiltration membrane: a parametric study to enhance the membrane performance. *Journal of Water Process Engineering* 36: 101289.

97 Tan, Y.Z., Velioglu, S., Han, L. et al. (2019). Effect of surfactant hydrophobicity and charge type on membrane distillation performance. *Journal of Membrane Science* 587: 117168.

98 Okamoto, Y. and Lienhard, J.H. (2019). How RO membrane permeability and other performance factors affect process cost and energy use: a review. *Desalination* 470: 114064.

99 Lazaridou, A. and Biliaderis, C. (2007). Molecular aspects of cereal β-glucan functionality: physical properties, technological applications and physiological effects. *Journal of Cereal Science* 46 (2): 101–118.

100 Bernhardsen, I.M., Ansaloni, L., Betten, H.K. et al. (2019). Effect of liquid viscosity on the performance of a non-porous membrane contactor for CO_2 capture. *Separation and Purification Technology* 222: 188–201.

101 Strathmann, H. (2001). Membrane separation processes: current relevance and future opportunities. *AIChE Journal* 47 (5): 1077–1087.

102 Rezakazemi, M., Sadrzadeh, M., and Matsuura, T. (2018). Thermally stable polymers for advanced high-performance gas separation membranes. *Progress in Energy and Combustion Science* 66: 1–41.

103 Vu, D.Q. (2003). Formation and characterization of asymmetric carbon molecular sieve and mixed-matrix membranes for natural gas purification. University of Texas at Austin.

104 Liu, Y. and Wang, G. (2016). Membranes: technology and applications. In: *Nanostructured Polymer Membranes: Applications*, vol. 2, 2, 27. Wiley.

105 Prabhakar, R.S. (2004). Low hydrocarbon solubility polymers: plasticization-resistant membranes for carbon dioxide removal from natural gas. Doctoral dissertation, The University of Texas at Austin.

106 Lee, M., Wu, Z., and Li, K. (2015). Advances in ceramic membranes for water treatment. In: *Advances in Membrane Technologies for Water Treatment*, 43–82. Woodhead Publishing.

107 Ghosal, K. and Freeman, B.D. (1994). Gas separation using polymer membranes: an overview. *Polymers for Advanced Technologies* 5 (11): 673–697.

108 Robeson, L.M. (1991). Correlation of separation factor versus permeability for polymeric membranes. *Journal of Membrane Science* 62 (2): 165–185.

109 Rezakazemi, M., Vatani, A., and Mohammadi, T. (2016). Synthesis and gas transport properties of crosslinked poly(dimethylsiloxane) nanocomposite membranes using octatrimethylsiloxy POSS nanoparticles. *Journal of Natural Gas Science and Engineering* 30: 10–18.

110 Rezakazemi, M., Vatani, A., and Mohammadi, T. (2015). Synergistic interactions between POSS and fumed silica and their effect on the properties of crosslinked PDMS nanocomposite membranes. *RSC Advances* 5 (100): 82460–82470.

111 Younas, M., Rezakazemi, M., Daud, M. et al. (2020). Recent progress and remaining challenges in post-combustion CO_2 capture using metal-organic frameworks (MOFs). *Progress in Energy and Combustion Science* 80: 100849.

112 Rezakazemi, M., Shahidi, K., and Mohammadi, T. (2012). Sorption properties of hydrogen-selective PDMS/zeolite 4A mixed matrix membrane. *International Journal of Hydrogen Energy* 37 (22): 17275–17284.

113 Rezakazemi, M., Shahidi, K., and Mohammadi, T. (2012). Hydrogen separation and purification using crosslinkable PDMS/zeolite a nanoparticles mixed matrix membranes. *International Journal of Hydrogen Energy* 37 (19): 14576–14589.

114 Robeson, L.M. (2008). The upper bound revisited. *Journal of Membrane Science* 320 (1–2): 390–400.

115 Rezakazemi, M. Dashti, A., Asghari, M. et al. (2017). H2-selective mixed matrix membranes modeling using ANFIS, PSO-ANFIS, GA-ANFIS. *International Journal of Hydrogen Energy* 42 (22): 15211–15225.

116 Rezakazemi, M. and Mohammadi, T. (2013). Gas sorption in H_2-selective mixed matrix membranes: experimental and neural network modeling. *International Journal of Hydrogen Energy* 38 (32): 14035–14041.

117 Shahverdi, M., Baheri, B., Rezakazemi, M. et al. (2012). Pervaporation study of ethylene glycol dehydration through synthesized (PVA–4A)/polypropylene mixed matrix composite membranes. *Polymer Engineering & Science* 53 (7): 1487–1493.

118 Mirqasemi, M.S., Homayoonfal, M., and Rezakazemi, M. (2020). Zeolitic imidazolate framework membranes for gas and water purification. *Environmental Chemistry Letters* 18 (1): 1–52.

119 Ahmadijokani, F., Tajahmadi, S., Rezakazemi, M. et al. (2021). Aluminum-based metal-organic frameworks for adsorptive removal of anti-cancer (methotrexate) drug from aqueous solutions. *Journal of Environmental Management* 277: 111448.

120 Ahmadijokani, F., Tajahmadi, S., Bahi, A. et al. (2021). Ethylenediamine-functionalized Zr-based MOF for efficient removal of heavy metal ions from water. *Chemosphere* 264: 128466.

121 Rezakazemi, M., Ebadi Amooghin, A., Montazer-Rahmati, M.M. et al. (2014). State-of-the-art membrane based CO_2 separation using mixed matrix membranes (MMMs): an overview on current status and future directions. *Progress in Polymer Science* 39 (5): 817–861.

122 Aroon, M., Ismail, A., Matsuura, T. et al. (2010). Performance studies of mixed matrix membranes for gas separation: a review. *Separation and Purification Technology* 75 (3): 229–242.

123 Iarikov, D.D. and Oyama, S.T. (2011). Review of CO_2/CH_4 separation membranes. In: *Membrane Science and Technology*, 91–115. Elsevier.

124 Xu, Z.-l., Yu, L.-y., and Han, L.-f. (2009). Polymer-nanoinorganic particles composite membranes: a brief overview. *Frontiers of Chemical Engineering in China* 3 (3): 318–329.

125 Goh, P., Ismail, A., Sanip, S. et al. (2011). Recent advances of inorganic fillers in mixed matrix membrane for gas separation. *Separation and Purification Technology* 81 (3): 243–264.

126 Bet-Moushoul, E., Mansourpanah, Y., Farhadi, K. et al. (2016). TiO_2 nanocomposite based polymeric membranes: a review on performance improvement for various applications in chemical engineering processes. *Chemical Engineering Journal* 283: 29–46.

127 Ng, L.Y., Mohammad, A.W., Leo, C.P. et al. (2013). Polymeric membranes incorporated with metal/metal oxide nanoparticles: a comprehensive review. *Desalination* 308: 15–33.

128 Nor, F.M., Karim, N.H.A., Abdullah, I. et al. (2016). Permeability of carbon dioxide and nitrogen gases through SiO_2 and MgO incorporated ENR/PVC membranes. *Journal of Elastomers & Plastics* 48 (6): 483–498.

129 Farnam, M., Mukhtar, H., and Shariff, A. (2016). Analysis of the influence of CMS variable percentages on pure PES membrane gas separation performance. *Procedia Engineering* 148: 1206–1212.

130 Fu, Y.-J., Hu, C.-C., Lin, D.-W. et al. (2017). Adjustable microstructure carbon molecular sieve membranes derived from thermally stable polyetherimide/polyimide blends for gas separation. *Carbon* 113: 10–17.

131 Dorosti, F., Omidkhah, M., Pedram, M. et al. (2011). Fabrication and characterization of polysulfone/polyimide–zeolite mixed matrix membrane for gas separation. *Chemical Engineering Journal* 171 (3): 1469–1476.

132 Li, Y., Chung, T.-S., Cao, C. et al. (2005). The effects of polymer chain rigidification, zeolite pore size and pore blockage on polyethersulfone (PES)-zeolite a mixed matrix membranes. *Journal of Membrane Science* 260 (1–2): 45–55.

133 Chen, Y., Wang, B., Zhao, L. et al. (2015). New Pebax®/zeolite Y composite membranes for CO_2 capture from flue gas. *Journal of Membrane Science* 495: 415–423.

134 Rafiq, S., Man, Z., Maulud, A. et al. (2012). Separation of CO_2 from CH_4 using polysulfone/polyimide silica nanocomposite membranes. *Separation and Purification Technology* 90: 162–172.

135 Panapitiya, N.P., Wijenayake, S.N., Huang, Y. et al. (2014). Stabilization of immiscible polymer blends using structure directing metal organic frameworks (MOFs). *Polymer* 55 (8): 2028–2034.

136 Shahid, S., Nijmeijer, K., Nehache, S. et al. (2015). MOF-mixed matrix membranes: precise dispersion of MOF particles with better compatibility via a particle fusion approach for enhanced gas separation properties. *Journal of Membrane Science* 492: 21–31.

137 Daramola, M.O., Hlanyane, P., Sadare, O.O. et al. (2017). Performance of carbon nanotube/polysulfone (CNT/Psf) composite membranes during oil–water mixture separation: effect of CNT dispersion method. *Membranes* 7 (1): 14.

138 Nejad, M.N., Asghari, M., and Afsari, M. (2016). Investigation of carbon nanotubes in mixed matrix membranes for gas separation: a review. *ChemBioEng Reviews* 3 (6): 276–298.

139 Zhang, H., Guo, R., Hou, J. et al. (2016). Mixed-matrix membranes containing carbon nanotubes composite with hydrogel for efficient CO_2 separation. *ACS Applied Materials & Interfaces* 8 (42): 29044–29051.

140 Shahid, S. and Nijmeijer, K. (2017). Matrimid®/polysulfone blend mixed matrix membranes containing ZIF-8 nanoparticles for high pressure stability in natural gas separation. *Separation and Purification Technology* 189: 90–100.

141 Ammar, A., Elzatahry, A., Al-Maadeed, M. et al. (2017). Nanoclay compatibilization of phase separated polysulfone/polyimide films for oxygen barrier. *Applied Clay Science* 137: 123–134.

142 Ebadi-Dehaghani, H., Barikani, M., Khonakdar, H.A. et al. (2015). On O2 gas permeability of PP/PLA/clay nanocomposites: a molecular dynamic simulation approach. *Polymer Testing* 45: 139–151.

143 Gryta, M. (2005). Osmotic MD and other membrane distillation variants. *Journal of Membrane Science* 246 (2): 145–156.

144 Gabino, F., Belleville, M.-P., Preziosi-Belloy, L. et al. (2007). Evaluation of the cleaning of a new hydrophobic membrane for osmotic evaporation. *Separation and Purification Technology* 55 (2): 191–197.

145 Romero, J., Draga, H., Belleville, M.-P. et al. (2006). New hydrophobic membranes for contactor processes—applications to isothermal concentration of solutions. *Desalination* 193 (1–3): 280–285.

146 Barbe, A., Bartley, J., Jacobs, A. et al. (1998). Retention of volatile organic flavour/fragrance components in the concentration of liquid foods by osmotic distillation. *Journal of Membrane Science* 145 (1): 67–75.

147 Hogan, P.A., Canning, R.P., Peterson, P.A. et al. (1998). A new option: osmotic distillation. *Chemical Engineering Progress* 94 (7): 49–61.

148 Michaels, A.S. (1999). Osmotic distillation process using a membrane laminate. *Google Patents.*

149 Mansouri, J. and Fane, A. (1999). Osmotic distillation of oily feeds. *Journal of Membrane Science* 153 (1): 103–120.

150 Gabelman, A. and Hwang, S.-T. (1999). Hollow fiber membrane contactors. *Journal of Membrane Science* 159 (1–2): 61–106.

151 Sakai, K., Muroi, T., Ozawa, K. et al. (1986). Extraction of solute-free water from blood by membrane distillation. *ASAIO Journal* 32 (1): 397–400.

152 Sadeghi, A., Nazem, H., Rezakazemi, M. et al. (2018). Predictive construction of phase diagram of ternary solutions containing polymer/solvent/nonsolvent using modified Flory-Huggins model. *Journal of Molecular Liquids* 263: 282–287.

153 Loloei, M., Moghadassi, A., Omidkhah, M. et al. (2015). Improved CO_2 separation performance of Matrimid® 5218 membrane by addition of low molecular weight polyethylene glycol. *Greenhouse Gases: Science and Technology* 5 (5): 530–544.

154 Ahn, J.-H., Kim, J.-K., Kim, H.-S. et al. (2009). Solubility of 1, 1-diamino-2, 2-dinitroethylene in N, N-dimethylformamide, dimethyl sulfoxide, and N-methyl-2-pyrrolidone. *Journal of Chemical & Engineering Data* 54 (12): 3259–3260.

155 Nasirzadeh, K., Neueder, R., and Kunz, W. (2005). Vapor pressures of propylene carbonate and *N,N*-dimethylacetamide. *Journal of Chemical & Engineering Data* 50 (1): 26–28.

156 Cequier-Sánchez, E., Rodriguez, C., Ravelo, A.n.G. et al. (2008). Dichloromethane as a solvent for lipid extraction and assessment of lipid classes and fatty acids from samples of different natures. *Journal of Agricultural and Food Chemistry* 56 (12): 4297–4303.

157 Li, Y., Chung, T.-S, Cao, C., and Kulprathipanja, S. (2005). A review on membrane fabrication: structure, properties and performance relationship. *Desalination* 326: 77–95.

158 Freeman, B., Yampolskii, Y., and Pinnau, I. (2006). *Materials Science of Membranes for Gas and Vapor Separation.* Wiley.

159 Ulbricht, M. (2020). Design and synthesis of organic polymers for molecular separation membranes. *Current Opinion in Chemical Engineering* 28: 60–65.

160 Merkel, T., Pinnau, I., Prabhakar, R., and Freeman, B.D. (2006). Gas and vapor transport properties of perfluoropolymers. In: *Materials Science of Membranes for Gas and Vapor Separation*, 251–270.

5

Mode of Operation in Membrane Contactors

Waheed Ur Rehman[1], Zarrar Salahuddin[2], Sarah Farrukh[2], Muhammad Younas[3], and Mashallah Rezakazemi[4]

[1] *PCSIR Laboratories Complex, Jamrud Road, Peshawar, 25120, Pakistan*
[2] *National University of Sciences and Technology, School of Chemical and Materials Engineering, Islamabad, Pakistan*
[3] *University of Engineering and Technology, Department of Chemical Engineering, Peshawar, Pakistan*
[4] *Shahrood University of Technology, Faculty of Chemical and Materials Engineering, Shahrood, Iran*

5.1 Membrane Distillation (MD)

5.1.1 Basic Principles of MD Process

Bodell was the first researcher to patent the technology of membrane distillation (MD) [1]. MD is a type of membrane separation process where water vapors permeate through a porous hydrophobic membrane and condense in liquid form on the other side of the membrane where permeate flows [2–4]. In MD process, mass transfer takes place due to the difference in vapor pressure induced by the temperature gradient that exists between the feed and permeate sides of the membrane [5, 6]. MD is not a purely thermal process because of the presence of a membrane as a separation barrier. Therefore, a much lower temperature can be used for causing separation using MD. Another important aspect of MD is the use of a hydrophobic membrane, which prevents the permeation of water through it. Membrane hydrophobicity averts the entry of water droplets into the pores of the membrane, due to the surface tension, which forms an interface at the surface of the membrane, where the liquid and vapor are to come in contact. Water vapor does not enter the membrane pores, until and unless the *transmembrane pressure* is lower than the membrane liquid entry pressure (LEP). Thus, it can be said that the pressure gradient is an important character in defining the separation efficiency of the membrane, in addition to the hydrophobicity of the membrane material. The stepwise transport of vapors occurs as following: (i) heat transfer from the bulk solution to the surface of the membrane, (ii) evaporation of liquid at the interface of hot feed membrane surface, (iii) vapor transport through membrane pores due to the temperature difference, and (iv) vapor condensation by coolants or cooling plates [7, 8].

Membrane Contactor Technology: Water Treatment, Food Processing, Gas Separation, and Carbon Capture, First Edition. Edited by Mohammad Younas and Mashallah Rezakazemi.

5.1.2 MD Configurations

MD process can be carried out in four main types of configurations. These are direct contact membrane distillation (DCMD), air gap membrane distillation (AGMD), vacuum membrane distillation (VMD), and sweeping gas membrane distillation (SGMD). The basic variance lies in the method to induce a difference in vapor pressure for driving the permeation flux. DCMD is the most commonly applied in desalination and concentration of liquid food products. In this process, the feed (hot solution) and permeate (the cold solution) are in straight connection with opposite surfaces of a hydrophobic membrane, as shown in Figure 5.1. Vapor transfer across the membrane surface occurs due to the difference in vapor pressure induced by temperature gradient. DCMD has low thermal efficiency because large conductive heat losses occur in this configuration. In AGMD, the water vapors pass across the membrane's pores and a stagnant air gap and consequently condense over the cold surface, as shown in Figure 5.1. Due to an air gap, there is a higher resistance to the vapors flow, which results in a lower vapor flux than DCMD. However, AGMD has smaller conduction losses and higher thermal resistance, making it an energy-efficient configuration. In VMD, a vacuum pump is used to produce suction on the cold permeate side of the membrane module. For inducing a vapor flow, the saturation pressure of the hot feed solution must be higher than the vacuum pressure applied on the permeate side. Although VMD enjoys high transmembrane flux (TMF) and lower conduction heat losses, yet it makes the membrane more likely to wetting and fouling. In SGMD, a cold and inert flowing gas stream is applied for sweeping the water vapors at the cold side of the membrane surface. The vapors are then subsequently condensed through a condenser, which is installed outside the module. Due to a non-stagnant gas barrier in SGMD, the heat loss reduces, and accordingly, the mass transfer coefficient improves. In addition, SGMD has a lower thermal polarization and best wetting resistance [7].

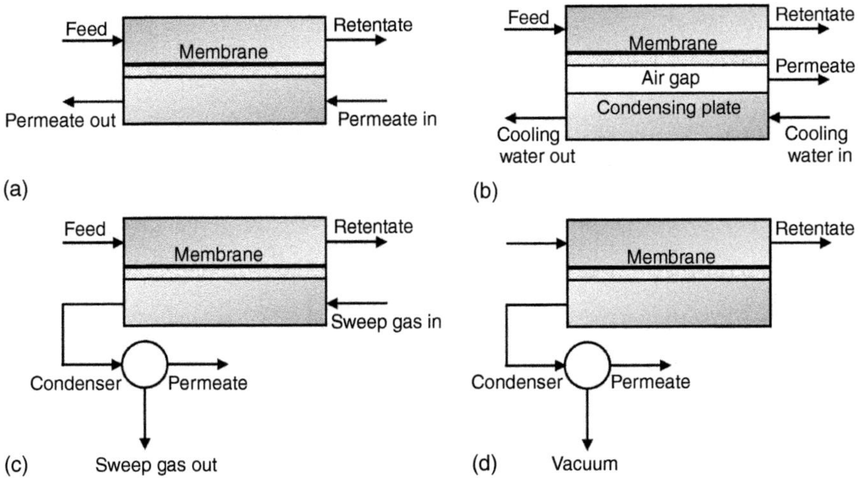

Figure 5.1 Four major MD configurations: (a) DCMD, (b) AGMD, (c) SGMD, and (d) VMD.

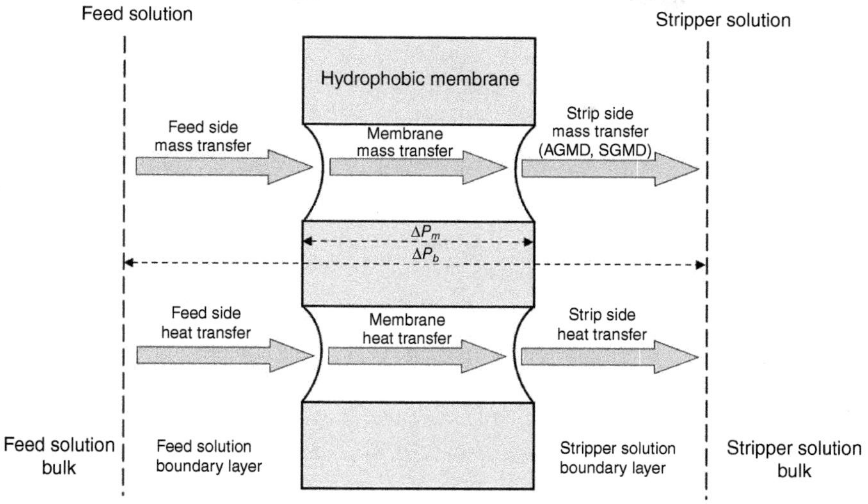

Figure 5.2 Mass and heat transfer in MD process.

The mass and heat transfer steps are shown in Figure 5.2. An overall representation of the MD process is provided by the following phenomenological equation (5.1) [9]:

$$J_w = \theta_v K \Delta P_b \tag{5.1}$$

where ΔP_b is overall driving force (Pa), K is overall mass transfer coefficient (kg/m^2 s Pa), and θ_v is the vapor pressure polarization coefficient of water. ΔP_b represents the vapor pressure gradient between the boundaries of the process, as defined by the bulk streams. K signifies the mass transfer coefficients applicable to the membrane and the concentration polarization boundary layers on either side of the membrane. θ_v is the fraction of the overall driving force that remains in place after losses due to the temperature polarization. In DCMD, for example, this equates to the ratio of the vapor pressure gradient between the membrane faces to that between the bulk streams, as given in Eq. (5.2) [9]:

$$\theta_v = \frac{P^F_{wm} - P^E_{wm}}{P^F_{wb} - P^E_{wb}} \tag{5.2}$$

where P^F_{wm} and P^E_{wm} are the vapor pressures at the feed–membrane and membrane–strip interfaces, respectively, and P^F_{wb} and P^E_{wb} are the vapor pressures of their corresponding bulk streams.

A systematic investigation of the various mass and heat transfer processes that mutually establish the MD process, as represented by Eq. (5.1), is discussed in detail by Johnson et al. [9]. The same approach will be summarized below to get a concise idea about the transport processes taking place in the MD process.

5.1.3 Overall Driving Force

After selecting a suitable feed temperature, the overall driving force can be assessed using a systematic three-step technique. In the first step, the vapor pressure of pure water at the bulk feed stream temperature is calculated. In the second step, an adjustment is made for lowering in vapor pressure of pure water due to the presence of solutes in the feed stream. In the third step, the vapor pressure of the bulk stripper medium is calculated according to the applicable MD configuration mode. This procedure permits ΔP_b to be determined as $(P^F_{wb} - P^E_{wb})$. The three-step technique is explained below.

Step 1: For calculating the pure water vapor pressure of the bulk feed stream, the Antoine equation can be used:

$$P^*_w = \ln\left(A - \frac{B}{C+T}\right) \tag{5.3}$$

In Eq. (5.3), P^*_w is the pure water vapor pressure (Pa), T is the temperature (K or °C), and A, B, and C are constants. It is clear that the vapor pressure varies exponentially with temperature. This fact has important implications for the selection of temperature in MD process. The driving force upsurges when the temperature of feed and stripper streams is increased while keeping the temperature gradient constant.

Step 2: Raoult's law (Eq. (5.4)) is applied to make adjustments for the lowering in vapor pressure of pure water, due to the presence of solutes:

$$P^F_{wb} = x^F_b \, P^{F*}_{wb} \tag{5.4}$$

where x^F_b is the mole fraction of water in the bulk of feed stream. Equation (5.4) is valid for dilute feed streams. In the case of concentrated feed solutions, like the liquid foods in MD, significant deviations may occur from Raoult's law. Under such circumstances, x^F_b is replaced by water activity, a^F_b, in Eq. (5.4):

$$P^F_{wb} = a^F_b \, P^{F*}_{wb} \tag{5.5}$$

where

$$a^F_b = x^F_b \, \gamma^F_b \tag{5.6}$$

where γ^F_b is the water activity coefficient at a given concentration of solutes. As the concentration of solutes increases, water activity coefficient decreases. Although many equations have been put forward for calculating the activity coefficient, yet the simplest one is Margules equation, presented as Eq. (5.7):

$$\gamma^F_b = x_s{}^2[\ln \gamma^\infty_b + 2(\ln \gamma^\infty_s + \ln \gamma^\infty_b)x^F_b] \tag{5.7}$$

where x_s is the mole fraction of the organic solute and γ^∞_b and γ^∞_s are the activity coefficients of water and solute, respectively, at infinite dilution. Tiegs et al. [10] tabulated the experimentally determined activity coefficients at infinite dilution for an extensive range of solute–solvent combinations.

Step 3: In this step, a separate strategy is used for determining pure water vapor pressure of stripper stream, for each type of MD configuration. In the case of DCMD, the Antoine equation is utilized for calculating the vapor pressure, as described for feed stream in *Step 2*. In the case of AGMD, the overall driving force is considered to be the difference in vapor pressures of the bulk feed and cooling water streams, while in the case of SGMD, the overall driving force is taken to be the difference in vapor pressures of the bulk feed and sweeping gas streams. The bulk sweeping gas stream is considered to be the incoming stream before permeate collection, as it flows through the module. The vapor pressure of the incoming stream is determined from its humidity ratio as given in Eq. (5.8):

$$w_b^E = \frac{0.622\phi\rho_{\text{wv}}}{\rho_{\text{ha}} - \rho_{\text{wv}}} \tag{5.8}$$

where w_b^E is the humidity ratio of the bulk strip stream, ϕ is the relative humidity expressed as a fraction, ρ_{wv} is the density of water vapor (kg/m^3), and ρ_{ha} is the density of the humid air (kg/m^3), all at a given temperature. The water vapor pressure of the sweeping gas stream bulk is then determined according to Eq. (5.9):

$$P_b^E = \frac{w_b^E P_t^E}{w_b^E + 0.622} \tag{5.9}$$

where $P_t^E$is the total strip-side gas pressure (Pa).

In the case of VMD, the overall driving force is taken to be the difference between the water vapor pressure of the bulk feed solution and the downstream pressure afforded by the application of a vacuum.

5.1.4 Overall Mass Transfer Coefficient, *K*

As shown in Figure 5.2, water flux perpendicular to the membrane surface must necessarily be the same in each stage of the process. This section investigates the determination of mass transfer coefficient for each individual stage as well as summing up these individual entities into the overall mass transfer coefficient. First of all, the overall resistance to mass transfer, $1/K$, is linked to the individual resistances according to Eqs. (5.10) and (5.11):

$$\frac{1}{K} = \frac{1}{k^F} + \frac{1}{k^m} + \frac{1}{k^E} \tag{5.10}$$

$$K = \frac{1}{\left(\frac{1}{k^F} + \frac{1}{k^m} + \frac{1}{k^E}\right)} \tag{5.11}$$

where k^F, k^m, k^E denote the feed-, membrane-, and stripper-side mass transfer coefficients, respectively. $\frac{1}{k^F}$ represents the resistance offered by the feed-side concentration polarization boundary layer, and $\frac{1}{k^m}$ denotes the resistance offered by the membrane, while $\frac{1}{k^E}$ depends on the form of MD configuration used. DCMD and VMD offer no resistance on the strip side; therefore, $\frac{1}{k^E}$ becomes zero. In AGMD, the air gap offers resistance, while in SGMD, the mass transfer resistance is offered by strip-side concentration polarization boundary layer.

5.1.4.1 Feed-Side Mass Transfer

Sherwood number, Sh, is used to calculate the feed-side mass transfer coefficient according to Eqs. (5.12) and (5.13):

$$k^F = Sh\frac{D_w}{d_h} \tag{5.12}$$

$$Sh = \frac{k^F d_h}{D_w} \tag{5.13}$$

where d_h is the hydraulic diameter of the conduit along which the feed stream flows (m) and D_w is the diffusion coefficient of the solute in the bulk solution (m^2/s). In the case of hollow fiber and tubular membrane modules, the hydraulic diameter is equal to the actual diameter for circular channels. In the case of rectangular channels such as those in plate-and-frame and spiral-wound membrane modules, the hydraulic diameter is equal to $\frac{4S}{P}$, where S denotes the cross-sectional area of the channel (m^2) while P represents the wetted perimeter (m) (portion of the cross-sectional area of the channel in contact with the liquid).

The Sherwood number is the ratio between the convective (turbulence) and diffusional mass transfers. An increase in the former entity lessens the thickness of boundary layer, and as a consequence the mass transfer resistance is decreased. A semiempirical relationship in the form of Eq. (5.14) is used to determine the Sherwood number.:

$$Sh = \alpha(Re^{\beta} Sc^{\gamma}) \tag{5.14}$$

where Re and Sc are the dimensionless Reynolds and Schmidt numbers, respectively, and α, β, and γ are the experimentally determined coefficients for a given module design and fluid velocity:

$$Re = \frac{d_h u \rho}{\mu} \tag{5.15}$$

$$Sc = \frac{\mu}{\rho D_w} \tag{5.16}$$

In Eq. (5.15), the Reynolds number gives the ratio between the inertial and viscous forces for a fluid flowing through a channel, while in Eq. (5.16), the Schmidt number portrays the ratio between momentum and mass diffusivities. Many forms of Eq. (5.14) have been proposed for use under specific conditions, prominent examples of which are listed in Table 5.1.

After obtaining the value of k^F, the concentration of solutes at the feed–membrane interface, C_m^F, and the concentration polarization coefficient, CMF^F, can be calculated using Eqs. (5.17) and (5.18), respectively:

$$C_m^F = C_b^F \exp\left(\frac{J_w}{\rho_w k^F}\right) \tag{5.17}$$

$$\mathrm{CMF}^F = \frac{C_m^F}{C_b^F} \tag{5.18}$$

Table 5.1 Mass transfer correlations used in MD.

Correlation	Flow type and limitations	References
$Sh = 1.615\left(Re^{0.33}Sc^{0.33}\left(\frac{d_h}{L}\right)^{0.33}\right)$	Laminar flow $Re < 2100$	[11]
$Sh = 1.5\left(Re^{0.33}Sc^{0.33}\left(\frac{d_h}{L}\right)^{0.33}\right)$	Laminar flow $Re < 2100$	[12]
$Sh = 0.023(Re^{0.83}Sc^{0.44})$	Turbulent flow $0.6 < Sc < 2.5$	[13]
$Sh = 0.023(Re^{0.8}Sc^{0.33})$	Turbulent flow $Re > 10^5$ $Sc > 0.5$	[14]
$Sh = 0.34(Re^{0.75}Sc^{0.33})$	Turbulent flow $104 < Re < 10^5$ $Sc > 0.5$	[14]
$Sh = 0.023(Re^{0.875}Sc^{0.25})$	Turbulent flow $300 < Sc < 700$	[14]
$Sh = 0.0149(Re^{0.88}Sc^{0.33})$	Turbulent flow $Sc > 100$	[14]
$Sh = 0.107(Re^{0.9}Sc^{0.5})$	Turbulent flow $0.5 < Sc < 10$	[14]
$Sh = 1.25\left(Re^{0.93}Sc^{0.33}\left(\frac{d_h}{L}\right)^{0.93}\right)$	Parallel flow over loosely packed fibers (void fraction 0.74–0.97) $0.5 < Re < 500$	[15]
$Sh = 1.38(Re^{0.34}Sc^{0.33})$	Transverse flow over loosely packed fibers (void fraction 0.93) $1 < Re < 25$	[15]
$Sh = 0.24$	Parallel flow over tightly packed fibers (void fraction 0.60)	[15]
$Sh = 0.90(Re^{0.4}Sc^{0.33})$	Transverse flow over tightly packed fibers (void fraction 0.30) $1 < Re < 25$	[15]
$Sh = 0.57(Re^{0.31}Sc^{0.33})$	Transverse flow over fibers present in net wrapped around central feeder tube	[16]
$Sh = 0.46(Re^{0.4}Sc^{0.33})$	Baffled cylindrical module with flow both countercurrent and perpendicular to fibers	[17]
$Sh = (0.53 - 0.58\Phi)(Re^{0.53}Sc^{0.33})$	General equation for different fiber packing densities, Φ	[18]

Source: Johnson and Nguyen [19].

where C_m^F and C_b^F are the concentrations of solute at the feed–membrane interface and in the bulk feed stream, respectively, and ρ_w is the bulk feed solution density (kg/m^3).

5.1.4.2 Membrane Mass Transfer

The water vapor transport through the pores of membrane can be explained either by molecular, Knudsen, or Poiseuille flow [20, 21]. The type of vapors flow is governed by the diameter of the pores and the pressure inside the pores.

In molecular diffusion, the membrane pores are large enough such that the smashing frequency among the water and air molecules is greater than that between the water molecules and the membrane pore walls. This transport mode can be neglected in DCMD, where deaerated feed and strip streams are used, and in VMD, where air inflowing into the pores is quickly removed. For molecular diffusion, the membrane permeability, k_{MD}^m, can be calculated using Eq. (5.19):

$$k_{\mathrm{MD}}^m = \frac{M_w}{RT}\frac{\varepsilon D_{\mathrm{wa}}}{\tau\delta}\frac{P}{(\mathrm{Pa})_{\mathrm{lm}}} \tag{5.19}$$

where M_w is the molecular weight of water (kg/mol), R is the universal gas constant (8.314 J/mol K), ε is the membrane porosity, τ is the membrane tortuosity, δ is the membrane thickness, P is the total pressure in membrane pores (Pa), and $(\mathrm{Pa})_{\mathrm{lm}}$ is the logarithmic mean of partial pressure of air (Pa). For determining the value of $(\mathrm{Pa})_{\mathrm{lm}}$, the interfacial water vapor pressure is firstly calculated from the interfacial temperature. Thereafter, the interfacial water vapor pressure is subtracted from the bulk stream pressure to obtain the value of the interfacial air pressure and hence $(\mathrm{Pa})_{\mathrm{lm}}$.

The geometry parameters like membrane thickness, porosity, and pore diameter are generally described by the manufacturer. These values can be incorporated in the respective equations for the calculation of mass transfer. Tortuosity can be estimated from porosity according to Eq. (5.20):

$$\delta = \frac{(2-\varepsilon)^2}{\varepsilon} \tag{5.20}$$

In Knudsen diffusion, the membrane pores are small enough, such that the smashing frequency among the water and air molecules is lower than that between the water molecules and the membrane pore walls. Equation (5.21) can be used to calculate the membrane permeability, k_{Kn}^m:

$$k_{Kn}^m = \frac{2}{3}\frac{\varepsilon r}{\tau\delta}\left(\frac{8}{\pi RTM_w}\right)^{0.50} \tag{5.21}$$

In Poiseuille flow, the water vapor transport occurs convectively through the deaerated membrane pores. Mass transfer is facilitated by the impact of a temperature-induced vapor pressure gradient or downstream vacuum. Therefore, this mode of flow is mainly restricted to deaerated DCMD and VMD. The resistance to mass transfer is mainly attributed to the momentum transfer to the pore walls through viscous drag. Equation (5.22) can be used to calculate the membrane permeability, $k^m_{\rm Pl}$ [20]:

$$k^m_{\rm Pl} = \frac{r^2\varepsilon}{8\delta\tau}\frac{M_w P_{\rm av}}{\eta R T_{\rm av}} \tag{5.22}$$

where $P_{\rm av}$ is the average gas pressure (air plus water vapor in the pores (Pa)).

After a brief discussion about each flow mode, the question arises, how to determine the applicable mass transfer mechanism? For this purpose, Knudsen number, *Kn*, is used according to Eqs. (5.23) and (5.24), which is the ratio between mean free path and the nominal pore diameter of the membrane [22]. Mean free path is the distance traveled by a water vapor molecule before colliding with an air molecule or the pore wall:

$$Kn = \frac{\lambda}{2r} \tag{5.23}$$

$$\lambda = \frac{Tk_B}{P\sqrt{2\pi\sigma^2}} \tag{5.24}$$

where λ is mean molecular free path (m), r is pore radius (m), k_B is Boltzmann constant (1.3807×10^{-23} J/K), and σ is collision diameter (m). A collision diameter of 0.27 nm is water characteristic. The calculated value of *Kn* determines the mode of flow and thus the type of mass transfer equation to be used. In the case of air-filled pores, *Kn* of 0.01 or less indicates molecular diffusion, while in the case of deaerated pores, like deaerated DCMD and VMD, a *Kn* of 0.01 or less indicates Poiseuille flow. A *Kn* of 10 or more represents the Knudsen diffusion, both in air-filled and deaerated pores. In the case of air-filled pores, a value of *Kn* between 0.01 and 10 shows a transition between molecular and Knudsen diffusion. Likewise, in the case of deaerated pores, a value of *Kn* between 0.01 and 10 shows a transition between Poiseuille flow and Knudsen diffusion. In the former case, Eq. (5.25) is applied for evaluating the membrane permeability [23], while in the latter case, Eq. (5.26) is applied [24]:

$$k^m = \frac{1.8 * 10^{-15}}{RT\delta}\left(\frac{3}{2}\frac{\tau}{\varepsilon r}\left(\frac{8}{\pi RTM_w}\right)^{-0.50} + \frac{P_a\tau}{\varepsilon(4.46 * 10^{-6}T^{2.334})}\right) \tag{5.25}$$

where P_a represents the air partial pressure in the membrane pore that is generally considered to be 101 325 Pa [25] and

$$k^m = \frac{\pi}{RT\delta\tau}\left(\frac{2}{3}\left(\frac{8RT}{\pi M_w}\right)^{0.50} r^3 + \frac{r^4}{8\eta}P_{\rm av}\right) \tag{5.26}$$

5.1.4.3 Strip-Side Mass Transfer

As discussed earlier, DCMD and VMD offer no resistance on the strip side; therefore, the resistance to mass transfer can be neglected in these cases. In SGMD, the concentration polarization boundary layer on the strip side offers resistance to mass transfer, while in AGMD, the air gap offers resistance. In the former case, the calculation of the mass transfer coefficient can be made using the same equations used to make calculations on the feed side. In the latter case, vapor transfer occurs by molecular diffusion through an air gap that exists between the membrane–stripper interface and the condensate film on the condenser plate surface. Therefore, Eq. (5.27) can be used to determine the mass transfer coefficient [26]:

$$k^E = \frac{P_{\text{av}}}{(\text{Pa})_{\text{lm}}} \frac{\varepsilon}{(\tau\delta + b)} \frac{D_{\text{wa}} M_w}{RT_{\text{av}}} \tag{5.27}$$

where b is the air gap thickness (m). In AGMD, the air gap thickness is usually 1–10 mm, which is almost 10–100 times larger than the membrane's thickness [24].

5.1.5 Vapor Pressure Polarization Coefficient, θ_v

The determination of the vapor pressure polarization coefficient, θ_v, depends upon the heat transfer processes occurring on the stripper side, and therefore, it varies according to four configuration modes of MD.

5.1.5.1 DCMD

In DCMD, the loss in driving force is mainly due to the temperature polarization at both sides of the membrane, which thus necessitates the knowledge of interfacial temperatures at each side, for the calculation of θ_v. The interfacial temperature cannot be measured directly; rather it requires its calculation through heat balance equations. Therefore, a stepwise procedure, just as elaborated in the case of mass transfer, is applied. These steps can be summarized below:

$$Q_w = U\Delta T_b = h^F (T_b^F - T_m^F) = J_w \Delta H_v + h^m (T_m^F - T_m^E) = h^E (T_m^E - T_b^E) \tag{5.28}$$

In Eq. (5.28), Q_w is the heat flux across the boundary layer (W/m^2), U is the overall heat transfer coefficient (W/m^2 K), ΔT_b is the difference between the feed and stripper bulk temperatures (K), T_b^F is the feed bulk temperature (K), T_m^F is the feed–membrane interface temperature (K), T_b^E is the stripper bulk temperature (K), T_b^E is the temperature at membrane–stripper interface (K), h^F is the coefficient of heat transfer at feed-side boundary layer (W/m^2 K), h^m is the coefficient of heat transfer of the membrane (W/m^2 K), h^E is the coefficient of heat transfer at stripper-side boundary layer (W/m^2 K), and ΔH_v is the latent heat of vaporization of water (J/mol).

The overall heat transfer resistance, $1/U$, is equal to the sum of the individual stage resistances and can be represented by Eq. (5.29):

$$\frac{1}{U} = \frac{1}{h^F} + \frac{1}{h^m + \frac{J_w \Delta H_v}{\Delta T_m}} + \frac{1}{h^E} \tag{5.29}$$

The interfacial temperatures at the feed–membrane and membrane–stripper can be evaluated through Eqs. (5.30) and (5.31), respectively [20, 22],

$$T_m^F = \frac{h^m \left(T_b^E + \left(\frac{h^F}{h^E}\right) T_b^F\right) + h^F T_b^F - J_w \Delta H_v}{h^m + h^F \left(1 + \left(\frac{h^m}{h^E}\right)\right)} \tag{5.30}$$

$$T_m^E = \frac{h^m \left(T_b^F + \left(\frac{h^E}{h^F}\right) T_b^E\right) + h^E T_b^E + J_w \Delta H_v}{h^m + h^E \left(1 + \left(\frac{h^m}{h^F}\right)\right)} \tag{5.31}$$

In Eqs. (5.28)–(5.31), the values of J_w, T_b^F, and T_b^E can be determined experimentally, while the values of each of h^F, h^m, and h^E require specific relationships, as described below.

5.1.5.2 Feed-Side and Strip-Side Heat Transfer

Just like the Sherwood number in mass transfer, we can use the analogous Nusselt number, Nu, for estimation of feed- and stripper-side heat transfer coefficients through Eqs. (5.32) and (5.33), respectively,

$$Nu = \frac{h^F d_h}{k_F^T} \tag{5.32}$$

$$Nu = \frac{h^E d_h}{k_E^T} \tag{5.33}$$

where k_F^T is the thermal conductivity of feed boundary layer (W/m K) and k_E^T is the thermal conductivity of stripper boundary layer (W/m K).

A semiempirical relationship in the form of Eq. (5.34) is used to determine the Nusselt number:

$$Nu = \alpha Re^{\beta} Pr^{\gamma} \tag{5.34}$$

where Pr is the Prandtl number and α, β, and γ are the experimentally determined coefficients for a given module design and fluid velocity. Pr is the ratio between momentum diffusivity and heat diffusivity. In other words, Pr can be regarded as the ratio between thickness of velocity and thermal boundary layers, representing the heat transfer equivalent of the Schmidt number in mass transfer. For the feed and stripper side, Pr can be calculated using Eqs. (5.35) and (5.36), respectively,

$$Pr = \frac{\mu C_p}{k_F^T} \tag{5.35}$$

$$Pr = \frac{\mu C_p}{k_E^T} \tag{5.36}$$

$$Gz = RePr\left(\frac{d_h}{z}\right) \tag{5.37}$$

where C_p is the specific heat (J/kg K), Gz is the Gratz number, and z is the module length (m). Many forms of Eq. (5.34) have been proposed for use under specific conditions, prominent examples of which are provided in Table 5.2.

5.1.5.3 Membrane Heat Transfer

The heat transfer coefficient for membrane can be estimated using Eqs. (5.38)–(5.40):

$$h^m = \frac{k_m^T}{\delta} \tag{5.38}$$

$$k_m^T = \varepsilon k_{\text{gas}}^T + (1-\varepsilon)k_{\text{polymer}}^T \tag{5.39}$$

$$h^m = \frac{\varepsilon k_{\text{gas}}^T + (1-\varepsilon)k_{\text{polymer}}^T}{\delta} \tag{5.40}$$

Table 5.2 Heat transfer correlations used in MD.

Correlation	Flow type and limitations	References
$Nu = 1.62\left(Re^{0.33}Pr^{0.33}\left(\frac{d_h}{L}\right)^{0.33}\right)$	General laminar flow equation	[27]
$Nu = 1.86\left(Re^{0.33}Pr^{0.33}\left(\frac{d_h}{L}\right)^{0.33}\right)$	Laminar flow Low values for $RePr\left(\frac{d_h}{L}\right)$ term Flat-sheet modules	[28, 29]
$Nu = 3.66 + \frac{0.0668\left(RePr\frac{d_h}{z}\right)}{1+0.0011\left(Re^{0.8}Pr^{0.8}\left(\frac{d_h}{L}\right)^{0.8}\right)}$	Laminar flow Recommended correlation	[27, 28]
$Nu = 0.298(Re^{0.646}Pr^{0.316})$	Laminar flow $150 < Re < 3500$	[30]
$Nu = 0.023(Re^{0.8}Pr^{n})$ $n = 0.4$ for heated fluid $n = 0.3$ for cooled fluid (feed stream in all forms of MD)	Turbulent flow $2500 < Re < 1.25 \times 10^5$ $0.6 < Pr < 100$	[31]
$Nu = 0.036\left(Re^{0.8}Pr^{0.33}\left(\frac{d_h}{L}\right)^{0.055}\right)$	Turbulent flow	[32]
$Nu = 0.036\left(Re^{0.96}Pr^{0.33}\left(\frac{d_h}{L}\right)^{0.055}\right)$	Turbulent flow in tube $10 \leq L/d \leq 400$	[33]
$Nu = 0.206((Re\cos\omega)^{0.63}Pr^{0.36})$ where ω is the yaw angle (ranges between 0° and 90° for pure cross flow and parallel flow, respectively)	Flow through shell side of hollow fiber and tubular membranes	[28, 34]

Source: Johnson and Nguyen [19].

where h^m is the heat transfer coefficient of membrane (W/m^2 K) and k_m^T, k_{gas}^T, and k_{polymer}^T are the thermal conductivities of membrane, air, and polymer (W/m K), respectively.

After the calculation of h^F, h^E, and h^m, Eqs. (5.30) and (5.31) can be used to calculate T_m^F and T_m^E, respectively. The water vapor pressures for pure water on opposite sides of the membrane interface can then be calculated using the Antoine equation. For feed side, the adjustment for solute's concentration can be made using CMF^F and Raoult's law if required, thus allowing to calculate θ_v using Eq. (5.2).

5.1.6 AGMD

In AGMD, heat transfer ensues progressively through the feed-side boundary layer, air gap, condensate layer, cold plate, and the cooling water boundary layer. As the resistance offered to heat transfer by the air-filled membrane is insignificant in relation to the air gap, therefore it can be neglected. Figure 5.3 shows the AGMD's

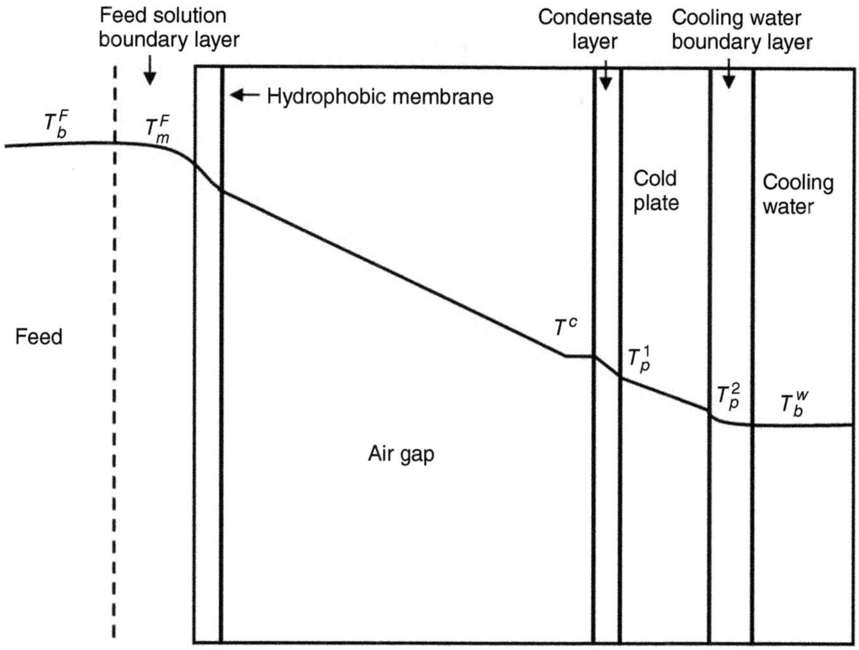

Figure 5.3 Temperature profile in AGMD.

temperature profile. The downstream interfacial temperature required for estimation of the vapor pressure polarization coefficient is taken as the temperature at the air gap–condensate layer interface, T^c. For the estimation of θ_v, the temperature of bulk stripper is taken as the temperature of cooling water, T_b^w.

After the condensation occurs, the final three heat transfer stages are grouped together as a collective heat removal stage. The heat transfer coefficient is then defined according to Eq. (5.41) [26]:

$$h^g = \frac{1}{\frac{1}{h^c} + \frac{1}{h^p} + \frac{1}{h^w}} \tag{5.41}$$

where g, c, p, and w represent the grouped stages, condensate layer, cold plate, and cooling water, respectively. The heat transfer in AGMD can then be summarized according to Eq. (5.42):

$$Q_w = U\Delta T_b = h^F(T_b^F - T_m^F) = J_w \Delta H_v + h^a(T_m^F - T^c) = h^g(T^c - T_b^w) \tag{5.42}$$

The feed–membrane and air gap–condensate interface temperatures can be evaluated through Eqs. (5.43) and (5.44), respectively,

$$T_m^F = \frac{h^a\left(T_b^w + \left(\frac{h^F}{h^g}\right)T_b^F\right) + h^F T_b^F - J_w \Delta H_v}{h^a + h^F\left(1 + \left(\frac{h^a}{h^g}\right)\right)} \tag{5.43}$$

$$T^c = \frac{h^a\left(T_b^F + \left(\frac{h^g}{h^F}\right)T_b^w\right) + h^g T_b^w + J_w \Delta H_v}{h^a + h^g\left(1 + \left(\frac{h^a}{h^F}\right)\right)} \tag{5.44}$$

In Eqs. (5.43) and (5.44), J_w, T_b^F, and T_b^w can measured experimentally. However, estimation of h^F, h^a, and h^g require suitable heat transfer correlations, as discussed below.

h^F can be estimated as per procedure outlined for DCMD. h^a is calculated according to Eq. (5.45), which is based on simple conduction through the air-filled gap:

$$h^a = \frac{k_{\text{gas}}^T}{b} \tag{5.45}$$

For condensation over a vertical plate, the heat transfer coefficient, h^c, can be determined through Eq. (5.46) [35]:

$$h^c = 0.943\left(\frac{\rho^2 g \Delta H_v k_{\text{film}}^3}{\eta L(T^c - T^{\text{pl}})}\right)^{0.25} \tag{5.46}$$

where k_{film} is the condensate film's thermal conductivity and L is the height of the air gap (m). In AGMD, T^c can be replaced by T_b^F, due to the small loss of conductive heat afforded by the air gap and the total recovery of latent heat at the air gap–cooling plate interface. Moreover, T^{pl} can be replaced by T_b^w because of the insignificant effect of the variation of cooling water temperature on water vapor flux. Therefore in Eq. (5.46), the unknown term $T^c - T^{\text{pl}}$ can be replaced by the known term $T_b^F - T_b^w$ [26].

The heat transfer coefficient applicable to conduction through the cold plate is calculated as through Eq. (5.47):

$$h^p = \frac{k_p}{\Delta z} \tag{5.47}$$

where k_p is the plate material's thermal conductivity and Δz is the plate thickness (m). As a boundary layer exists in the stream of cooling water, therefore h^w should be calculated using Eq. (5.32) and a suitable equation for *Nu* from Table 5.2. h^g can then be determined by Eq. (5.41), thus permitting to calculate T_m^F and T^c using Eqs. (5.43) and (5.44), respectively.

5.1.6.1 SGMD

Just like in DCMD, the heat transfer in the case of SGMD can also be calculated using Eq. (5.42). Hence, the same procedure, as discussed previously, is adopted to determine all of the heat transfer coefficients. Likewise, the water vapor pressure at the feed–membrane interface can be estimated from temperature at interface obtained from Eq. (5.43). However, it should be remembered that the water stripper stream does not exist in SGMD; therefore, the water vapor pressure cannot be calculated from the temperature at the membrane–stripper interface. Instead, humidity ratio, w_b^E, before entering the membrane module, can be applied to determine the water vapor pressure of the bulk sweeping gas stream, according to Eq. (5.9). Nevertheless, there is a gradual rise in the humidity ratio at the membrane's surface along the module length, as the permeate arises from the membrane–stripper interface. This results in a gradual decline in the driving force. Therefore, the water vapor pressure at the stripper side of the membrane surface can be calculated by adjusting the value of w_b^E for the effect of water permeate flux according to Eqs. (5.48) and (5.49):

$$w_m^E = w_b^E + \frac{J_w A}{m_{\text{fa}}} \tag{5.48}$$

$$P_m^E = \frac{w_m^E P_t^S}{w_m^E + 0.622} \tag{5.49}$$

where w_m^E is the humidity ratio at the membrane–stripper interface, A is area of the membrane (m^2), and m_{fa} is the mass flow rate of air (kg/s) in the membrane module.

5.1.7 VMD

As the pressure inside the membrane pores is very low in VMD, therefore, the conductive heat transfer across the membrane can be neglected [28, 34]. Likewise, because of external condensation phenomenon and the lack of temperature polarization boundary layer, heat transfer on stripper side can also be ignored. Thus, in VMD, the heat transfer occurs in two steps. In the first step, evaporation occurs due to the heat transfer across the feed-side boundary layer. In the second step, latent heat transfer occurs over the membrane due to water vapors. Therefore, the heat transfer in this process can be presented in Eq. (5.50). Accordingly, the temperature at feed–membrane interface can be determined using Eq. (5.51), which thus facilitates the calculation of water vapor pressure from Antoine equation:

$$h^F(T_b^F - T_m^F) = J_w \Delta H_v \tag{5.50}$$

$$T_m^F = T_b^F - \frac{J_w \Delta H_v}{h^F} \tag{5.51}$$

As only the temperature polarization is responsible for the driving force loss on the feed side, therefore, θ_v can be determined using Eq. (5.52):

$$\theta_v = \frac{P_{wb}^F - P_{wm}^F}{P_{wb}^F - P_{wb}^E} \tag{5.52}$$

5.1.8 Membranes for MD Process

The main component of the MD unit is the membrane itself. The membrane merely acts as a physical support mechanism for the separation of vapor and liquid, but some properties must be met. Therefore, the membrane's characteristics must be carefully analyzed, and a membrane should only be made of the material with the required properties. Effective MD must take place that the pores must only be filled with vapor, and no liquid should enter them. Therefore, the utmost requirement for a suitable MD membrane is to be sufficiently hydrophobic and permeable. Moreover, it should have a low mass transfer resistance, high heat transfer resistance, high thermal stability, and good chemical compatibility.

The pore size and porosity of an MD membrane should typically lie in the range of 0.1–1 μm and 40–90%, respectively. Likewise, the thickness of the membrane should not exceed 30–60 μm. For this purpose, the commercially available microfiltration membranes have proved to be a good choice. However, conventional microfiltration membranes are prone to wetting and fouling when dealing with high-saline and low surface tension liquids. Therefore, these membranes still need to improve performance in terms of wetting and fouling resistance and high water vapor flux. The antiwetting properties of MD membranes can be improved by adjustment of the membrane surface roughness. Moreover, the chemical composition of the MD membrane can also be manipulated to get the desired antiwetting and antifouling properties for specific applications [8, 36].

The commercially available MD membranes are mostly made of polypropylene (PP), polyvinylidene difluoride (PVDF), and polytetrafluoroethylene (PTFE). Out of these three membranes, PTFE exhibits the best chemical resistance, good thermal stability, and, most importantly, the highest hydrophobicity. Likewise, PVDF membrane also enjoys better chemical and thermal resistance. On the other hand, PP membrane displays high crystallinity and remarkable solvent resistance [36]. Table 5.3 shows the physical details of commercial MD membrane supplied by various manufacturers.

The recent works in MD are mostly concerned with the fabrication of innovative membrane materials with superior properties to enhance the system performance further. An example of this is developing an asymmetric membrane for MD and subsequent thermo-osmotic conversion done by Shaulsky et al. [38]. They produce the asymmetric membranes by treating cellulose ultrafiltration membranes with perfluorodecyltrichlorosilane. The resulting membranes gave a water vapor permeability coefficient of 1.7×10^{-7} kg/m^2 s Pa. Another major problem encountered in MD is a higher risk of membrane fouling. This phenomenon has been put to study for the past decade, but only recently could such materials be developed to withstand membrane fouling and scaling to an appreciable extent. In a recent work carried out by Liao et al. superhydrophobic membranes were produced by electrospraying a fumed solution of polydimethylsiloxane (PDMS), PVDF, and silica, followed by coating this membrane on an electrospun nanofibrous PVDF support membrane. The resulting membrane has better abrasion resistant and antiwetting properties, which was shown to have ultimately retarded the formation of scales and fouling on the membrane. These membranes not only gave better antiwetting properties but also gave much higher flux values of up to 28 kg/m^2 h with a sustained salt rejection of 99.99% [39]. Xu et al. [40] produced poly(ethylene chlorotrifluroethylene) (PECTFE) by treating it with acetyl tributyl citrate (ATBC). This membrane was produced by the TIPS (thermally induced phase separation) method. The resulting membranes had a superior hydrophobicity and were used in VMD. The membranes gave a salt rejection of 99.9% and a permeate flux of 22.3 l/m^2 h. This showed that the treatment had indeed helped in improving the performance of VMD.

5.1.9 Pros and Cons of MD Process

MD, as a separation process, has several advantages over other conventional processes. The main advantage that it serves is its moderate energy consumption, which

Table 5.3 Typical commercial membranes used in MD applications.

Manufacturer	Configuration	Material	Thickness (μm)	Porosity (%)	Average pore size (μm)
Gelman	Flat sheet	PTFE/PP	178	80	0.20
		PTFE/PP	178	80	0.45
		PTFE/PP	178	80	1.00
		PTFE	60	50	0.80
		PP	90	55	0.10
Millipore	Flat sheet	PTFE/PE	130	70	0.2
		PTFE/PE	175	85	0.5
		PTFE/PE	150	85	1.0
		PVDF	110	75	0.22
		PVDF	140	75	0.45
Hoechst Celanese Co.	Flat sheet	PP	28	45	0.07
	Flat sheet	PP	25	38	0.05
	Capillary	PP	25	35	0.03
Gore	Flat sheet	PTFE/PP	184	44	0.2
		PTFE	64	90	0.2
		PTFE	77	89	0.45
GoreTex Sep.	Flat sheet	PTFE	100	80	0.20
		PP	91	66	0.29
3M Corporation	Flat sheet	PP	81	76	0.40
		PP	76	79	0.51
		PP	86	80	0.58
		PP	79	85	0.73
Teknorama	Flat sheet	PTFE	—	80	0.2
		PTFE	—	80	0.5
		PTFE	—	80	1.0
Akzo Nobel Microdyn	Capillary	PP	450	70	0.2
Enka Microdyn	Capillary	PP	1550	75	0.2
Enka A.G. Euro-Sep	Capillary	PP	200	70	0.2

Source: Cassano et al. [37].

makes it an energy-saving technology. Moreover, it requires minimum land and capital to operate, as compared with conventional distillation [41]. Also, because of the lower operating temperatures required, MD can be used in conjunction with alternate energy sources such as solar energy [42]. Another key benefit of MD is that it produces a very high purity product as opposed to conventional membrane separation processes such as reverse osmosis (RO), nanofiltration, or ultrafiltration (theoretically 100% rejection of ion, macromolecules, and colloids is possible) [43]. Some other advantages of MD over RO include lower pressures leading to thinner pipelines, the provision to treat higher-saline feedstocks, less chance of leakage, no necessity for pretreatment of feed, and system efficiency independent of wastewater salinity [44].

The operational setup required for MD process is compact, which means lower space is required for installation. The feed stream for MD does not need to be heated above its boiling point. Therefore, the process temperature is lower, typically in the range of 35–85 °C, thus minimizing the loss of heat to the surrounding environment. For the purpose of attaining operating temperature, waste heat streams can be utilized along with the option of alternative energy sources like wind, solar, or geothermal. This makes MD process further economical and an attractive choice over conventional desalination and wastewater treatment technologies. The operation of MD plant at ambient atmospheric pressure is another added advantage. This ensures the safety of the process, meanwhile lowering the fouling propensity of MD membranes.

Some of the disadvantages of MD process are lower water vapor flux than the established technologies of desalination and wastewater treatment. MD process also suffers from the temperature polarization phenomena. The process has a high specific thermal energy consumption. Other downsides of MD process are risks of membrane wetting, sensitivity to low surface tension feeds, and a large amount of heat dissipation due to the conduction [45].

As discussed earlier, there are four basic configurations of MD, each representing its benefits and drawbacks. Therefore, after looking at the advantages and limitations of MD as a whole, we will now discuss the pros and cons of each configuration separately.

The advantages of DCMD include the fact that it gives a very high permeate flux and is simple in its design and operation. Moreover, it can incorporate internal heat recovery into the system. However, DCMD has some major limitations as well, including low thermal efficiency, high sensitivity to temperature and concentration polarization, and a very strong likelihood of the contamination of permeating stream.

AGMD was devised initially to address the limitations of DCMD. Therefore, the main advantages of AGMD are that ocean water can be applied as a coolant on the permeate side and the system has a remarkable thermal efficiency. Also, interior heat recovery can easily be done for this system as well. However, it faces some disadvantages while tackling the limitations of DCMD. AGMD has a very high resistance to water vapor, and this results in very low permeate flux. Furthermore, AGMD also has a very large carbon footprint.

VMD enjoys some key advantages, such as high permeate flux and low conductive heat losses. It also gives us another big advantage, which is that there is a very small formation of temperature and concentration boundary layer on each side of the membrane. Therefore, temperature and concentration polarization do not affect VMD as much as it affects MD configurations. However, in providing these exceptional advantages, it suffers from some major drawbacks, which include a high degree of pore wetting and a very low chance and possibility of heat recovery.

Lastly, SGMD offers advantages, which include exceptional mass transfer rate and very minimal conductive heat losses. However, it also has some drawbacks of its own, including the fact that there is considerable difficulty in heat recovery, dealing with sweeping gas. It needs much more stringent and delicate controls, and a much larger condenser is required for cooling purposes within SGMD [46].

5.1.10 Future Prospects of MD Process

Among the many future prospects of MD, the one thing that seems to be advancing at an appreciable rate is the novel membranes and membrane materials. Ceramic membranes have their advantages in being chemically resistant and tough materials; however, they are highly hydrophilic, making them unsuitable for MD. For this purpose, coatings are being developed to make ceramic membranes hydrophobic while keeping a high flux and salt rejection. Mixed matrix membranes formed by the combination of polymeric matrix embedded with nano- and microfiller particles may also give an advantage by forming stiffer membranes, which have lower heat transfer rates and reduced wettability. Moreover, the production of nanofibrous materials and the development of the electrospinning technique has created new opportunities for the development of even more improved membrane materials in the future [46]. One very promising solution to the problem of membrane fouling can be the use of nanoparticle-incorporated membranes, which will produce a superhydrophobic material having contact angles in excess of 150°. Moreover, nanoparticle-embedded materials also give self-cleaning properties to the fabricated membrane through the lotus effect. Thus, nanoparticle-embedded membranes could be the future of MD materials [47].

In the end, as stated above, the fact that MD can be operated at mild conditions means that it can use solar energy and other alternative energy resources to provide heat, meaning more work needs to be done to incorporate solar energy heating into the MD system for better performance and energy utilization as well as environmental protection. Finally, as is the case with current trends, we must continue to utilize hybrid separation techniques and conventional distillation with MD so that the disadvantages and limitations of the different separation techniques can be overcome [48].

5.2 Osmotic Membrane Distillation (OMD)

The concept and process for OMD were introduced by Lefebvre [49]. As opposed to MD, OMD was operated at ambient temperature. Therefore, OMD had a lower risk of volatile loss and heat degradation [50].

5.2.1 Basic Principles of OMD Process

In the OMD process, water vapors are evaporated due to the higher water activity of the feed solution, which are then transported by means of diffusion across the pores, and finally condensed at the stripper side due to the lower water activity of

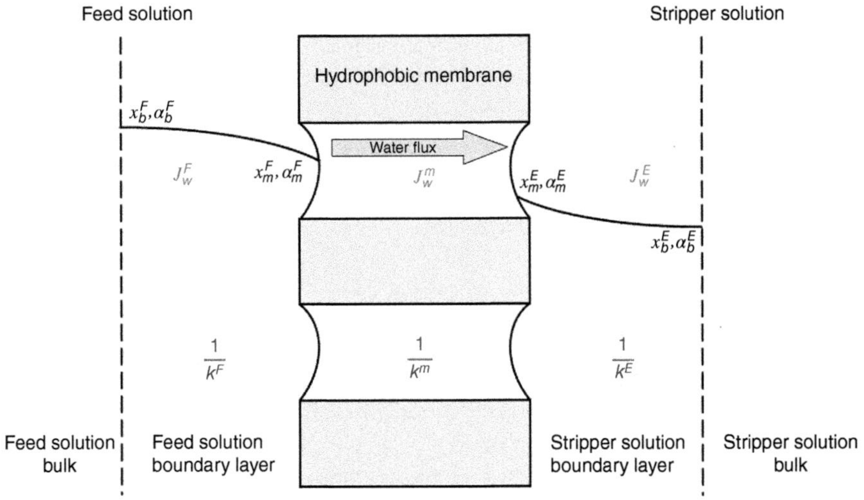

Figure 5.4 Schematic of mass transfer in osmotic membrane distillation.

the solution [51]. In the case of OMD, the stripper is in the form of a concentrated brine solution, which has a lower water activity compared with the feed solution. The difference between water activity of the feed and stripper solution works as a driving force for the OMD process. As opposed to MD, such a driving force eliminates the need for heating of feed solution, making the OMD process further economical. Also, ambient temperature makes OMD process significantly beneficial for concentration of liquid foods like fruit juices, which are otherwise vulnerable to high temperature and may lose their nutritional properties upon heating. Figure 5.4 shows an overview of the OMD process. The stepwise process of mass transfer is (i) water vapor evaporation near the membrane pores on the feed side, (ii) water vapor transport through the air-filled pore, and (iii) water vapor condensation near the membrane pores on the stripper side [52–55]. Under ambient process conditions, the difference in temperature of feed and stripper liquids is pretty minor, and the process may be well thought-out isothermal [52, 56].

The water vapor flux in OMD process is calculated according to Eq. (5.53) [57]:

$$J = \frac{1}{A}\frac{\Delta W}{\Delta t} \tag{5.53}$$

where J is water vapor flux (kg/m^2 h), ΔW is variation in weight of feed (kg), Δt is period of time during which the variation in weight is noted (hours), and A is membrane surface area (m^2).

Now, the mass transfer steps in OMD will be briefly described in terms of mathematical expressions. Most of the equations are almost the same as previously discussed in detail in the MD process.

As OMD can be considered as an isothermal process, therefore, the heat transfer is negligible and thus generally neglected.

5.2.2 Overall Mass Transfer

At steady state, the overall mass transfer in OMD can be expressed in the form of Eq. (5.54) for hollow fiber membrane contactor (HFMC) module:

$$J_w^F \cdot A_{OD} = J_w^m \cdot A_{av} = J_w^E \cdot A_{ID} \tag{5.54}$$

where A_{OD} is the contact surface area (m^2) outside the HFMC fiber, A_{av} is the average area (m^2) between inside and outside the fiber, and A_{ID} is the contact surface area (m^2) inside the fiber.

In the context of water vapor pressure and water activity, the overall mass transfer flux (J_w) can be expressed according to Eqs. (5.55) and (5.56), respectively [22, 58],

$$J_w = K(P_{wb}^F - P_{wb}^E) \tag{5.55}$$

$$J_w = K(a_b^F - a_b^E) \tag{5.56}$$

where overall mass transfer coefficient (K) is expressed as

$$K = \frac{1}{\left(\frac{1}{k^F} + \frac{1}{k^m} + \frac{1}{k^E}\right)}$$

5.2.2.1 Mass Transfer Across Feed Boundary Layer

In terms of mole fraction, the water vapor flux along the feed-side boundary layer is given by Eq. (5.57) [59, 60]:

$$J_w^F = k^F(x_b^F - x_m^F) \tag{5.57}$$

In terms of vapor pressure, the water vapor flux is given by Eq. (5.58):

$$J_w^F = k^F(P_{wb}^F - P_{wm}^F) \tag{5.58}$$

In terms of water activity, the water vapor flux is given by Eq. (5.59) [53, 58]:

$$J_w^F = k^F(a_b^F - a_m^F) \tag{5.59}$$

where P_{wb}^F and P_{wm}^F can be further elaborated mathematically through Eqs. (5.60) and (5.61), respectively,

$$P_{wb}^F = a_b^F P_{wb}^{F*} = x_b^F \gamma_b^F P_{wb}^{F*} \tag{5.60}$$

$$P_{wm}^F = a_m^F P_{wm}^{F*} = x_m^F \gamma_m^F P_{wm}^{F*} \tag{5.61}$$

5.2.2.2 Mass Transfer Across Stripper Boundary Layer

In terms of mole fraction, the water vapor flux along the stripper-side boundary layer is given by Eq. (5.62) [59, 60]:

$$J_w^E = k^E(x_m^E - x_b^E) \tag{5.62}$$

In terms of vapor pressure, the water vapor flux is given by Eq. (5.63) [22, 23, 61]:

$$J_w^E = k^E(P_{wm}^E - P_{wb}^E) \tag{5.63}$$

In the context of water activity, the water vapor flux is given by Eq. (5.64) [53]:

$$J_w^E = k^E(a_m^E - a_b^E) \tag{5.64}$$

where P_{wb}^{E} and P_{wm}^{E} can be further elaborated mathematically through Eqs. (5.65) and (5.66), respectively,

$$P_{wb}^{E} = a_{b}^{E}P_{wb}^{E*} = x_{b}^{E}\gamma_{b}^{E}P_{wb}^{E*} \tag{5.65}$$

$$P_{wm}^{E} = a_{m}^{E}P_{wm}^{E*} = x_{m}^{E}\gamma_{m}^{E}P_{wm}^{E*} \tag{5.66}$$

5.2.2.3 Mass Transfer Across Membrane

The mass transfer across the pores of hydrophobic membrane in OMD can be presented by Eq. (5.67) [23, 52, 62]:

$$J_{w}^{m} = k^{m}(P_{wm}^{F} - P_{wm}^{E}) \tag{5.67}$$

5.2.2.4 Mass Transfer Coefficient for Feed and Stripper Side

The mass transfer coefficient on the feed side can be assessed according to Eqs. (5.68) and (5.69) while the mass transfer coefficient on the stripper side can be worked out through Eqs. (5.70) and (5.71):

$$k^{F} = Sh\frac{D_{w}}{d_{h}} \tag{5.68}$$

$$Sh = \frac{k^{F}d_{h}}{D_{w}} \tag{5.69}$$

$$k^{E} = Sh\frac{D_{w}}{d_{h}} \tag{5.70}$$

$$Sh = \frac{k^{E}d_{h}}{D_{w}} \tag{5.71}$$

A suitable equation for the estimation of Sherwood number on feed side and stripper side can be used from Table 5.4, where Φ is the packing fraction, n is the total fibers in HFMC module, d_0 is the fiber external diameter, and D_i is the shell inner diameter. All the remaining entities used in the equations for determination of Sh have been previously described in the MD process.

5.2.2.5 Mass Transfer Coefficient Across Membrane

The mass transfer coefficient for membrane can be found using Eq. (5.72), according to the dusty gas model [23, 58, 62]:

$$k^{m} = \frac{1.8 * 10^{-15}}{RT\delta}\left(\frac{1}{D_{we}^{Kn}} + \frac{P_{a}}{D_{wa}^{\mathrm{MD}}}\right)^{-1} \tag{5.72}$$

where

$$D_{we}^{Kn} = K_{0}\left(\frac{8}{\pi RTM_{w}}\right)^{0.50} \tag{5.73}$$

$$K_{0} = \frac{2}{3}\frac{\varepsilon r}{\tau} \tag{5.74}$$

$$D_{wa}^{\mathrm{MD}} = K_{1}PD_{wa} \tag{5.75}$$

$$K_{1} = \frac{\varepsilon}{\tau} \tag{5.76}$$

$$PD_{wa} = 4.46 * 10^{-6}T^{2.334} \tag{5.77}$$

Table 5.4 Mass transfer coefficient for feed and stripper side in OMD.

Feed side	Stripper side	References
$Sh = 0.082(Re^{0.69}Sc^{0.33})$	$Sh = 0.082(Re^{0.69}Sc^{0.33})$	[60]
$Sh = 0.66(Re^{0.50}Sc^{0.33})$	$Sh = 0.66(Re^{0.50}Sc^{0.33})$	[23, 58]
$Sh = 0.09\left(Re^{0.33}Sc^{0.33}\left(\frac{d_h}{z}\right)^{0.33}\right)$	$Sh = 0.0149(Re^{0.88}Sc^{0.33})$ (For $Sc > 100$), $Sc = \frac{\mu}{\rho D_w}$	[53, 61]
$Sh = (0.53 - 0.58\Phi)(Re^{0.53}Sc^{0.33})$	(For $Gz > 400$) $Sh = 1.615\left(Re^{0.33}Sc^{0.33}\left(\frac{d_h}{z}\right)^{0.33}\right)$ (For $Re < 34$ and $Gz < 65$) $Sh = 0.20\left(Re^{1.01}Sc^{0.33}\left(\frac{d_h}{z}\right)^{0.33}\right)$ $Gz = ReSc\left(\frac{d_h}{z}\right)$	[52]
$Sh = 15.40\left(Re^{0.92}Sc^{0.33}\left(\frac{d_h}{z}\right)\right)$	$Sh = 2.66\left(Re^{0.25}Sc^{0.33}\left(\frac{d_h}{z}\right)^{0.33}\right)$	[62]
$Sh = 5.85\left((1-F)\left(\frac{d_h}{z}\right)\right)(Re^{0.60}Sc^{0.30})$	$Sh = 1.64\left(Re^{0.33}Sc^{0.33}\left(\frac{d_h}{z}\right)^{0.33}\right)$	[63]

So, incorporating Eqs. (5.73)–(5.77) in Eq. (5.72) yields Eq. (5.78) [23]:

$$k^m = \frac{1.8 * 10^{-15}}{RT\delta}\left(\frac{3}{2}\frac{\tau}{\varepsilon r}\left(\frac{8}{\pi RTM_w}\right)^{-0.50} + \frac{P_a\tau}{\varepsilon(4.46 * 10^{-6}T^{2.334})}\right) \tag{5.78}$$

where P_a is the air partial pressure in the membrane pore (assuming $P_a = 101\,325$ Pa) [25].

Membrane permeability for Knudsen diffusion and molecular diffusion can be worked out using Eqs. (5.79) and (5.80), respectively [22, 53, 59–61],

$$k^m_{Kn} = \frac{2}{3}\frac{\varepsilon r}{\tau\delta}\left(\frac{8}{\pi RTM_w}\right)^{0.50} \tag{5.79}$$

$$k^m_{\text{MD}} = \frac{M_w}{RT}\frac{\varepsilon D_{wa}}{\tau\delta}\frac{P}{(\text{Pa})_{\text{lm}}} \tag{5.80}$$

$$k^m_{\text{MD}} = \frac{P}{RT}\frac{\varepsilon D_{wa}}{\tau\delta}\frac{1}{P^F_{wb} - P^E_{wb}}\ln\left[\frac{P - P^E_{wb}}{P - P^F_{wb}}\right] \tag{5.81}$$

Water-in-air diffusion coefficient, D_{wa}, can be estimated using Eq. (5.82) [25]:

$$D_{wa} = -2.775 * 10^{-6} + 4.479 * 10^{-8}T + 1.656 * 10^{-10}T^2 \tag{5.82}$$

5.2.3 Stripping Solutions for OMD

Various inorganic salts like $CaCl_2$, NaCl, $MgCl_2$, $MgSO_4$, K_2HPO_4, KH_2PO_4, or organic solvents like polyglycerol and glycerol have been tested in a number of

OMD applications [51, 54, 64]. As a general rule, a stripping solution offering the lowest water activity and facilitating highest water flux is the most favorable option for OMD process. However, certain other features must also be considered while choosing a particular stripping solution. The selected stripping solution should be non-wetting toward the hydrophobic membrane, nonvolatile, thermally stable, nontoxic, food grade quality, and noncorrosive toward the ferrous alloys. Stripping solutions having these properties can safely be reconcentrated through thermal evaporation without the danger of material loss or crystallization in the evaporator. However, too high a concentration of stripper may result in salt crystallization and increase the corrosion risk [19].

During the OMD operation, the draw solution continuously gets diluted, which leads to a fall in the driving force for water vapor transport across the membrane contactor. To make the process economical, some way must be sorted out to regenerate the draw solution [65]. For this purpose, waste or natural heat can be utilized to regenerate the draw solution. Wingara Wine Group utilized a wastewater at 70 °C to regenerate the $CaCl_2$ stripper solution [19]. As suggested by Gryta [66], another method recommends the continuous revival of stripping solution by means of natural evaporation.

5.2.4 Membranes for OMD Process

OMD process generally employs the same membranes as used in the MD process. PP, PTFE, and PVDF membranes have been extensively used for OMD operations. The pore size ranging from 0.1 to 0.45 μm is usually employed. Depending on the existence or nonexistence of the support layer, membranes that have porosity and thickness in the range of 60–80% and 80–250 μm can be employed in the OMD process. As compared with flat-sheet and spiral-wound membranes, the HFMC module has proved to be more efficient because of the higher effective surface area [19, 37].

5.2.5 Pros and Cons of OMD Process

Water evaporation consumes a large amount of energy, reaching a power requirement of as high as 628 kWh/m^3 of pure water produced [19]. As opposed to the traditional thermal evaporation processes for the concentration of aqueous feeds, OMD process can be performed at ambient pressure and temperature. With such operating parameters, the natural attributes of fruit juices like the taste, color, and fragrance are efficiently preserved. Moreover, a higher concentration can be achieved with OMD process as compared with the forward osmosis (FO) and RO systems. The attractive features of OMD process are further enhanced by a lower investment and operational costs [5, 67].

Some of the disadvantages of OMD are lower water vapor flux and wetting of the membrane pores. The use of membranes having thickness of more than 30 μm results in a longer diffusion path for water vapors, leading to a reduced water permeability. Moreover, membrane thickness significantly affects the thermal insulation properties of the membrane. Thick membranes result in the temperature polarization phenomenon, which is undesirable as it leads to low water vapor flux [68]. There are two modes of membrane wetting in OMD, i.e. partial and full wetting. In the former case, the liquid feed infiltrates almost half of the pore length, while, in the latter case, the liquid feed travels the full length of the membrane pores. Therefore, the full wetting scenario badly damages the quality of feed [69]. The wetting process is also related to the nature of membrane material, operating conditions, and surface tension and viscosity of the feed and brine solutions. A fluid having low surface tension may readily wet the membrane surface and pores. Conversely, a high surface tension fluid has a lower membrane wetting propensity [70]. Therefore, the assessment of membrane wetting effects in osmotic distillation (OD) concentration of fruit juices requires devoted attention.

5.3 Forward Osmosis

5.3.1 Basic Principles of FO Process

FO works on the principle of difference in osmotic pressure between a draw solution and a feed solution. The draw solution is concentrated, having a high salinity and osmotic pressure but a low water chemical potential. Conversely, the feed solution is dilute, having a low salinity and osmotic pressure during a high water chemical potential. Due to the difference in osmotic pressure, transfer of water (permeate) occurs from the feed to the draw solution side through a hydrophilic membrane contactor, which is semipermeable in nature [71, 72]. During the FO process, the draw solution dilutes while the feed solution concentrates until an equilibrium is attained between the osmotic pressures of draw and feed solutions. Thus, FO process can be viewed as a natural osmosis process, as opposed to RO, where a hydrostatic pressure (higher than the osmotic pressure) is applied over the concentrated solution side to transfer water to the dilute solution side [73, 74].

In FO process, the mass transfer can be expressed according to Eq. (5.83) [72]:

$$J_w = \sigma A \Delta\pi_m = \sigma A(\Delta\pi_{Dm} - \Delta\pi_{Fm}) \tag{5.83}$$

where σ denotes the reflection coefficient; A represents the membrane's intrinsic water permeance; $\Delta\pi_D$ and $\Delta\pi_F$ are the corresponding osmotic pressures on feed and draw side, respectively; and $\Delta\pi_m$ denotes the effective osmotic pressure difference athwart the dense layer of the membrane, where subscript m refers to the interface between the dense layer and the solution.

5.3.2 Calculation of the Osmotic Pressures

As, FO process depends on the osmotic pressure difference; therefore its calculation is of utmost importance. For determining the osmotic pressure, π, of a dilute solution from the concentration of the solute, van't Hoff equation according to Eq. (5.84) can be used:

$$\pi = nRTc \tag{5.84}$$

where n shows the number of dissolved entities formed by the solute. As an example, NaCl has two and $MgCl_2$ has three dissolved entities. R, T, c represent the ideal gas constant, absolute temperature, and the solute's molar concentration, respectively. As discussed, the draw solution in FO is extremely concentrated, which may not allow the direct use of Eq. (5.84), which is applicable to dilute solutions. To address this issue, an osmotic pressure coefficient φ is incorporated in Eq. (5.84), which yields Eq. (5.85):

$$\pi = \varphi nRTc \tag{5.85}$$

Pitzer equations can be used to evaluate φ can be determined theoretically by or using the three-characteristic-parameter correlation model [75, 76].

FO membranes are generally asymmetric or composite, comprising a dense separating and a porous support layer. In the FO process, two modes of membrane orientation can be applied. In the first orientation, the feed solution faces the membrane active layer. This mode is called AL-FS or FO mode. In the second orientation, the draw solution is in contact with the membrane active layer. This mode is called AL-DS or PRO mode. Although, AL-DS mode enjoys the advantage of higher water flux, the fouling risk in this mode is increased because of the transport and clogging of foulants within the porous membrane support. Therefore, in applications like water treatment and desalination, the AL-FS mode is preferred as it poses lower fouling threats compared with the AL-DS mode. However, the AL-FS mode suffers from external concentration polarization (ECP) and internal concentration polarization (ICP). ECP can be further subclassified into concentrative ECP and dilutive ECP. Concentrative ECP occurs when the water permeates to the draw solution side, and as a consequence, the solute accumulates at the interface of the active layer on the feed side. Dilutive ECP happens when the draw solution at the porous support layer interface dilutes up because of the incoming permeate water from the feed side. ICP occurs when the draw solution inside the porous support layer dilutes up because of the incoming permeate water from the feed side. Both ECP and ICP cause a deterioration in the effective osmotic driving force for FO process [72]. Figures 5.5 and 5.6 show the illustration of FO and PRO modes, respectively.

McCutcheon and Elimelech [77] proposed a mathematical model for the FO process that considered the effects of concentration polarization on water flux. For the AL-FS (FO) mode, the water flux is predicted by Eq. (5.86):

$$J_w^{\mathrm{FO}} = A\left(\Delta\pi_{Db}\exp(J_w K) - \Delta\pi_{Fb}\exp\left(-\frac{J_w}{k}\right)\right) \tag{5.86}$$

For the AL-DS (PRO) mode, the water flux is predicted by Eq. (5.87)):

$$J_w^{\mathrm{PRO}} = A\left(\Delta\pi_{Db}\exp\left(-\frac{J_w}{k}\right) - \Delta\pi_{Fb}\exp(J_w K)\right) \tag{5.87}$$

where k represents the mass transfer coefficient, while K shows the quickness of the solute to diffuse into and out of the membrane support layer. It can be termed as the measurement of the severity of ICP. A downside of the McCutcheon and Elimelech model is that it ignores the reverse salt diffusion effects in the FO membrane.

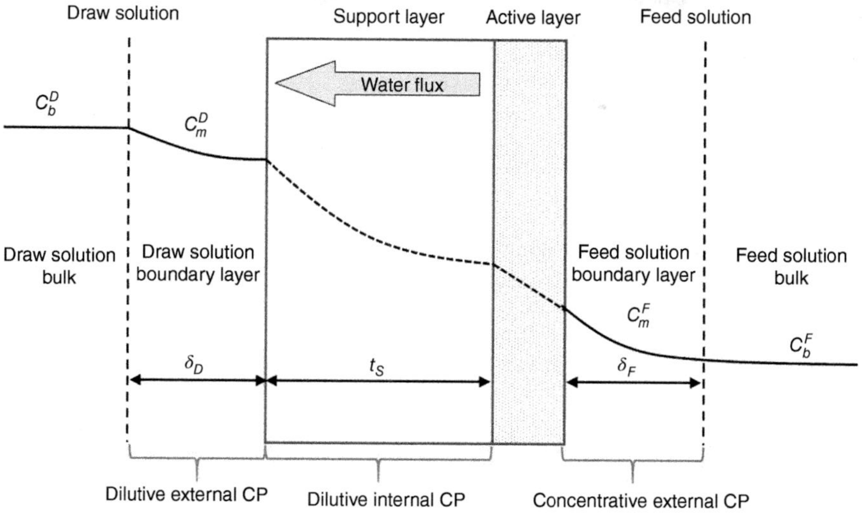

Figure 5.5 Mass transfer in FO mode.

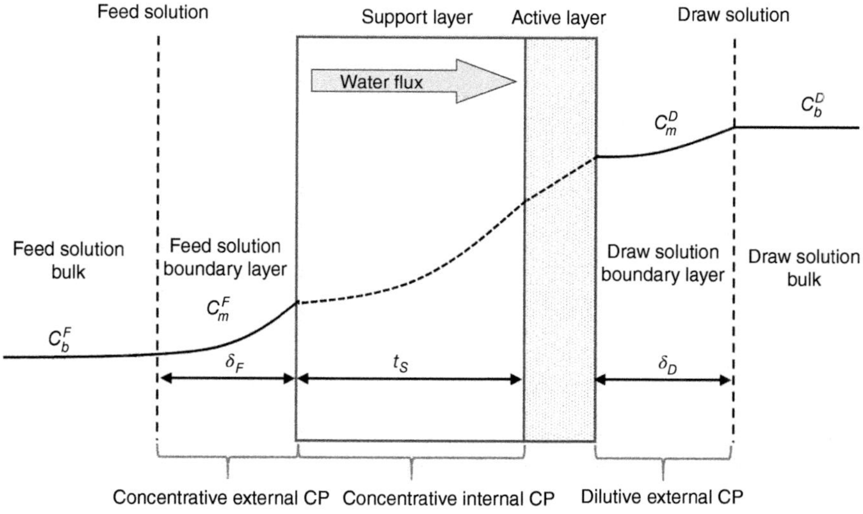

Figure 5.6 Mass transfer in PRO mode.

To address the shortcomings of the previous model, Nagy [78] proposed an improved mathematical model expressed according to Eqs. (5.88) and (5.89) for AL-FS (FO) and AL-DS (PRO) modes, respectively,

$$J_w^{\mathrm{FO}} = A\left(\frac{\Delta\pi_{Db}\exp\left(-J_w\left(\frac{S}{D}+\frac{1}{k_D}\right)\right) - \Delta\pi_{Fb}\exp\left(\frac{J_w}{k_F}\right)}{1+\frac{B}{J_w}\left\{\exp\left(\frac{J_w}{k_F}\right) - \exp\left(-J_w\left(\frac{S}{D}+\frac{1}{k_D}\right)\right)\right\}}\right) \tag{5.88}$$

$$J_w^{PRO} = A\left(\frac{\Delta\pi_{Db}\exp\left(-\frac{J_w}{k_D}\right) - \Delta\pi_{Fb}\exp\left(\frac{S}{D}+\frac{1}{k_F}\right)}{1+\frac{B}{J_w}\left\{\exp\left(\frac{S}{D}+\frac{1}{k_F}\right) - \exp\left(-\frac{J_w}{k_D}\right)\right\}}\right) \tag{5.89}$$

where D is the draw solute's diffusion coefficient in water and $S = t_S \frac{\tau}{\varepsilon}$ is the structural parameter of support layer, the third inherent transport parameter of the membrane, along with A and B. The parameters $k_F = \frac{D}{\delta_F}$ and $k_D = \frac{D}{\delta_D}$ are the coefficients of mass transfer at the active layer–feed solution and support layer–draw solution boundary layers, respectively. $\exp\left(\frac{J_w}{k_F}\right)$ and $-\exp\left(\frac{J_w}{k_D}\right)$ are therefore the relative rectifications for the concentrative and the dilutive ECPs. $-\exp\left(\frac{J_w S}{D}\right)$ is the rectification for dilutive ICP while the denominator of the equation explains the leakage of draw solute through the active layer [72].

5.3.3 Reverse Solute Flux in FO

An adverse feature of the FO process is the reverse solute flux (RSF), denoted by J_s. RSF occurs due to difference in concentration of the solute across the membrane active layer. Some of the negative impacts of RSF are decline in driving force across the FO membrane, higher draw solution loss, deterioration of feed quality, and higher risk of membrane fouling [79].

In the FO process, RSF occurs when an individual solute moves from the draw solution side to the feed side across the membrane active layer. J_s can be determined according to Eq. (5.90) [72]:

$$J_s = B\left(\frac{c_{Db}\exp\left[-J_w\left(\frac{S}{D}+\frac{1}{k_D}\right)\right] - c_{Fb}\exp\left(\frac{J_w}{k_F}\right)}{1+\frac{B}{J_w}\left\{\exp\left(\frac{J_w}{k_F}\right) - \exp\left[-J_w\left(\frac{S}{D}+\frac{1}{k_D}\right)\right]\right\}}\right) \tag{5.90}$$

where B is the membrane's inherent coefficient of the solute permeability. The value of B is unique for each combination of the solute and membrane. c_{Db} and c_{Fb} are the bulk concentrations of the draw and feed solutions, respectively.

5.3.4 Membranes for FO Process

In general, the essential features of an ideal FO membrane are an adequate water flux, a sufficient resistance toward RSF, and a good mechanical strength. The water permeability of the FO membrane should be greater than 1 l/m^2 h bar [73]. FO membranes are typically asymmetric in characteristics. The active layer provides selectivity and has a thickness of 100–200 nm, while the porous support layer provides mechanical strength and has a thickness of 100–200 μm [80]. The pore size range of the dense layer is usually 0.4–1.0 nm for an efficient FO membrane. Moreover, the membrane should have a contact angle less than 60° [81].

The physical properties of the active and support layers typically dictate the FO membrane performance. For instance, the water coefficient and salt permeability depend upon the active layer while the resistance to the solute diffusion depends upon the structural parameter (S) of the support layer. The most commonly used hydrophilic polymers for the synthesis of the support layer of FO membrane are cellulose triacetate (CTA), polyethersulfone (PES), polysulfone (PSf), polyacrylonitrile (PAN), and hydrophobic PVDF [82], while the additives used for enhancing the properties of support layer are graphene [83], zeolites [84–86], metal–organic frameworks (MOFs) [87–89], zeolitic imidazolate frameworks (ZIFs) [90], ZnO nanoparticles [91], and titania [81, 92]. Other than the additives, surface modification techniques like grafting and coating can also uplift the membrane properties. For example, the layer-by-layer technique is widely used to synthesize a nanofilm over the FO membrane's support layer, which helps minimize the hydraulic resistance and maximize the water flux [80].

A foremost challenge in the successful commercialization of the FO process is the cost of membrane. When a cost comparison of FO and RO membranes is made, it turns out that the previous ones are almost 10 times expensive than the later ones. To make FO technology a success story, exhaustive efforts are needed to bring the cost of membranes down. Only then, the FO process will compete with the well-established RO technology [79].

5.3.5 Draw Solutes for FO Process

A carefully selected draw solute for FO process can enhance the process efficiency to a significant level as a major source of driving force. An ideal draw solute must be well suited for an FO membrane and produce an osmotic pressure larger than the feed solute. Other requisites for the best draw solute are low molecular weight, small size and charge, high diffusion coefficient for minimizing ICP, low viscosity, chemical stability, low reverse flux, ease of regeneration, easier and cheaper availability, and no toxicity. In the earlier periods of FO technology, volatile compound-based draw solutions like SO_2 were employed. However, appreciable water flux could not be achieved with these draw solutions due to a high biofouling propensity. Moreover, the draw solute had a high rate of diffusion, which promoted RSF. The most widely tested draw solutes for FO process are inorganic salts like NaCl, $MgCl_2$, Na_2SO_4, $Ca(NO_3)_2$, $CaCl_2$, $MgSO_4$, NH_4HCO_3, $(NH_4)_2SO_4$, KNO_3, KSO_4, KCl, etc. Organic compounds like glucose and fructose have also been used as draw solutes in applications like desalination of seawater. Other novel draw solutes for FO process are magnetic nanoparticles, hydrogels, and inorganic fertilizers.

No single draw solute can be recommended as an ideal candidate for every application. Rather, the choice of draw solute mainly depends upon the desired application. For example, NaCl is the cheapest draw solute having a large osmotic pressure, low molecular weight, and large diffusion coefficient. However, on the other hand, it suffers from the drawback of high reverse flux. In comparison, salts such as $MgCl_2$ and $MgSO_4$ enjoy a low reverse flux, but unfortunately, these solutes are responsible for organic fouling. Likewise, hydrogels exhibit high water flux, but the regeneration costs are too high. Magnetic nanoparticles can be easily regenerated through a magnetic field; however, agglomeration of particles is a severe problem in these types of solutes. Although inorganic fertilizers do not need the regeneration step, the concentrated end stream cannot be used directly as a plant feed. It must be diluted 10–100 times before its agricultural use [72, 79, 80].

5.3.6 Pros and Cons of FO Process

FO process needs no hydraulic pressure; so, the fouling propensity is lower. A lower fouling tendency means that the FO membranes require less frequent cleaning cycles, which ultimately positively affects the membrane life. A low-cost or waste heat can be used to regenerate the draw solute. FO process can demonstrate its capability to recover clean water from complex streams without extensive pretreatment needs. Moreover, FO process is more selective compared with RO process [73].

As it is clear from the previous discussion, the end product of the FO process is a diluted draw solution and not a purified clean water. Therefore, a secondary step like RO process is added with the FO process to restore the draw solution and recover clean water. This increases the cost of production. However, in some applications like fertilization, there is no need to draw solution regeneration. The need for recovery of draw solute is eliminated, making the FO process further economical. The selection of a suitable draw solute that can offer ideal solute features is a challenging issue in FO process. The selection of an appropriate draw solute for a particular application is usually a trade-off, and some essential features must therefore be compromised [73].

5.4 Pressure-Retarded Osmosis

5.4.1 Basic Principles of PRO Process

Pressure-retarded osmosis (PRO) works on the principle of salinity gradient difference. Salinity gradient difference is the driving power for PRO, which is the difference in the osmotic pressure of high-saline draw solution and low-saline feed solution. In the PRO process, a hydraulic pressure is exerted over the draw solution side. When the water permeates from the feed side to the draw solution side across the membrane, it is rapidly pressurized. This pressure energy can be used to drive a turbine or some other device for energy recovery. As compared with the traditional fuels, the salinity gradient energy generated in PRO is a clean and renewable form of energy [93–95]. An illustration of PRO process is shown in Figure 5.7.

As the pure water permeates from the feed side to the pressurized draw solution side, the draw solution gets expanded and diluted, resulting in a brackish water stream. This stream is split into two separate streams, one of which drives a hydro turbine for electricity generation, while the other one is fed to an energy recovery device (ERD), such as a pressure exchanger. The pressure exchanger pressurizes the arriving stream of draw solution. Suppose an efficient ERD is installed in the PRO system. In that case, the operational cost reduces because mere the turbine energy is not enough to meet the energy demands for pretreatment of feed stream, driving the feed pumps, and pressurizing the draw solution [96].

A comparative illustration of the FO, PRO, and RO is shown in Figure 5.8. As discussed earlier, the feed solution in FO process faces the membrane active layer. Here, hydraulic pressure needs not to be applied over the side of the draw solution. So, in the FO process, $\Delta P = 0$ and $\Delta\pi > 0$ (where $\Delta P = P_d - P_f$ and $\Delta\pi = \pi_d - \pi_f$). In the PRO process, the draw solution flows over the membrane active layer. Here,

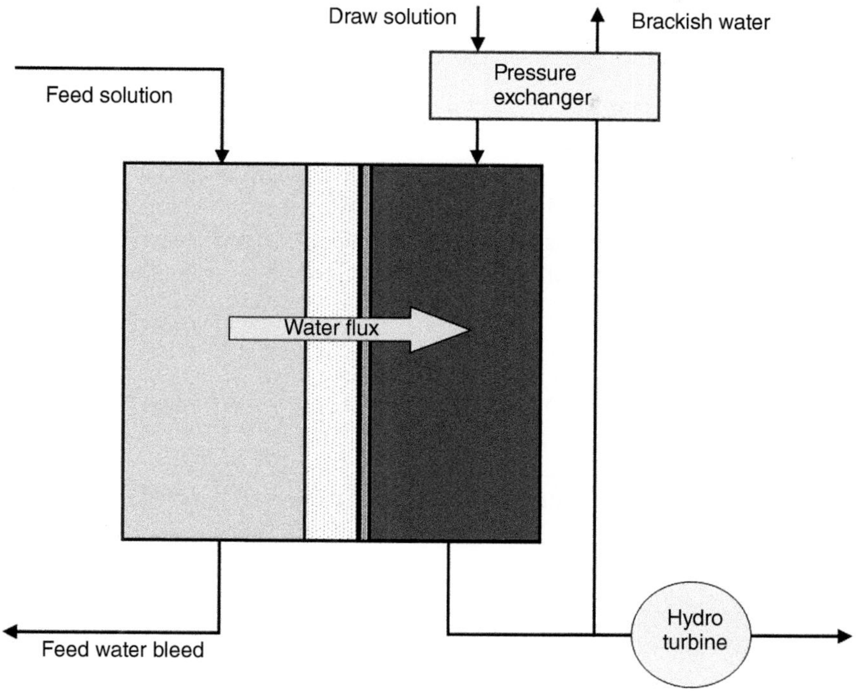

Figure 5.7 Illustration of PRO process.

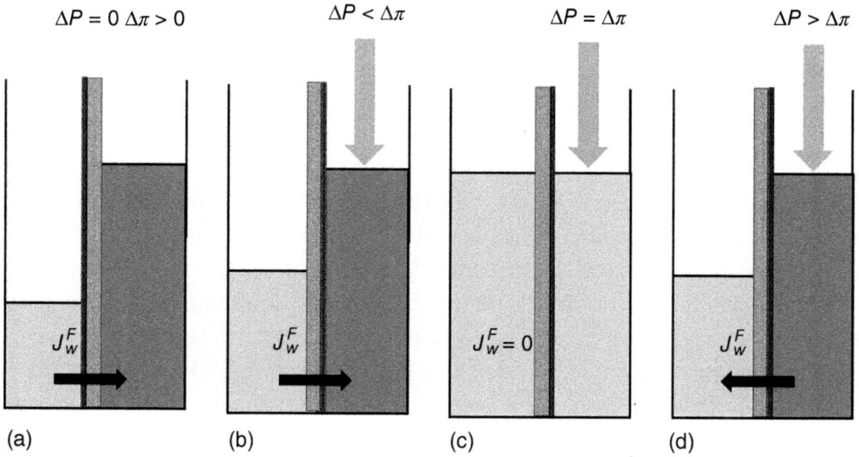

Figure 5.8 Comparison of (a) FO, (b) PRO, (c) equilibrium, and (d) RO processes.

a hydraulic pressure lesser than the osmotic pressure difference is applied over the draw solution side. So, in the PRO process, $\Delta P < \Delta\pi$. Although the water flux has decreased because of the exerted hydraulic pressure, it still remains in the same direction. A hydraulic pressure higher than the osmotic pressure difference is exerted over the draw solution side in the RO process. So, in the RO process, $\Delta P > \Delta\pi$, and the water flux occurs in the opposite direction [96, 97].

In the PRO process, the volumetric flow rate (ΔV) of permeate from feed to the draw solution side can be given by Eq. (5.91) [96]:

$$\Delta V = J_w A_m \tag{5.91}$$

where A_m is the membrane area and J_w is the water flux.

Power density (W) is an important entity in PRO, which is used to analyze the membrane performance. It is the osmotic energy output per unit area of the membrane (W/m^2). Power density is calculated using Eq. (5.92), which is the product of hydraulic pressure (ΔP) applied over the draw solution and the water flux (J_w) across the membrane [96, 98]:

$$W = J_w \Delta P \tag{5.92}$$

The water flux across a PRO membrane can be determined through Eq. (5.93):

$$J_w = A_w(\Delta\pi - \Delta P) \tag{5.93}$$

where A_w is the membrane's intrinsic coefficient of the water permeability (m/s bar), ΔP is the differential feed pressure across the membrane (bar), and $\Delta\pi$ is the difference in osmotic pressure between the feed and draw solutions (bar).

Substitution of Eq. (5.93) in Eq. (5.92) gives Eq. (5.94):

$$W = A_w(\Delta\pi - \Delta P)\Delta P \tag{5.94}$$

If the value of ΔP is half of $\Delta\pi$, then W is anticipated to attain a maximum theoretical value, $W_{\max}$ expressed according to Eq. (5.95):

$$W_{\max} = A_w \frac{\Delta\pi^2}{4} \tag{5.95}$$

Therefore, for achieving a maximum power output, the PRO process should be run at a pressure close to $\frac{\Delta\pi}{2}$ [96].

Just like in the FO process, PRO also suffers from the detrimental effects of ECP and ICP upon the water flux. Figure 5.9 shows the mass transfer in the PRO process. The concentration polarization occurring on each side of the membrane causes the water flux to decline. In Figure 5.9, C_s and C_f are the concentrations of salt in the bulk of draw and feed solutions, respectively. C_1, C_2, and C_3 are the salt concentrations at the membrane interface on draw solution side, membrane porous support layer, and membrane interface on feed solution side, respectively. It should be noted that as compared with ECP, ICP poses a higher risk toward reduction in water flux and power density. As ICP occurs inside the pores of support layer, the detrimental effects of ICP cannot be forfeited by merely increasing the feed flowrate and promoting turbulence. ECP exhibits its negative effects when a higher concentration draw solution is used [99].

For evaluating the effects of ICP, Eq. (5.96) defines the resistivity of solute in the pores of support layer [96]:

$$K = \frac{S}{D_s} \tag{5.96}$$

where K is the resistivity of the solute in the pores of the support layer (s/m), $S = \frac{t\tau}{\varepsilon}$, is a structural parameter (m), t is the support layer thickness (m), τ is the support

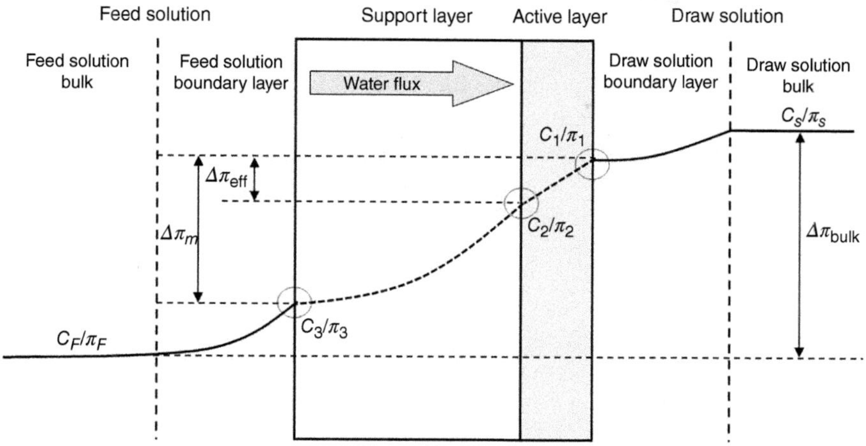

Figure 5.9 Mass transfer in PRO process.

layer tortuosity, ε is the porosity of the support layer, and D_s is the solute's diffusion coefficient in support layer (m^2/s).

5.4.2 Membranes for PRO Process

As discussed previously in the FO process, asymmetric membranes are most commonly employed in the PRO process. The membrane has two parts. One of these is the dense layer, which can also be termed as skin, active, or selective layer, having a thickness of 0.1–1.0 μm, and the other one is the porous support layer, having a thickness of 100–200 μm. The pore size of the active layer is generally in the range of 0.4–1.0 nm. The separation characteristics of the PRO membrane are dictated by the chemical nature, pore size, and active layer thickness. On the other hand, the porous support layer is responsible for providing mechanical support to the membrane [97].

Cellulose acetate (CA) membrane is widely used in PRO application for water treatment. Hydration Technology Innovations Llc (HTI) has commercialized both flat-sheet and spiral-wound modules. CA membrane has turned out to be the favorite choice for PRO because it has superior hydrophilicity, lower fouling propensity, better water flux, lower cost, higher resistance toward chlorine attack, and longer service life. However, the CA membrane is recognized to have a lower tolerance toward alteration in pH.

Other than the CA membrane, the thin film composite (TFC) membranes have also been tested in the PRO process. Although these membranes can show higher water flux and durability and can work on a broader pH range, they suffer from the disadvantages of low tolerance toward oxidants and chlorine [96]. The polyamide TFC membrane, manufactured by HTI, has shown best performance in the PRO process. Until now, flat-sheet PRO membranes, having a capability to tolerate pressure of 22 bar and a power density of 18 W/m^2, have been fabricated. Similarly, hollow fiber PRO membranes, having a capability to tolerate pressure of 20 bar and a power density of 27 W/m^2, have been synthesized [100, 101].

An innovation in the TFC membranes is thin film nanocomposite (TFN) membrane. In TFN membranes, nanomaterials like zeolite and carbon nanotubes (CNT) are used to achieve desired properties like water flux, hydrophilicity, antimicrobial features, mechanical strength, and power density. These membranes have been tested experimentally and have shown a good performance [96].

5.4.3 Pros and Cons of PRO Process

The biggest benefit of the PRO process is its ability to generate clean and sustainable energy. Generally, the seawater and freshwater are sources of draw and feed solutions, respectively, which depicts the economics of the process. Suppose an efficient ERD is installed in the PRO system. In that case, the operational cost reduces because the energy demands for pretreatment of the feed stream, driving the feed pumps, and pressurizing the draw solution are met through the same system.

Although a sustainable energy generation is the winning point for the PRO process, it faces some drawbacks. For instance, no specialized membranes have been produced for the PRO process so far. The requirement of feed pretreatment and the need to efficiently run an ERD also pose difficulties in PRO operation. However, as the FO and PRO are closely connected, a rapid development in the membranes for the FO process reflects a positive impact on PRO. A further improvement in PRO process economics can be achieved through optimization and efficient plant design, which would ultimately result in a decrease in total energy requirement [96, 102].

5.5 Conclusions

In this chapter, the non-pressure-driven membrane separation techniques like MD, OMD, forward osmosis, and PRO were discussed in detail. The membrane contactors used in each mode and the relevant transport equations were briefly elaborated. As compared with the conventional pressure-driven processes, the techniques that expoit the vapor pressure or osmotic pressure difference are energy efficient and less prone to membrane fouling. However, a considerable research is still required to make the water flux in these techniques comparable with the pressure-driven processes. For this purpose, the transport properties of existing membrane contactors must be upgrdaded through physical or chemical surface modification. Moreover, dedicated efforts should be made to develop and synthesize novel membrane contactors, having remarkable water transport and antifouling properties.

References

1 Bodell, B.R. (1968). Distillation of saline water using silicone rubber membrane. *Google Patents.*

2 Shokrollahi, M., Rezakazemi, M., and Younas, M. (2020). Producing water from saline streams using membrane distillation: modeling and optimization using CFD and design expert. *International Journal of Energy Research* 44: 8841–8853.

3 Rehman, W.U., Khan, A., Mushtaq, N. et al. (2020). Electrospun hierarchical fibrous composite membrane for pomegranate juice concentration using osmotic membrane distillation. *Journal of Environmental Chemical Engineering* 8: 104475.
4 Ahmad, S., Marson, G.V., Zeb, W. et al. (2020). Mass transfer modelling of hollow fiber membrane contactor for apple juice concentration using osmotic membrane distillation. *Separation and Purification Technology* 250: 117209.
5 Rehman, W.U., Muhammad, A., Khan, Q.A. et al. (2019). Pomegranate juice concentration using osmotic distillation with membrane contactor. *Separation and Purification Technology* 224: 481–489.
6 Esfandiari, A., Hosseini Monjezi, A., Rezakazemi, M., and Younas, M. (2019). Computational fluid dynamic modeling of water desalination using low-energy continuous direct contact membrane distillation process. *Applied Thermal Engineering* 163: 114391.
7 Alsebaeai, M.K., Otitoju, T.A., Makhtar, M.b.M.Z., and Ahmad, A.L. (2020). Membrane distillation: progress in the improvement of dedicated membranes for enhanced hydrophobicity and desalination performance. *Journal of Industrial and Engineering Chemistry* 86: 13–34.
8 Yao, M., Tijing, L.D., Naidu, G. et al. (2020). A review of membrane wettability for the treatment of saline water deploying membrane distillation. *Desalination* 479: 114312.
9 Johnson, D.J., Oatley-Radcliffe, D.L., and Hilal, N. (2018). State of the art review on membrane surface characterisation: visualisation, verification and quantification of membrane properties. *Desalination* 434: 12–36.
10 Tiegs, D., Gmehling, J., Medina, A. et al. (1986). *Activity Coefficients at Infinite Dilution*, vol. IX, Part 2. Dechema.
11 Gawroński, R. and Wrzesińska, B. (2000). Kinetics of solvent extraction in hollow-fiber contactors. *Journal of Membrane Science* 168: 213–222.
12 Dahuron, L. and Cussler, E. (1988). Protein extractions with hollow fibers. *AIChE Journal* 34: 130–136.
13 Gilliland, E.R. and Sherwood, T.K. (1934). Diffusion of vapors into air streams. *Industrial & Engineering Chemistry* 26: 516–523.
14 Gekas, V. and Hallström, B. (1987). Mass transfer in the membrane concentration polarization layer under turbulent cross flow: I. Critical literature review and adaptation of existing sherwood correlations to membrane operations. *Journal of Membrane Science* 30: 153–170.
15 Yang, M.C. and Cussler, E. (1986). Designing hollow-fiber contactors. *AlChE Journal* 32: 1910–1916.
16 Bhaumik, D., Majumdar, S., and Sirkar, K. (1998). Absorption of CO_2 in a transverse flow hollow fiber membrane module having a few wraps of the fiber mat. *Journal of Membrane Science* 138: 77–82.
17 Wang, K.L. and Cussler, E. (1993). Baffled membrane modules made with hollow fiber fabric. *Journal of Membrane Science* 85: 265–278.

18 Costello, M., Fane, A., Hogan, P., and Schofield, R. (1993). The effect of shell side hydrodynamics on the performance of axial flow hollow fibre modules. *Journal of Membrane Science* 80: 1–11.

19 Johnson, R.A. and Nguyen, M.H. (2017). *Understanding Membrane Distillation and Osmotic Distillation*. Wiley.

20 Schofield, R., Fane, A., and Fell, C. (1990). Gas and vapour transport through microporous membranes. I. Knudsen-Poiseuille transition. *Journal of Membrane Science* 53: 159–171.

21 Schofield, R., Fane, A., and Fell, C. (1990). Gas and vapour transport through microporous membranes. II. Membrane distillation. *Journal of Membrane Science* 53: 173–185.

22 Courel, M., Dornier, M., Rios, G.M., and Reynes, M. (2000). Modelling of water transport in osmotic distillation using asymmetric membrane. *Journal of Membrane Science* 173: 107–122.

23 Babu, B.R., Rastogi, N., and Raghavarao, K. (2008). Concentration and temperature polarization effects during osmotic membrane distillation. *Journal of Membrane Science* 322: 146–153.

24 Alkhudhiri, A., Darwish, N., and Hilal, N. (2012). Membrane distillation: a comprehensive review. *Desalination* 287: 2–18.

25 Salmón, I.R., Janssens, R., and Luis, P. (2017). Mass and heat transfer study in osmotic membrane distillation-crystallization for CO_2 valorization as sodium carbonate. *Separation and Purification Technology* 176: 173–183.

26 Banat, F.A. and Simandl, J. (1998). Desalination by membrane distillation: a parametric study. *Separation Science and Technology* 33: 201–226.

27 Bandini, S., Gostoli, C., and Sarti, G. (1992). Separation efficiency in vacuum membrane distillation. *Journal of Membrane Science* 73: 217–229.

28 Mengual, J., Khayet, M., and Godino, M. (2004). Heat and mass transfer in vacuum membrane distillation. *International Journal of Heat Mass Transfer* 47: 865–875.

29 Martínez-Díez, L. and Vazquez-Gonzalez, M.I. (1999). Temperature and concentration polarization in membrane distillation of aqueous salt solutions. *Journal of Membrane Science* 156: 265–273.

30 Gryta, M., Tomaszewska, M., and Morawski, A. (1997). Membrane distillation with laminar flow. *Separation and Purification Technology* 11: 93–101.

31 Srisurichan, S., Jiraratananon, R., and Fane, A. (2006). Mass transfer mechanisms and transport resistances in direct contact membrane distillation process. *Journal of Membrane Science* 277: 186–194.

32 Curcio, E. and Drioli, E. (2005). Membrane distillation and related operations—a review. *Separation and Purification Reviews* 34: 35–86.

33 Gryta, M. and Tomaszewska, M. (1998). Heat transport in the membrane distillation process. *Journal of Membrane Science* 144: 211–222.

34 Pangarkar, B.L., Sane, M., Parjane, S.B. et al. (2010). The heat and mass transfer phenomena in vacuum membrane distillation for desalination. *International Scholarly and Scientific Research & Innovation* 10: 818–823.

35 Guijt, C.M., Racz, I.G., van Heuven, J.W. et al. (1999). Modelling of a trans-membrane evaporation module for desalination of seawater. *Desalination* 126: 119–125.

36 González, D., Amigo, J., and Suárez, F. (2017). Membrane distillation: perspectives for sustainable and improved desalination. *Renewable and Sustainable Energy Reviews* 80: 238–259.

37 Cassano, A., Conidi, C., and Drioli, E. (2020). A comprehensive review of membrane distillation and osmotic distillation in agro-food applications. *Journal of Membrane Science and Research* 6: 304–318.

38 Shaulsky, E., Karanikola, V., Straub, A.P. et al. (2019). Asymmetric membranes for membrane distillation and thermo-osmotic energy conversion. *Desalination* 452: 141–148.

39 Liao, Y., Zheng, G., Huang, J.J. et al. (2020). Development of robust and superhydrophobic membranes to mitigate membrane scaling and fouling in membrane distillation. *Journal of Membrane Science* 601: 117962.

40 Xu, K., Cai, Y., Hassankiadeh, N.T. et al. (2019). ECTFE membrane fabrication via TIPS method using ATBC diluent for vacuum membrane distillation. *Desalination* 456: 13–22.

41 Weyl, P.K. (1967). Recovery of demineralized water from saline waters. *Google Patents.*

42 Hogan, P., Fane, A., and Morrison, G.J.D. (1991). Desalination by solar heated membrane distillation. *Desalination* 81: 81–90.

43 Rabiee, H., Khalilpour, K.R., Betts, J.M., and Tapper, N. (2019). Energy-water nexus: renewable-integrated hybridized desalination systems. In: *Polygeneration with Polystorage for Chemical and Energy Hubs*, 409–458. Elsevier.

44 Zheng, H. (2017). Chapter 1. General problems in seawater desalination. In: *Solar Energy Desalination Technology* (ed. Hongfei Zheng), pp. 1–46. Elsevier.

45 Ghaffour, N., Soukane, S., Lee, J.-G. et al. (2019). Membrane distillation hybrids for water production and energy efficiency enhancement: a critical review. *Applied Energy* 254: 113698.

46 Dumée, L.F., Smart, S., Duke, M., and Gray, S. (2015). Next generation membranes for membrane distillation and future prospects. In: *Pervaporation, Vapour Permeation and Membrane Distillation* (eds. A. Basile, A. Figoli and M. Khayet), 415–447. Elsevier.

47 Nthunya, L.N., Gutierrez, L., Derese, S. et al. (2019). A review of nanoparticle-enhanced membrane distillation membranes: membrane synthesis and applications in water treatment. *Journal of Chemical Technology & Biotechnology* 94: 2757–2771.

48 Onsekizoglu, P. (2012). Membrane distillation: principle, advances, limitations and future prospects in food industry. In: *Distillation – Advances from Modeling to Applications* (ed. S. Zereshki), 282. IntechOpen.

49 Lefebvre, M.S. (1988). Method of performing osmetic distillation. *Google Patents.*

50 Sheng, J., Johnson, R., and Lefebvre, M. (1991). Mass and heat transfer mechanisms in the osmotic distillation process. *Desalination* 80: 113–121.

51 Alves, V., Koroknai, B., Bélafi-Bakó, K., and Coelhoso, I. (2004). Using membrane contactors for fruit juice concentration. *Desalination* 162: 263–270.
52 Thanedgunbaworn, R., Jiraratananon, R., and Nguyen, M. (2007). Mass and heat transfer analysis in fructose concentration by osmotic distillation process using hollow fibre module. *Journal of Food Engineering* 78: 126–135.
53 Zambra, C., Romero, J., Pino, L. et al. (2015). Concentration of cranberry juice by osmotic distillation process. *Journal of Food Engineering* 144: 58–65.
54 Jiao, B., Cassano, A., and Drioli, E. (2004). Recent advances on membrane processes for the concentration of fruit juices: a review. *Journal of Food Engineering* 63: 303–324.
55 Bánvölgyi, S., Horváth, S., Stefanovits-Bányai, É. et al. (2009). Integrated membrane process for blackcurrant (*Ribes nigrum* L.) juice concentration. *Desalination* 241: 281–287.
56 Nagaraj, N., Patil, G., Babu, B.R. et al. (2006). Mass transfer in osmotic membrane distillation. *Journal of Membrane Science* 268: 48–56.
57 Rehman, W.U., Zeb, W., Muhammad, A. et al. (2016). Osmotic distillation and quality evaluation of sucrose, apple and orange juices in hollow fiber membrane contactor. *Chemica Industry and Chemical Engineering Quarterly*: 35–35.
58 Babu, B.R., Rastogi, N., and Raghavarao, K. (2006). Mass transfer in osmotic membrane distillation of phycocyanin colorant and sweet-lime juice. *Journal of Membrane Science* 272: 58–69.
59 Romero, J., Rios, G., Sanchez, J. et al. (2003). Modeling heat and mass transfer in osmotic evaporation process. *AIChE Journal* 49: 300–308.
60 Romero, J., Rios, G., Sanchez, J., and Saavedra, A. (2003). Analysis of boundary layer and solute transport in osmotic evaporation. *AIChE Journal* 49: 2783–2792.
61 Valdés, H., Romero, J., Saavedra, A. et al. (2009). Concentration of noni juice by means of osmotic distillation. *Journal of Membrane Science* 330: 205–213.
62 Peñaranda-López, A., Martínez-Alvarado, J., Muvdi-Nova, C., and Torrestiana-Sánchez, B. (2016). Development and validation of a theoretical model for osmotic evaporation. *Desalination* 384: 52–59.
63 Prasad, R. and Sirkar, K. (1990). Hollow fiber solvent extraction: performances and design. *Journal of Membrane Science* 50: 153–175.
64 Conidi, C., Drioli, E., and Cassano, A. (2020). Perspective of membrane technology in pomegranate juice processing: a review. *Foods* 9: 889.
65 Belafi-Bako, K., Petrinić, I., Hélix-Nielsen, C. et al. (2019). Osmotic driven membrane processes for separation of special food compounds. In: *Separation of Functional Molecules in Food by Membrane Technology* (ed. C.M. Galanakis), 383–401. Elsevier.
66 Gryta, M. (2018). The long-term studies of osmotic membrane distillation. *Chemical Papers* 72: 99–107.
67 Rehman, W.-U., Muhammad, A., Younas, M. et al. (2019). Effect of membrane wetting on the performance of PVDF and PTFE membranes in the

concentration of pomegranate juice through osmotic distillation. *Journal of Membrane Science* 584: 66–78.

68 Lee, S. and Straub, A.P. (2020). Opportunities for high productivity and selectivity desalination via osmotic distillation with improved membrane design. *Journal of Membrane Science.* 611: 118309.

69 Rezaei, M., Warsinger, D.M., Duke, M.C. et al. (2018). Wetting phenomena in membrane distillation: mechanisms, reversal, and prevention. *Water Research* 139: 329–352.

70 Mosadegh-Sedghi, S., Rodrigue, D., Brisson, J., and Iliuta, M.C. (2014). Wetting phenomenon in membrane contactors–causes and prevention. *Journal of Membrane Science* 452: 332–353.

71 Awad, A.M., Jalab, R., Minier-Matar, J. et al. (2019). The status of forward osmosis technology implementation. *Desalination* 461: 10–21.

72 Tiraferri, A. (2020). Forward osmosis for water treatment and desalination. In: *Current Trends and Future Developments on (Bio-) Membranes* (eds. A. Comite and A. Basi), 53–81. Elsevier.

73 Mohammadifakhr, M., de Grooth, J., Roesink, H.D., and Kemperman, A.J. (2020). Forward osmosis: a critical review. *Processes* 8: 404.

74 Francis, L., Ogunbiyi, O., Saththasivam, J. et al. (2020). A comprehensive review of forward osmosis and niche applications. *Environmental Science: Water Research & Technology*.

75 Pitzer, K.S. (1973). Thermodynamics of electrolytes. I. Theoretical basis and general equations. *The Journal of Physical Chemistry* 77: 268–277.

76 Pitzer, K.S. and Mayorga, G. (1973). Thermodynamics of electrolytes. II. Activity and osmotic coefficients for strong electrolytes with one or both ions univalent. *The Journal of Physical Chemistry* 77: 2300–2308.

77 McCutcheon, J.R. and Elimelech, M. (2006). Influence of concentrative and dilutive internal concentration polarization on flux behavior in forward osmosis. *Journal of Membrane Science* 284: 237–247.

78 Nagy, E. (2014). A general, resistance-in-series, salt-and water flux models for forward osmosis and pressure-retarded osmosis for energy generation. *Journal of Membrane Science* 460: 71–81.

79 Ibrar, I., Altaee, A., Zhou, J.L. et al. (2020). Challenges and potentials of forward osmosis process in the treatment of wastewater. *Critical Reviews in Environmental Science and Technology* 50: 1339–1383.

80 Suwaileh, W., Pathak, N., Shon, H., and Hilal, N. (2020). Forward osmosis membranes and processes: a comprehensive review of research trends and future outlook. *Desalination* 485: 114455.

81 Yadav, S., Saleem, H., Ibrar, I. et al. (2020). Recent developments in forward osmosis membranes using carbon-based nanomaterials. *Desalination* 482: 114375.

82 Rezakazemi, M., Sadrzadeh, M., and Matsuura, T. (2018). Thermally stable polymers for advanced high-performance gas separation membranes. *Progress in Energy and Combustion Science* 66: 1–41.

83 Parsamehr, P.S., Zahed, M., Tofighy, M.A. et al. (2019). Preparation of novel cross-linked graphene oxide membrane for desalination applications using (EDC and NHS)-activated graphene oxide and PEI. *Desalination* 468: 114079.

84 Sodeifian, G., Raji, M., Asghari, M. et al. (2019). Polyurethane-SAPO-34 mixed matrix membrane for CO_2/CH_4 and CO_2/N_2 separation. *Chinese Journal of Chemical Engineering* 27: 322–334.

85 Riasat Harami, H., Riazi Fini, F., Rezakazemi, M., and Shirazian, S. (2019). Sorption in mixed matrix membranes: experimental and molecular dynamic simulation and grand canonical Monte Carlo method. *Journal of Molecular Liquids* 282: 566–576.

86 Riasat Harami, H., Amirkhani, F., Khadem, S.A. et al. (2019). Mass transfer through PDMS/zeolite 4A MMMs for hydrogen separation: molecular dynamics and grand canonical Monte Carlo simulations. *International Communications in Heat and Mass Transfer* 108: 104259.

87 Ahmadijokani, F., Tajahmadi, S., Rezakazemi, M. et al. (2021). Aluminum-based metal-organic frameworks for adsorptive removal of anti-cancer (methotrexate) drug from aqueous solutions. *Journal of Environmental Management* 277: 111448.

88 Ahmadijokani, F., Tajahmadi, S., Bahi, A. et al. (2021). Ethylenediamine-functionalized Zr-based MOF for efficient removal of heavy metal ions from water. *Chemosphere* 264: 128466.

89 Younas, M., Rezakazemi, M., Daud, M. et al. (2020). Recent progress and remaining challenges in post-combustion CO_2 capture using metal-organic frameworks (MOFs). *Progress in Energy and Combustion Science* 80: 100849.

90 Mirqasemi, M.S., Homayoonfal, M., and Rezakazemi, M. (2020). Zeolitic imidazolate framework membranes for gas and water purification. *Environmental Chemistry Letters* 18: 1–52.

91 Sheikh, M., Pazirofteh, M., Dehghani, M. et al. (2020). Application of ZnO nanostructures in ceramic and polymeric membranes for water and wastewater technologies: a review. *Chemical Engineering Journal* 391: 123475.

92 Rezakazemi, M., Ebadi Amooghin, A., Montazer-Rahmati, M.M. et al. (2014). State-of-the-art membrane based CO_2 separation using mixed matrix membranes (MMMs): an overview on current status and future directions. *Progress in Polymer Science* 39: 817–861.

93 Sarp, S. (2019). Fundamentals of pressure retarded osmosis. In: *Current Trends and Future Developments on (Bio-) Membranes* (eds. A. Basile, E. Curcio and Inamuddin), 271–283. Elsevier.

94 Rahimpour, M.R. and Mohsenpour, S. (2019). Integrating pressure-retarded osmosis and membrane distillation. In: *Current Trends and Future Developments on (Bio-) Membranes* (eds. A. Basile, E. Curcio and Inamuddin), 385–402. Elsevier.

95 Wan, C.F., Cui, Y., Gai, W.X. et al. (2020). Nanostructured membranes for enhanced forward osmosis and pressure-retarded osmosis. In: *Sustainable Nanoscale Engineering* (ed. G.S.A. Livingston), 373–394. Elsevier.

96 Ripin, N.H.M., Goh, P.S., Lau, W.J. et al. (2019). Pressure-retarded osmosis. In: *Membrane Separation Principles and Applications* (eds. A.F. Ismail, M.A. Rahman, M.H. Dzarfan and T.M. Othman), 339–359. Elsevier.

97 Nagy, E. (2018). *Basic Equations of Mass Transport through a Membrane Layer*. Elsevier.

98 Han, G., Liu, J.T., Lu, K.J., and Chung, T.-S. (2018). Advanced anti-fouling membranes for osmotic power generation from wastewater via pressure retarded osmosis (PRO). *Environmental Science & Technology* 52: 6686–6694.

99 Han, G., Zhang, S., Li, X., and Chung, T.-S. (2015). Progress in pressure retarded osmosis (PRO) membranes for osmotic power generation. *Progress in Polymer Science* 51: 1–27.

100 Hickenbottom, K.L., Vanneste, J., Elimelech, M., and Cath, T.Y. (2016). Assessing the current state of commercially available membranes and spacers for energy production with pressure retarded osmosis. *Desalination* 389: 108–118.

101 Chung, T.-S., Luo, L., Wan, C.F. et al. (2015). What is next for forward osmosis (FO) and pressure retarded osmosis (PRO). *Separation and Purification Technology* 156: 856–860.

102 Kim, J., Lee, J., and Kim, J.H. (2012). Overview of pressure-retarded osmosis (PRO) process and hybrid application to sea water reverse osmosis process. *Desalination and Water Treatment* 43: 193–200.

6

Applications of Membrane Contactor Technology in Wastewater Treatment

Ayesha Rehman[1], Xianhui Li[2], Sarah Farrukh[1], Mohammad Younas[3], and Mashallah Rezakazemi[4]

[1]*National University of Sciences and Technology, School of Chemical and Materials Engineering, Islamabad, Pakistan*
[2]*Guangdong University of Technology, Institute of Environmental and Ecological Engineering, Guangzhou 510006, PR China*
[3]*University of Engineering and Technology, Department of Chemical Engineering, Peshawar, Pakistan*
[4]*Shahrood University of Technology, Faculty of Chemical and Materials Engineering, Daneshgah Street, Box 3619995161, Shahrood, Iran*

6.1 Introduction

Water is a basic ingredient of life. About 70% of the terrestrial region is covered with water. A small portion of approximately 2.5% is freshwater, well suited for terrestrial life [1, 2]. However, the availability of this basic ingredient has been disturbed seriously due to global issues like the speedy growth of inhabitants, fast industrialization, climate changes, maltreatment of water assets, and water pollution. In addition, water pollution has been increased due to pesticides, minerals, drugs, or other residues. Nearly 1 billion people worldwide are still short of safe water, and approximately 2 billion people do not possess safe sanitation. Another alarming issue is that more than 3 million people die each year due to water-borne diseases. Fifty percent of the world population is thought to be lived in water-deficient regions by 2025. These estimates cast a light on the urgent precautions needed to manage better water resources and wastewater treatment [3, 4]. Therefore, an important need of the day is to develop cost-effective technologies to save existing water resources and address water pollution problems. Among many other techniques, membrane technology is considered to be a potential candidate for solving the issues like clean potable water supply, desalination of brackish and seawater, and wastewater treatment, the reasons being simple in concept, easy operation, no phase changes, no need for chemical additives, and easy scale-up [5–9]. Membrane technology has played a well-defined role in process sustainability and intensification [10]. On the industrial scale, it has produced significant results like a decrease in equipment size and environmental impacts while increasing process efficiency, energy savings, safety, and remote control and automation. Economic and technology play a vital role in forcing industrial entrepreneurs to choose membrane technology over conventional treatment techniques [11, 12].

Membrane Contactor Technology: Water Treatment, Food Processing, Gas Separation, and Carbon Capture, First Edition. Edited by Mohammad Younas and Mashallah Rezakazemi.

In the wastewater treatment plants (WWTPs), the integrated membrane systems (IMSs) can produce high-quality water for applications such as boiler feed, irrigation, and indirect potable water supply. The IMSs, however, have a higher capital cost as compared with conventional wastewater treatment techniques. Anyhow, a positive trend is that a good deal of research is being carried out on optimizing the costs associated with the construction, operation, and maintenance of IMS-based WWTPs. Moreover, the higher capital costs of membrane separation processes as compared with conventional techniques are outweighed by several advantages like lower space requirement, less chemical usage, higher flux and recovery, and production of a stable and higher-quality feed for further processes [13, 14]. Energy cost is another important factor in the operation of membrane-based systems. Energy required for desalinating seawater through reverse osmosis (RO) lies in the range of 2.5–4 kWh/m^3 while the same costs in the range of 0.5–1.0 USD/m^3 that has only become possible due to the development of improved membrane materials and energy recovery devices. The production cost of wastewater recovery and reuse lies in the range of 0.40–1.26 USD/m^3, which depends on the required treatment level, i.e. primary or secondary, and the purpose of reuse, i.e. direct/indirect potable or non-potable reuse, industrial water, irrigation, etc. [15].

Nowadays, anthropogenic activities are the major source of water pollution. These include municipal wastewater, industrial wastewater, and domestic wastewater. Industrial or agro-based wastewater is generated as the conversion of raw material into a valuable product. The wastewater effluent arises from various manufacturing steps such as washing, mining, and separation. Household water effluent may contain domestic sewage. However, various industries are responsible for wastewater effluents:

- Food processing industries
- Pharmaceutical industries
- Textile wastewater
- Coal processing industries

Coke is a very important material utilized in metallurgical amenities. The pyrolysis process is done for producing coke as a product from coal. In pyrolysis, coal breakdown into various products such as volatile compounds, liquor, and solid-organic complexes. Wastewater produced in coal processing industries has a large amount of NH_3, C_6H_6O, polyaromatic hydrocarbons, and other toxic compounds. The amount of pollutants produced is mainly dependant on coal quality. These pollutants are toxic and carcinogenic in nature. Treatment of wastewater is needed for the removal of such venomous pollutants.

In the textile industry, pigments have been used for fabric dyeing purpose. A variety of textile pigments are available that are utilized in different textile fabrics. Wastewater effluent originating from textile industries usually contains dyes [14]. These dyes and pigments may damage the quality of water [15, 16]. Dyes may contain harmful and carcinogenic components such as heavy metal complexes, Cr(III), and various others. Thus, it must be necessary to treat water before its discharge or reuse.

The pharmaceutical industries manufacture a large variety of medicines that are being utilized for both humans and animals. Wastewater effluent collected from pharmaceutical industries is difficult to treat. Due to the large diversity in products, erratic wastewater composition is generated. The materials produced in the pharmaceutical industry are generally complex organic compounds. Another important source of wastewater is the food processing industries. Wastewater produced in food processing industries usually has a high concentration of nutrients, fats, oil, grease, sugar, or other organic components. Wastewater effluent is released without any appropriate treatment, causing serious environmental pollution [17]. The availability of clean water is a primary concern for maintaining the diversity of life. Water released from industries often has numerous noxious wastes that needed to be recovered and demolished to lessen the ecological impact [18].

6.2 Common Toxic Substances in Wastewater

Water released from the industries often has various wastes. These contaminants must be removed to reclaim the cleaned water. Membrane contactors (MC) technology has been utilized for the elimination of impurities from water, such as volatile organic compounds (VOCs), phenols, toluene, and heavy toxic metals (Cu, Cr, Co) [19]. There are different types of contaminants, such as inorganic pollutants, organic pollutants, and biological pollutants. All of these contaminants may get substantial interest because of their detrimental effects on exposure to the environment [18]. It can be observed that the increase in the concentration of potentially noxious substances when discharged into the environment is alarming to human health [11, 13]. Biological contaminants may contain bacteria, viruses, mites, etc. Wastewater also contains some organic contaminants, which can be classified based on their biological degradation capability. These organic contaminants having a good affinity for water are easily decomposed in the atmosphere. Some common organic contaminants are, for example, carbohydrates, methanol, and polycyclic aromatic hydrocarbons, which can be simply deteriorated by the microorganisms [13].

Nevertheless, pollutants such as ethanol and methanol can be highly toxic when present in a higher amount. Other chemicals such as polycyclic aromatic hydrocarbons, polychlorinated biphenyl, and similar compounds take much time to decompose. Pesticides were extensively utilized for agriculture purposes. These pesticides are generally present in a low amount and less toxic than soluble-organic contaminant in wastewater. On the other hand, persistent organic contaminants have got more attention because of their harmful and highly carcinogenic nature. Some common pollutants present in water effluent are listed below.

6.2.1 Phenols

Phenols are the most common organic contaminants, generally found in wastewater. The main sources responsible for phenol are chemical industries, pharmaceuticals, and petroleum industries. Phenols and their related compound have negative impacts on human health. In addition, the presence of phenol in water stream greatly reduces the growth of marine life. Even at a small amount of phenol can cause an adverse effect on the ecosystem [19, 20].

6.2.2 Heavy Metals

Metallurgy and textile industrial effluents usually produce wastewater that contains heavy metals that pose a serious danger to human health and ecosystem. Heavy metal pollutants have gained more attention, owing to their toxic nature. Generally, in natural water, some metals are already present in trace amount. However, various metals such as Co, As, Cd, Ni, and Cr are poisonous even at small concentrations [21, 22].

6.2.3 Ammonia

The detrimental consequence of ammonia and derived compounds present in the effluent stream has encouraged the development of an effective process for its elimination. In general, ammonia is found in both urban and industrial effluent. The excess concentration of ammonia/N-compounds in the receiving water can cause a major contribution to pollution. The possible disadvantage of these N-compounds present in the effluent stream can decrease the dissolved oxygen (DO) concentration. This reduced amount of DO in water may affect the aquatic life and the disinfection effectiveness and increase the rate of metal rusting [23, 24].

6.2.4 Hydrogen Sulfide

Wastewater stream coming from gas and oil plants contains toxic H_2S gas. Hydrogen sulfide is known as sour gas and is found as corrosive in nature. In oil and gas refineries, a significant amount of water is produced as a by-product [25, 26].

6.2.5 Carbon Dioxide

WWTPs are one of the possible resources of anthropogenic greenhouse gas (GHG) emissions. The GHG emissions result in weather changes and also contaminate the air [27, 28]. Carbon dioxide (CO_2) is considered as major impurity found in water. Organic carbon in WWTPs will react with oxygen and produce carbon dioxide or incorporated into biomass. Biomass is then transformed into methane and carbon dioxide to some extent. The oxidation of methane also produces CO_2 [29, 30].

6.2.6 Petroleum Hydrocarbons

The major source of petroleum hydrocarbons present in wastewater is industrial or municipal water effluent. The transportation industry, petroleum, and oil refineries generate various harmful hydrocarbons in water. These hydrocarbon-based water effluents have a detrimental effect on aquatic life. Petroleum hydrocarbons may worsen the water condition by creating an oily layer that reduces the oxygen exchange of the air and water body.

6.2.7 Polycyclic Aromatic Hydrocarbons

Polycyclic aromatic hydrocarbons are complex organic contaminants that contain more than one benzene ring in their structure, having various geometries. These polycyclic aromatic hydrocarbons are present in grease, petroleum, and coal. In the marine system, they may enter because of accidentally leakage of oil. The accumulation of polycyclic aromatic hydrocarbons is toxic to the environment and may cause cancer or other fatal diseases [31].

6.2.8 Nitrobenzene

Nitrobenzene is an aromatic organic contaminant that is produced on a large level. In the lab scale, sometimes nitrobenzene utilizes as a solvent. On larger exposure to nitrobenzene, this might result in a severe breakdown of the central nervous system weakening vision and affect the kidneys [32].

6.3 Environmental Risks of Wastewater

Wastewater contains many impurities such as organic matter, salts, and other harmful chemicals that worsen the water quality. The unwanted impurities in water have a negative effect on health and the environment. Wastewater contains a high concentration of salts and heavy metals that affect soil and crop irrigation system properties. When discharge in sea or river, the industrial effluent utilizes a large amount of dissolve oxygen due to the organic contaminants' presence. The higher amount of consumption of DO in water leads to damage the aquatic life. In Table 6.1, various sources of wastewater, pollutants, and their detrimental effects have been mentioned. However, the most important negative environmental issues of permitting unprocessed wastewater to be released in lands are as follows [36]:

- The breakdown of the organic matter present in wastewater can cause the formation of large amount of malodorous gases.
- Unprocessed raw wastewater containing a huge quantity of organic substance mixed in river streams. This organic matter utilizes the DO present within river and reduces its concentration. Depletion of oxygen causes suffocation, which results in fish death and other unpleasant effects.
- Wastewater usually consists of nutrients that play an important part in the growth of aquatic plants. Too much quantity of these nutrients in wastewater, consequently, leads to the eutrophication process of the lakes.
- Wastewater generally contains various pathogens and noxious compounds. These may pollute the earth or wherever such sewage is disposed off.

Likewise, the pretreatment of industrial water effluent before its release is the only solution to efficiently reduce its detrimental environmental consequences. The developments of proper technologies are needed for treating wastewater. In the agricultural area, establishing a suitable irrigation system can be greatly effective for its

Table 6.1 Various sources of wastewater, pollutants, and their detrimental effects [33–35].

Sources of wastewater	Pollutants	Effects of pollutants
Domestic waste		
• Municipal • Hospitals • School • Offices	Carbohydrates, fats, oil, surfactants, H_2S, CH_4, O_2, alkalinity, chlorides, heavy metals	Effects on human, plants, and animals Disruption of food chain Fatal diseases such as typhoid fever, cancer, etc.
Industrial waste		
• Food processing industries • Pharmaceutical industries • Textile industries • Coal processing industries • Chemical industries • Petroleum processing	Volatile fatty acids, sugar, antibiotics, phenol, butyl-hydroxyl, toluene, BOD, COD, TOC, ammonia-chlorinated compound, nitrophenol, polynuclear aromatic hydrocarbons	Algae blooms Eutrophication Devastation of ecosystems
Agricultural waste		
• Animal feedlots • Irrigation • Cultivation • Dairy farming • Orchards • Aquaculture	Biphenyls, nitrogen, fertilizers, nitrates, sulfates, phosphates, fluorides, chlorides, oxalates	Unpalatable water Death of aquatic life Reduces oxygen level

Source: Based on Diban et al. [33], Djenouhat et al. [34], and Dlamini et al. [35].

safe usage. Environmental pollution can be highly managed and controlled using applying appropriate irrigation techniques [37]. There are choices of different alternatives that may be used in the treating, recycling, and reclaiming of wastewater.

6.4 Membrane Technology for Wastewater Treatment

Several reasons make it necessary to treat wastewater before going to its final disposal [36]. Some key characteristics for selecting a proper technology for treating wastewater that obtained from various sources are as follows [25, 38]:

- The process must be efficient, economical, and eco-friendly.
- The process should need less energy intensive and should have less possibility of malfunctioning.
- The process should be easy to operate, be maintained, and require less expensive chemicals. The process must be capable of recycling all resources.
- The technology is supposed to be able of being upgraded as per needed.
- The process must need easy maintenance and control.

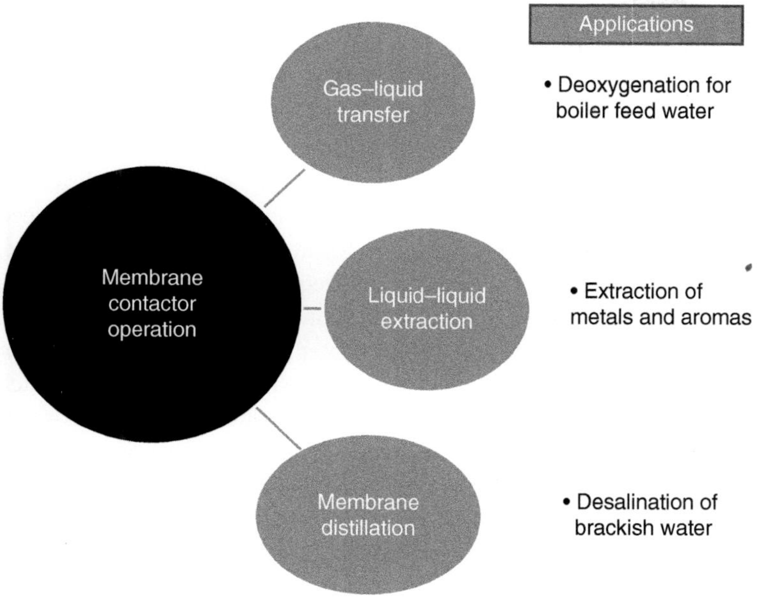

Figure 6.1 Comparison between various physicochemical techniques applied in wastewater treatment. Source: Based on Devold [36] and Dlamini et al. [39].

The impurities present in wastewater effluent have currently been paid attention to because of their possible risks and negative health effects. Public interest over the environmental effect of contaminated water has risen. Some traditional physicochemical wastewater treatment technologies are coagulation, adsorption process, and membrane processes. The comparison between various physicochemical techniques applied in wastewater treatment and their limitation is mentioned in Figure 6.1 [36, 39].

There are several traditional methods used for the treatment of wastewater polluted by organic materials. Commercially activated carbon as adsorbent material is considered as the most efficient adsorbent for removing the organic matter. Nevertheless, because of high expenditure and loss in some quantity of activated carbon during its regeneration, it is regarded as uneconomical. Other successful, efficient, and economical alternatives of known adsorbents are saw dust, fly ash, and many more [40]. Ionic liquids (ILs) have possible potential to be the best substitute for noxious and harmful solvents [41]. These conventional processes have been applied to eliminate contaminants. Nevertheless, these still have some constrictions, particularly of high operation costs.

The major types of membrane separation processes commonly employed in water treatment are microfiltration (MF), ultrafiltration (UF), nanofiltration (NF), and RO. Each of these processes is used to separate dissolved or finely dispersed particles from liquids. All these technologies depend on transport through an organic or inorganic porous membrane.

Two common modes of operation are dead-end filtration and crossflow filtration. In crossflow filtration, the movement of feed is parallel to the filter medium. Shear stress is thus generated, which causes a rubbing effect on the surface. Although more energy is required for crossflow filtration, yet cake layer thickness is controllable. If a balance is established between the shearing effect and particle settlement, the cake layer growth almost stops. Crossflow filtration is most effective in feeds containing higher amounts of suspended solids and macromolecules. Crossflow modes are generally adopted in all MBR processes and most of the wastewater filtrations. On the other hand, dead-end filtration refers to filtration at one end where the feed moves toward the filter medium. The target particles are settled on the filter surface. Backwashing and filter medium replacement are frequently required in this case. Dead-end filtration is most effective for feed that contains a low level of foulants. Dead-end modes are mostly adopted in surface water filtrations, pretreatment for seawater RO, and tertiary filtrations [19]. One of the problems associated with such a system is the vulnerability to easy clogging of the membrane. This problem is surmounted in a crossflow filtration system that is generally used in water and wastewater treatment. The polymeric membranes have conical-shaped pores that converge at the top, where contact is made with wastewater. The material passes unhindered with high flow rates through the membrane surface, and therefore, the crossflow filters do not plug easily. However, during routine operation, a thin fouling layer is formed over the membrane surface due to the accumulation of contaminants, resulting in a decrease in water flux. This layer can be easily cleaned in place with soap and water. Crossflow membrane filters can be repetitively cleaned for years of useful service before a replacement is needed [20].

The distinction between the pressure-driven membrane separation processes (MF, UF, and RO) can be made according to the particle or molecule size that the membrane can retain or pass, which is somewhat the pore size of the membrane. The most available MF membranes retain particles in the size range of 3–12 μm, while the tighter MF membranes can retain particles up to 0.2 μm in size. The dividing line between MF and UF membranes at 0.1 μm is, to some extent, arbitrary. The most available UF membranes have a molecular weight cutoff of about 0.08 μm. Again, the dividing line between UF and RO membranes at 0.001 μm is semantic. The pores in RO membranes are the intermolecular spacing between the polymer chains in the membrane. The narrowest size UF membranes can retain simple sugars but allow salt molecules to pass through, while RO membranes can retain both of these. RO processes typically require operating pressure in the range of 13–80 bars, the reason being high osmotic pressure of salt solutions. As the MF and UF membranes freely allow salt molecules to pass through and no considerable osmotic pressure has to be overcome, therefore, operating pressure of these membranes is generally below 7 bar. In the process of osmosis, the water transport across a membrane takes place due to the difference in chemical potential. Pure water has a higher chemical potential, while the salt solution has a lower chemical potential, which causes an increase in pressure on the salt solution side. The rise above the atmospheric pressure is known as osmotic pressure. To get pure water from the salty one, the osmotic pressure must be exceeded, and so it is termed as "reverse osmosis." The osmotic pressure varies with the nature of the salts and their concentration. Desalination of seawater through RO is carried out at pressures of around 1000 psi where the raw water has an osmotic pressure of nearly 25 bars [21, 22]. The detailed comparison of the four common membrane processes can be found in Chapter 1.

Membrane technology is now the most widely used process and played an important role in removing these impurities from wastewater effluent. It has been mainly categorized into filtration, membrane bioreactors, and membrane contactors to exclude these noxious contaminants. However, the performance of conventional pressure-based membrane technologies (pressure as a driving force) is restricted. The reasons behind the limitation of these traditional technologies are the occurrence of fouling, polarization process, and hindrance. Membrane contactor technology is reliable and easy to scale up and shows potential for growth in wastewater treatment or other applications [36, 39]. Membrane-based treatment processes have been paid remarkable attention by environmental management. Membrane contactor technology is more feasible due to its less costs and simple maintenance [42].

Membrane contactors have advantages over traditional contactors because of its high interfacial area/volume ratio and less operational cost. Membrane contactor technology has been used for the removal of toxic organic pollutants present in wastewater effluent. Numerous other important applications of membrane contactor technology related to industrial wastewater effluent are the following:

- Elimination of DO from food and beverage industry.
- Elimination of carbon dioxide from wastewater in electrodeionization technique.
- Elimination of dissolved nitrogen from wastewater in blanketed storage tanks.
- Elimination of VOCs from wastewater effluent.
- Humidification of air/gases.
- Elimination of ammonia (NH_3) compound from wastewater stream.

6.5 Membrane Contactor Technology for Removal of Organic Contaminants from Wastewater

There are several pollutants found in the wastewater stream, as organic contaminants play a key role. Industrial, agro-based wastewater, and domestic wastewater could be the main source of producing organic contaminants. This causes the scarcity of water sources. Agricultural wastewater may hold large amount of pesticides and herbicides. Wastewater obtained from coke plant may consist of different polycyclic aromatic hydrocarbons. The water received from chemical industries has a variety of compounds PCB and PBDE. Food industries consist of mainly complex organic contaminants with a large quantity of suspended solids and biochemical oxygen demand. Municipal wastewater may have a variety of organic impurities as oil, surfactants, food-based contaminants, etc. These organic matter contaminants can be hazardous to humans and cause danger to the environment.

The main methods of wastewater treatment include biological, physical, and chemical treatments. These methods have been widely applied in the industry. However, these wastewater treatment technologies suffer from some inevitable downsides. More reliable solutions are needed to remove hazardous substances from wastewater. Membrane contactor is a convenient and simple tool that holds the potential to address environmental issues, especially in the removal and extraction of organic matters [43–45]. A membrane contactor is a process that combines absorption by a liquid absorbent and membrane as a mass transfer facilitator [46–49]. Therefore, in membrane contactor, the membrane is not contributing to the selectivity of the separation. It either is hosting a liquid within its pores that provide selective media or provides a controlled contact within its pores with a selective liquid. The key to membrane contactor focuses on the mass transfer between two phases through a porous membrane. The pore size of the contacting membranes generally ranges from 0.001 to 0.100 μm to prevent the phases' direct mixing. In this section, the removal of organic foulant has been studied based on the membrane separation processes, and for each, the separation capability and mass transfer rate were investigated.

Membrane contactor technology is a novel process that shows potential in the field of wastewater treatment. This technology proposes several advantages over conventional techniques. To further commercialize membrane contactor technology, many challenges needed to be addressed; appropriate knowledge of shell-side flow is important to upgrade systems, high ease of utilizing this technology for maximum possible users. Also, good materials of construction are needed that present an extensive choice of solvent compatibility. The various advantages of membrane contactors will make the process pervasive as more development is done in these areas [50].

Various researchers have used membrane contactors for the VOCs' elimination. Some important applications of membrane contactors in wastewater treatment are given in Figure 6.2. For example, a polypropylene (PP) hollow fiber module was utilized for the elimination of $CHCl_3$ and aromatic hydrocarbon compound C_7H_8 from water. Thus, C_7H_8, an aromatic compound, is absorbed more preferentially as compared with $CHCl_3$ [4].

Phenol and its related compounds are the most common organic contaminants present in wastewater. Phenol is a major chemical compound utilized in various industries as a raw material [51, 52]. However, it has a detrimental effect on ecosystems due to its toxic nature. Only about 1 ppb or less amount of phenol is allowed in surface water according to EPA standard. Eliminating phenol from industrial wastewater is an imperative challenge. Conventional processes used to eliminate phenol are distillation, chemical process, adsorption technique, etc. [19]. These conventional techniques are not feasible because they utilized a large amount of organic solvents and energy-intensive processes. These hollow fiber membrane contactor technology has been preferred to eliminate various compounds from wastewater containing phenol [19, 53]. González-Muñoz et al. studied the recycling of C_6H_6O organic compound using hollow fiber contractor. A phenolic compound was recovered and sent back to the reactor again [54].

Odorant or flavored compound recovery from wastewater is an important application in the food industry. Recovery of these odorant is also done to control the flavor in food. In literature methyl ketones, removing odorant compounds from oil using

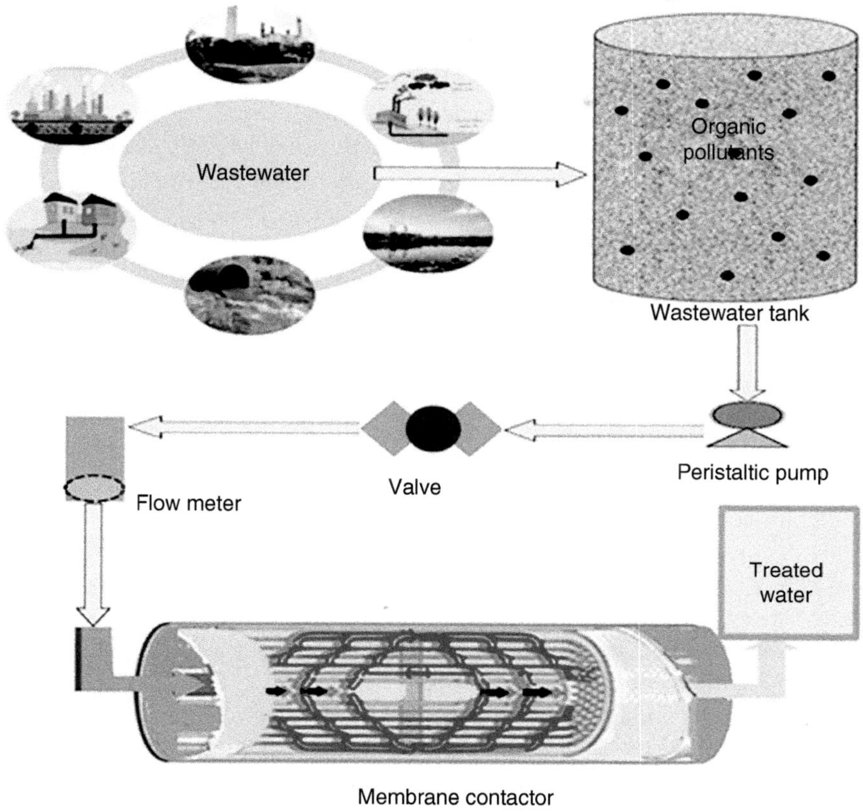

Figure 6.2 Membrane contactors in wastewater treatment.

Liqui-Cel hollow fiber membrane was reported [55]. Sulfur-containing odorant compounds present in food industry water effluent removed by a hollow fiber contactor were also studied. The water effluent stream was passed through contactor fibers, and n-C_6H_{14} was used as an extractant [56].

The performance of the commercial hollow fiber membrane contactor was investigated. Two contactors were used in series for the removal of $C_6H_{14}O$ from water. The wastewater sent in shell side 1-Octanol and n-$C_8H_{18}O$ in fibers. The regeneration of n-$C_8H_{18}O$ was done by using distillation column or recovered in the contactor. The efficiency of the membrane contactor was 10 times greater as compared with the packed column [40]. The elimination of $C_5H_{10}O_2$ from wastewater obtained from polymer industries has also been done using membrane contactor Amberlite LA-2. In this extraction process, toluene (C_7H_8) was used as an extractant. The regeneration of C_7H_8 was done with sodium hydroxide solution [57].

A hollow fiber membrane contactor was also utilized for the separation of isomers. Cyclodextrin-based liquid membrane was kept in hollow fiber. Isomers of nitroaniline such as o-$C_6H_6N_2O_2$ and p-$C_6H_6N_2O_2$ were present in organic solvents such as 80% $C_8H_{18}O$ and 20% n-C_7H_{16} as a feed [58].

Elimination of benzoic acid ($C_7H_6O_2$) from wastewater was done through a hollow fiber membrane contactor. The $C_7H_6O_2$ is utilized as an additive in several foodstuffs such as juices, jams, and many others. Benzoic acid helps to retard microbial and chemical actions that deteriorate food items. It keeps hold of the essential nutritional rate of foodstuffs and also maintains their shelf life [59]. The wastewater released from these food processing industries may include dissolved $C_7H_6O_2$. The high amount of $C_7H_6O_2$ has a toxic effect on human and animals health [60, 61]. Commercially, a hydrophobic polypropylene hollow fiber module membrane contactor was utilized for the elimination of dissolved $C_7H_6O_2$ from water effluent. In this case, liquid–liquid extraction approach was applied using a membrane contactor. Using benzoic acid-containing wastewater was sent on the lumen side and trioctylamine ($C_6H_{15}N$) mixed in 1-$C_8H_{18}O$ as extractant on the shell side. Benzoic acid reacts with trioctylamine at the boundary line of the wastewater and membrane. This causes the formation of a complex that was eliminated with the organic extractant. About 95% benzoic acid removal was achieved by the hollow fiber module membrane contactor [1].

MD has been applied to treat radioactive wastewater produced from nuclear sources such as nuclear plants. The elimination of these pollutants involves methods to remove radioisotopes. To diminish the amount of pollutants in wastewater, traditionally, various methods were used such as chemical precipitation and cation–anion exchange process [62].

MD can be done under mild conditions (T&P), appreciably cutting down the limitation of previous technologies. Moreover, MD facilitates plastics' utilization resulting in the exclusion of corrosion issues and decreases its installation cost [63]. It was effectively used to clean textile industrial water effluent that mainly consists of dyes/pigments [26]. The reliance of concentrate flux, membrane rejection, and polarization process on the retentate amount and operating temperature provided the chance of combining MD technique in the manufacturing phase with RO preconcentration phase.

Another proposed application of membrane contactor is treating the water effluent containing oil impurities using the combination of UF and MD processes. Permeate comes after the UF operation, usually holding oil not more than 5 ppm. More decontamination of obtained permeate was done by using MF. The final product obtained from MD column contains oil-free water stream. This reduces the total organic carbon of up to 99.5% and 99.9% of TDS concentration [64].

Currently, the primary use of membrane contactors at a commercial scale is associated with eliminating dissolved gases from wastewater. The removal of soluble gases in wastewater is essential in manufacturing ultrapure water, treating boiler feed water, and other related processes. The dissolved gases at elevated temperatures present in wastewater produce bubbles that cause fractional wetting of surfaces and can break them. Membrane contactors are an appropriate way to control the dissolved gas content [42].

Liquid membranes have also been widely applied in water treatment as a separation technique in membrane contactor [65]. They usually contain a liquid phase that exists in supported form. The liquid phase is made of liquid solutions and acts as a

Table 6.2 Summary of liquid membrane process during the extraction of organic materials.

Extractant	Diluent	Surfactant	Internal phase regent	Application	Efficiency	References
Soy lecithin	Dodecane	Abil EM 90	Hydrochloric acid	Chlorpheniramine recovery	99	[66]
Aliquat 336	Bran oil	Span 80	NaOH	Extraction of lactic acid	95	[67]
Cyanex-923	Kerosene	Span 80	NaOH	Phenol removal	97	[68]
TOMAC	Bran oil	Span 80	NaOH	Extraction of lactic acid	90	[69]
TBP	Heptane	Span 80	NaOH	4-Chlorophenol removal	99	[70]
Alamine300	Toluene	Span 80	NaOH	Recovery of acetic acid	86	[71]
Alamine336	Chloroform	Span 80	NaOH	Recovery of citric acid	97	[72]
—	Heptane	Span 80	NaOH	Dye removal	99	[73]
—	Hexane	Span 80	Na_2CO_3	Extraction of anionic dye	99	[74]
—	Heptane	Span 80	KCl	Extraction of gibberellic acid	88	[75]
—	Octanol	Span 80	NaOH	Extraction of succinic acid	98	[58]

membrane barrier between the two aqueous phases, which provides a highly selective barrier. The selection of extractant is important because it will affect both the liquid membrane stability and extraction efficiency. In the liquid process, organic foulants diffuse through the membrane, and then it reaches the interface between the internal and membrane phases. The solute molecule reacts with the agent by the stripping reaction to form the salt. Table 6.2 briefly describes some literature on the removal of organic acids. Generally, kerosene and span 80 have been widely applied as surfactant and diluent. However, the composition would change with the type of solute to be treated. For instance, for the lactic acid extraction, heptane, NaOH, and TOA are used as a diluent, an internal reagent, and an extractant, respectively (Table 6.2). The organic acid includes lactic acid, citric acid, gibberellic acid, and succinic acid. The other organic pollutants, such as phenol, chlorophenol, and dye, could also be removed using liquid membrane technology. It can be seen that the removal efficiency can reach above 85%. However, the solute concentration is extremely low, usually 100–1000 ppm. The solute-loaded emulsion was moved out from the feed phase after the extraction process. Afterward, the emulsion is demulsified to reuse. Finally, the recovery of reused resources from the membrane phase is completed for their further use again.

The most attractive application of membrane contact reactor is the ammonia removal and recovery from organic wastewater treatment, as illustrated in Figure 6.3. Membrane contactor is more prospective because of its ambient temperature and lower energy consumption. The PVDF hollow fiber membranes could be applied to remove ammonia from the ammonium chloride wastewater using sulfuric acid as a stripping solution [77]. The mass transfer coefficient was enhanced from 0.1×10^{-5} to 1.39×10^{-5} m/s as the pH was enhanced from 8 to 10, respectively. Similarly, PVDF membrane contactor was used to remove ammonia traces using 0.02 M sulfuric acid solution as a stripping phase [78]. The maximum efficiency obtained was 78% at the flow rate of 15 mg/l at pH of 10. To gain more insights into the transport mechanism and concentration distribution and transport mechanism of ammonia in hollow fiber membrane system, a mathematical model was established on the basis of the mass and momentum balances [76]. It was found that the Knudsen diffusion mechanism governed the transportation of ammonia through the membrane pores. The high feed velocity was not suitable for the stripping of ammonia because of the reduced feed residence time and lowers the mass transfer rate. The combination of a membrane contactor with the solid slurry separation process is an alternative for ammonia extraction [79]. The mass transfer coefficient was found to be increased from 17 to 34 mm/h with the increase of the operating temperature from 30 to 50 °C. However, protein is a major kind of foulant to limit the sustainable operation of membrane contactor. Thus, a suitable pretreatment process is desired for the practical application of membrane contactor.

In addition to the volatile solvent, the IL is an environmentally friendly solvent because of its nonvolatile nature and nonflammability. ILs have been widely applied for the selective recovery of many organics such as amines, dyes, toluene, and phenolic compounds from aqueous solutions. Therefore, the IL-based membrane has been widely used to extract organic pollutants from aqueous effluents. The separation of organics between IL and aqueous phase follows the same principle in the liquid membrane systems. The pH as a significant extraction condition can be adjusted to enhance the selectivity and affinity. The support material is of great significance to the extraction efficiency. For example, the porous polypropylene membrane can recover 68% of acetaminophen, whereas only 28% was recovered [59]. Thus, the physically and chemically stable membrane should be used to prevent liquid leakage. Meanwhile, the low solute concentration and high IL viscosity in the feed negatively influence both membrane filtration flux and recovery efficiency.

The extraction capacity of the IL is constantly reused because the extracted solute was removed from the stripping phase into the receiving phase. Long-term extraction duration such as 20–40 hours is enough to confirm the system's stability since the leaking of solvents poses is still harmful to the system. The feasibility of the extraction of the organic has been investigated by using the supported membrane system. The bulk membrane could extract up to 96% of phenol into the receiving phase, whereas 7.4% of $[C_4mim][NTf_2]$ was left in a aqueous phase [59]. IL was used as co-carriers and co-extractants in the emulsion liquid membrane. It could improve the transport rate of phenol into an internal phase of emulsion, and almost 99% of

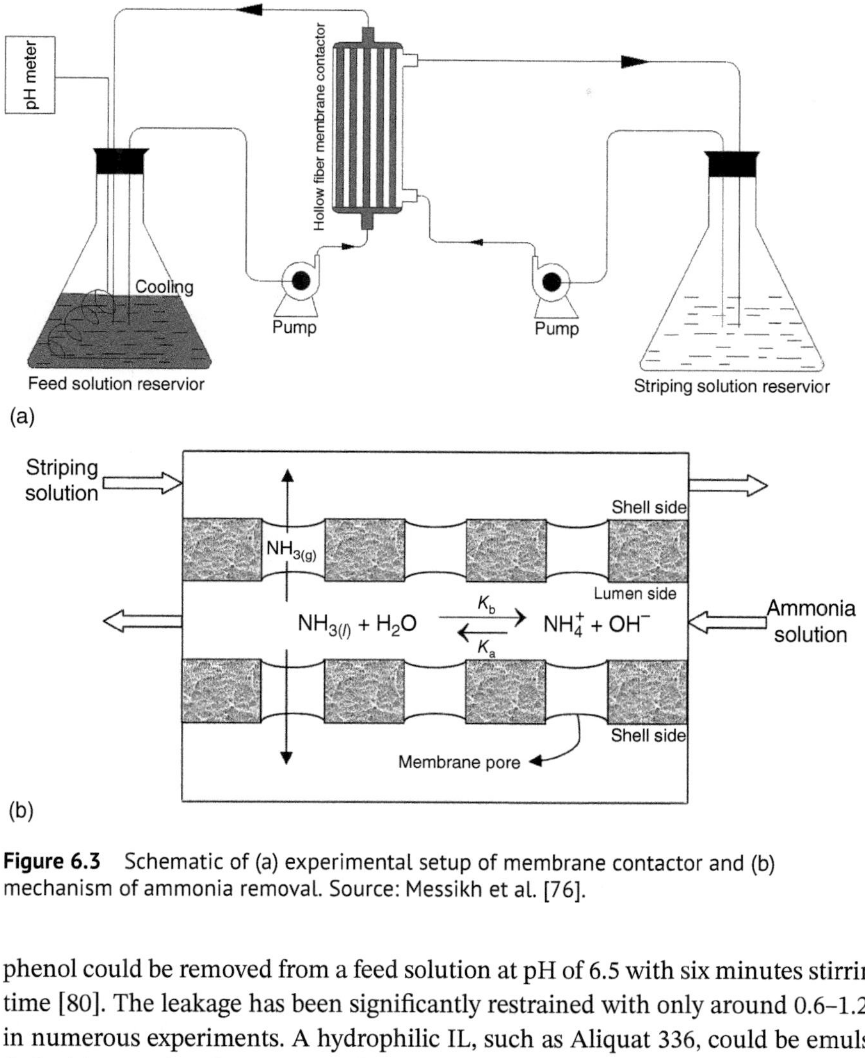

Figure 6.3 Schematic of (a) experimental setup of membrane contactor and (b) mechanism of ammonia removal. Source: Messikh et al. [76].

phenol could be removed from a feed solution at pH of 6.5 with six minutes stirring time [80]. The leakage has been significantly restrained with only around 0.6–1.2% in numerous experiments. A hydrophilic IL, such as Aliquat 336, could be emulsified with organic solvents in another emulsion liquid membrane study [81]. Nearly 100% of IL was recovered from distilled water.

The supported liquid membrane has been investigated for the removal of pharmaceuticals, pesticide, and phenol with Aliquat 336, $[C_4DMIm][PF_6]$, and $[C_4MIm][HSO_4]$, respectively [82]. The membrane permeability during removal of bisphenol A was higher with $[TBP][PF_6]$ than that with ammonium- and imidazolium-based ILs (Figure 6.4) [83]. The $[TBP][PF_6]$ was a stable IL with only 0.35% loss during the experiment. The tributyl phosphate (TBP) as a carrier and NaOH solution as stripping were used for levulinic removal [56]. It showed that the removal rate of levulinic was increased with the increase of concentrations of NaOH and TBP.

OMD-based membrane contactors have also been applied in wastewater treatment. The major advantage of OMD lies in its ability to work at low pressure

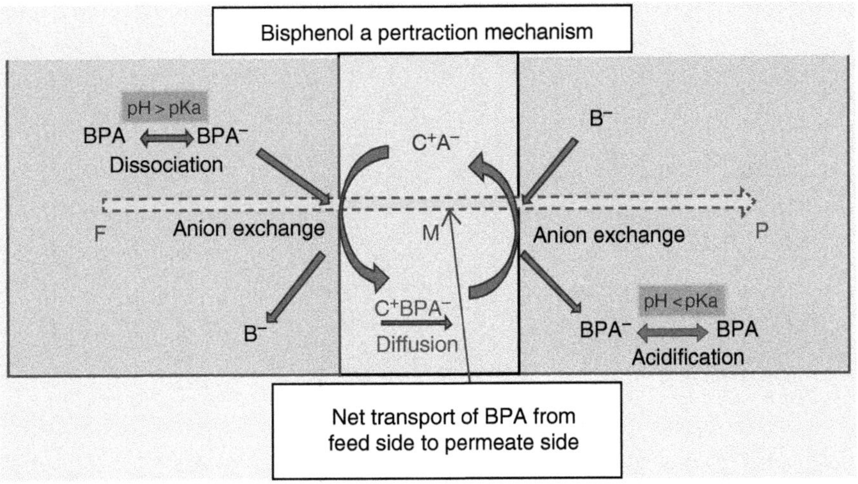

Figure 6.4 Separation mechanism of bisphenol A through the SILM (M = membrane, P = permeate phase, F = feed phase, B^- = anion, C^+ = cation, BPA^- = bisphenol A anion, BPA = bisphenol A). Source: Panigrahi et al. [83].

and temperature, thus avoiding thermal decomposition and mechanical damage. The feasibility of the application of hydrophobic hollow fiber membrane contactor was investigated during the dealcoholization of the wine solution [84]. The membrane resistance was found to be the major transportation resistance of aromatics and ethanol in the stripping and feed phases. The volatile compounds such as aroma compound were almost removed when the feed's residence time is too high. The OMD de-alcoholization performance was compared using three different kinds of stripping agents (pure water, 40% (w/w) $CaCl_2$, and 50% (w/w) glycerol) [43]. The results indicated that pure water is the best stripping agent. The removal rate of ethanol and flux were enhanced with increasing operating temperature, feed, and stripping solution rates. The removal rate of ethanol in the wine could be reached 66%. The laminar flow pattern was found to be an optimal condition for ethanol removal [57]. On the other hand, an increase in temperature would accelerate the de-alcoholization process.

6.6 Removal of Inorganic Contaminants from Wastewater

Heavy metal pollution is a major problem in the environment. Some toxic heavy metals, such as cadmium, nickel, lead, chromium, zinc, and arsenic, would enter the food chain and then the human body. As the heavy metals accumulate in an organ to more than standard limits, they would result in a serious health-related disease. More seriously, these heavy metals would remain in the human body for a long time because of their nonbiodegradable features. Especially, the concentration of heavy metal ions is getting high, owing to the increased industrial wastes.

Table 6.3 The maximum admissible limit set by WHO for safe drinking water [86].

Heavy metal	Toxicity	Maximum concentration (mg/l)
Cr	Headache, nausea, carcinogenic to human	0.050
Pb	Liver damage, irritability, memory problems	0.010
Cu	Insomnia, Wilson disease	2.00
Cd	Human carcinogen, kidney damage	0.003
Ni	Chronic asthma, nausea, dermatitis	0.070

Source: Based on Wasana et al. [86].

The major metal pollution comes from industries, including electroplating, tannery, fertilizer, mining, and metal processing industries [85]. The standard concentration limits of heavy metals regulated by the World Health Organization (WHO) in drinking water are shown in Table 6.3.

There are numerous and effective approaches based on the physicochemical or electrochemical technologies to achieve heavy metal removal. Physicochemical processes mainly include adoption, membrane filtration, and ion exchange. Electrodeposition and electrocoagulation are categorized as electrochemical techniques. Among these methods, membrane technology has made significant progress in applying heavy metal removal over the past decades because it is space saving and highly efficient. Especially, membrane contactor is an attractive technique for extracting heavy metal because of its small footprint and less extractant needed. Therefore, numerous studies concerning the practical application of membrane contactor have been carried out for the heavy metal recovery from hydrometallurgical leach solutions. Juang et al. [65] investigated the extraction of copper using di(2-ethylhexyl)phosphoric acid in both hollow fiber and flat-sheet membrane contactor. They confirmed the applicability of membrane contactors for copper removal, and a mass transfer model for prediction of membrane performance was developed on the basis of the resistance in-series models. Trtic-Petrovic et al. [87] used membrane contactor to separate thallium (TI) from chloride-containing acidic solutions using butyl acetate as stripping solution. Simulation results demonstrated that TI transport is mainly controlled by the concentration boundary layer's existence inside the membrane. Younas et al. [88] developed another mass transfer model for simulation of copper recovery from aqueous solutions using 2-hydroxy-5-nonylacetephenone oxime as an extracting agent in a hollow fiber membrane contactor. They solved kinetic extraction data. The diffusion in the aqueous phase dominates the integrated extraction process except for the partition coefficient equal to or less than 1 [88].

The membrane used in membrane contactor technologies for heavy metal removal from the wastewater are MF, UF, and IE membranes and LMs. Leakage of extractant and the instability of membrane material are key issues. Many approaches were tested to address these challenges, such as interfacial polymerization coating with hydrophobic and hydrophilic layers, immobilization of extraction molecules into the membrane matrix, and IE membranes. Therefore, the in-depth analysis of varied membrane material and processes' features and properties is of the great significance to the enhanced membrane application. Special attention will be paid to the varied polymeric materials and their properties and the optimization of the advanced polymeric membrane fabrication and structures to offer high removal efficiency and separation performance.

6.7 Polymer-Based Adsorption Membranes

The adsorption process has been widely accepted as the affordable available technology to remove pollutants from aqueous solution due to the simplicity to operate, suitability for small-scale use, and high efficiency for concentrated wastewater. Giles et al. [55] showed that the preferred adsorbents for heavy metal removal from aqueous solution are Al_2O_3, carbon, Fe_2O_3, modified iron oxides, layered double hydroxides, and even organic polymers. The forms of available adsorbents that are used in water treatment are generally powders or granules. However, the composite membranes with various types of functionalized nanomaterials have been tested as adsorbents to remove heavy metals and organic pollutants from aqueous solutions. This study highlights an exciting research scope that has not been gained significant attention from the research community. The membrane-based adsorbents would be compared with the commonly used powdered or granular adsorbents.

Membrane adsorbents in the form of polymeric composites have attracted a great deal of interest recently because of easy enhancements in membrane properties, e.g. the addition of small amounts of nanoparticles into the polymer matrix. Polymer nanocomposites are consist of a nanomaterial as the filler and a polymer as the matrix. The selection of the filler and polymer are based on the application scope of the nanocomposite. The preparation strategies of composited include melt blending, in situ intercalative polymerization, and solution blending [89].

The application of membranes in adsorption technology is relatively new. Dlamini et al. [38] investigated *Lantana camara*-reinforced ethylene vinyl acetate (EVA) composites' adsorptive properties. The experiments were carried out with an initial concentration of 100 mg/l *p*-nitrophenol (PNP) at pH 5.0. Predictably, *L. camara* reached equilibrium more rapidly in comparison with the organic composites. This observation results from the easy adsorption sites in the powder adsorbents. The active sites should be enveloped by the composite matrix's polymer matrix, thereby hindering accessibility by adsorption. The composite membrane could remove 88% of the PNP in about 75 minutes. The X-ray diffraction (XRD) spectroscopy images presented that *L. camara* contains various organic compounds such as cellulose, phenol, and pectins. The composite's adsorption capacity was approximately 21 mg/g, which was much lower than the record of 96 mg/g with *L. camara* powder at the initial concentration of 100 mg/l.

Zwane et al. [90] fabricated polyethersulfone/*Chromolaena odorata* (PES/CO) membrane for adsorption of dyes from water using phase inversion. The results showed that the adsorption capacity increased with an increase in CO loading during the Congo red removal with the maximum adsorption capacity of 85%. However, the adsorption capacity decreased marginally as the loading rate was increased beyond 20%.

The pH as an important factor has a significant influence on the adsorption process of pollutants from wastewater since it affects compounds' ionization [91]. From pH 2 to 4, PNP uptake was reported to decrease from 92% to 87% and then remained almost constant at the pH in the range of 4–7. High adsorption recorded at lower pH resulted from the ionization of PNP and the protonation of the active sites. The organic composite's protonated functional groups mainly include the hydroxyl and carboxylic groups of *L. camara*. The composite could be expected to interact with PNP molecules through hydrogen bonding mechanism.

The adsorption behavior of metal ions by using composites of EVA and lignocellulosic fiber has been reported [92]. The study described the applicability of lignocellulosic natural fiber dispersed into EVA solution to adsorb chromium and lead in an aqueous solution. It demonstrated that metal ion uptake takes place on the surface of the adsorbent and diffuses into the composite. Adsorption equilibrium could be reached after three-hours operation, and the maximum adsorption capacity was showed to be above 80%.

The clinoptilolite/polypropylene blend membrane was fabricated using the melt-blending method and tested for heavy metals' adsorption [80]. The optimal pH was reported to be in the range of pH 6–8. The composite removed 95% of the ions from the aqueous solution. The pretreatment of clinoptilolite with NaCl and HCl was observed to make a minor difference to the composites' performance. The authors concluded that the findings suggested the potential applicability of the composite in wastewater and water purification works. Similarly, Dlamini et al. [39] investigated the adsorption characteristics of EVA/bentonite clay composite using Pb^{2+} as a model pollutant. Polyurethane was applied to enhance the water permeation. The maximum adsorption capacity was obtained at pH 5 because the Pb^{2+} adsorption as hydroxide ions came closer to balancing the hydronium ions [92].

The kinetics and isotherms of removing Pb^{2+} from aqueous solutions were studied by Dlamini et al. [35]. The ability of clays and their minerals like montmorillonite to adsorb Pb^{2+} results from the presence of exchangeable ions, silanol, and aluminol groups [86]. The Si-OH and Al-OH have active donor atoms (O) on the surface of the adsorbants [93], leading to enough active adsorption sites. The possible adsorption mechanism of Pb^{2+} onto clay was proposed by Yang et al. [94]. It demonstrated that neat EVA and its composites could remove about 5% of Pb^{2+} ions in 10 hours. The reason was that the poor water uptake resulted from the nanocomposite's hydrophobic nature and the compounding and molding processes during synthesis using the melt-blending method. The compounding processes are regarded as an attractive strategy to suppress porosity in the nanocomposite, thus hindering the accessibility of the active sites inside the membrane. Therefore, the nanocomposite's sorption capacity was improved by the use of Na_2SO_4 as a pore former.

In another study, a C20A/polycaprolactone (PCL) nanocomposite was reported to have remarkable performance in terms of its adsorption affinity for Pb^{2+} [95]. It was reported that the nanocomposite could remove 87% of Pb^{2+} from the water with

3% filler. An artificial neural network model was developed to predict the adsorption efficiency of Pb^{2+} by using a clay–polymer nanocomposite fabricated via the melt-blending technique. The model was then used to simulate the influences of contact time, pH, and temperature on the removal of Pb^{2+} from aqueous solution by a C20A/PCL nanocomposite.

6.8 Ion-Exchange Nanoporous Membrane

The ion-exchange material, such as sulfonated polyether ether ketone (SPEEK), is resistant to the organic extractant, which can block the leakage of organic extractant after coating onto the UF membrane. This developed membrane has been successfully used as NF membrane for mono- and multi-valent ion separation, the concentration of oil emulsion due to its strong surface change [94]. As the SPEEK-coated UF membrane was utilized in the membrane contactor processes, a long-term stable operation for 65 days was displayed [96]. However, the organic extractant was in close with the porous UF support membrane, leading to the decomposition of the UF membrane and loss of mechanical strength. Even though the SPEEK separation layer was stable, the weakened structure of the support membrane could not retain the long-term stable extraction operation. To solve this issue, a new nanoporous SPEEK/PES ion-exchange membrane was displayed for Li^+ separation from salt lake brine by using a support liquid membrane contactor [88]. Unlike a composite SPEEK-coated membrane, the SPEEK/PES blend ion-exchange membrane could avoid the impregnation of organic extractant into the porous support so that the membrane stability has been largely improved.

To further investigate the possibility of the SPPESK/PES blend membrane for the separation of Ni and Cu, the mixed synthetic water containing Ni and Cu was used as the feed solution, 2-hydroxy-5-nonylbenzophenone was used as extractant, and the H_2SO_4 solution was used as the strip [97]. As feed pH was adjusted to 2.9, a complete separation of Ni and Cu was achieved using a membrane contactor. The permeate flow rate of copper ions of 1.10×10^{-8} mol/cm^2 s was observed. This attractive performance was attributed to the integration of the extraction and stripping process. The ion-exchange membrane-based contactor shows advantages in treating the wastewater containing heavy metals for rejecting the organic pollutants and recovering the valuable metals. However, the ion-exchange membranes have not been extensively investigated in the application of membrane contactor due to the high fabrication costs.

6.9 Micellar-Enhanced Ultrafiltration Membrane

Micellar-enhanced ultrafiltration (MEUF) membrane is principally applied to restrain heavy metal ions from passing through membranes via the formation of the larger molecular weight of micelles between surfactants and metal ions. The process forms micellar compounds with metal ions using charged surfactants such as cetyl trimethylammonium bromide (CTAB), cetylpyrinidium chloride (CPC), ethylenediaminetetraacetic acid (EDTA), and sodium dodecyl sulfate (SDS), to mention just a few. The surfactants are known as micelle agents. In this case the surfactant cannot be a polymer ligand. The surfactant's important property is electrical charge, at the point where the surfactant concentration exceeds the critical micelle concentration (CMC). Besides, the surfactant monomers aggregate and assemble with heavy metal ions to form amphiphilic micelles. The extensive application of the cationic/anionic surfactants in MEUF membrane was resulted from their larger molecular weight and lower CMC level that leads to the fast formation of micelle with cationic metal ions. Some significant parameters of MEUF process include pressure, pH, additives, temperature, and surfactant concentration [98]. The technique has been mostly tested at the experimental level due to critical concerns in commercial applications. One such utmost critical concern is the generation of toxic pollutants and membrane pre-fouling tendency, owing to the surfactants accumulation on the membrane surface [60]. Another concern is that complete heavy metal removal is not guaranteed. For example, if the surfactant has a high CMC level, the chemical reaction with metal ions is difficult so that none or very low heavy metal removal efficiency is obtained. For that reason, this technique requires large quantities of surfactants, rendering high operating cost.

Additionally, high surfactant dosage could form an insulating surfactant layer, which eventually leads to reduced DO [99]. Fortunately, high dosage may be avoided by selecting surfactants with low CMC, which is critical to provide the fast formation of micelles with metal ions. Alternatively, another possible solution could be selecting biosurfactants or mixing ionic and non-ionic surfactants [90].

6.10 Membrane Materials for Water Treatment

Ceramic and polymeric membranes are the two important types of membrane materials. Ceramic membranes are swiftly finding their way in applications like water and wastewater treatment, food processing, bioreactions, petrochemical materials separation, and radioactive materials enrichment. These membranes are remarked by a higher chemical, mechanical, and thermal stability and resistance to erosion and corrosion [100, 101]. Therefore, these membranes can be easily cleaned by physical and chemical approaches to remove the irreversible fouling. The newly developed silicon carbide ceramic membrane has a high hydrophilicity and a low isoelectric point (pH 2.6), making them highly resistant to fouling compared with other ceramic membranes. The higher reliability of ceramic membranes leads to low operational costs due to the reduced repetitive testing, repair, and replacement of membrane modules [23–26].

Conventional ceramic membranes are prone to fouling. Coating of the membrane surface with a functional layer like TiO_2 may reduce the risk of fouling. TiO_2 exhibits an excellent self-cleaning behavior, which stops the sticking of contaminants on the TiO_2 surface. However, the conventional TiO_2 coating has a low permeability due to higher density and specific platelike or flake-like membrane structure. Therefore, the flux eventually decreases as the pore size is reduced. Instead of using the platelike or flake-like structures, the mesh structure is more effective for the pressure-driven membrane processes. Since porosity is greatly enhanced with mesh structure, a high membrane flux can therefore be achieved [29]. Polymeric membranes enjoy a higher selectivity than ceramic membranes, although ceramic membranes can effectively withstand harsh working conditions [30].

6.11 Membrane Materials for Microfiltration (MF) and Ultrafiltration (UF)

MF and UF are well known in terms of low fouling and cost factors. These can be made from both ceramic and polymer materials. However, ceramic materials suffer from the disadvantage of brittleness, due to which it is difficult to fabricate these on a large scale. Sealing also becomes a significant issue when the modules are to be integrated. Polymeric membranes on the other hand do not face these problems. Therefore, these membranes have dominated the market. Poly(vinylidene fluoride), poly(ether sulfone), polypropylene, polyethylene, and poly(tetrafluoroethylene) are some of the polymers that are widely used in the fabrication of MF and UF membranes. All these polymers are very hydrophobic except for PES. The hydrophobicity of these membranes makes fouling one of the biggest issues for NF and MF membranes. However, fouling can be addressed by either hydrophilical modification before fabrication, hydrophilic agent blend, or hydrophilic polymer graft on the membrane surface. In some cases, the downside effects of these techniques are reduced pore size and blockage of the MF and UF membrane surfaces' pores. In other cases, the membrane pore size may expand, leading to lower salt rejection. Also, many of the coating layers do not have ample chemical and mechanical properties and so cannot uphold a prolonged operation [30].

Due to the diversity and ease of processability, polymers are regarded as the most promising class of materials for the synthesis of UF membranes. The features of the wastewater are main determining factor in the selection of suitable material for membrane fabrication. Higher chemical stability and mechanical are the major objectives in the synthesis of UF membranes [30].

6.12 Membrane Materials for Nanofiltration (NF)

Recent developments in membrane fabrication techniques have led to NF membranes with high flux, high salt rejection, and low fouling tendency. The efficiency of NF membranes depends on several factors:

- Membrane surface charge
- Module configuration
- Operation mode
- Operating pressure
- Temperature
- Product recovery rate
- Feed pH, concentration, and flow rate

Commercial NF membranes are mostly made of polymers like polyamide that contain full or slight charge on the surface, possibly because of the dissociation of functional groups on the surface of the charged solutes. These membranes are therefore more efficient in heavy metal removal from aqueous solution. Other polymers like polypiperazineamide, melamine polyamine, polysulfonamide, polybenzimidazole, sulfonated polyethersulfone, and polyvinyl alcohol have also been applied for synthesis of a variety of NF membranes [30–32].

6.13 Membrane Materials for Reverse Osmosis (RO)

Polymeric membranes lead the market for RO applications due to lower fabrication cost, easy handling, and better selectivity and permeability. Thin film composite (TFC) membrane and asymmetric cellulose acetate (CA) membrane are two broad classes of commercial polymeric RO membranes. Asymmetric CA membranes were the first to be developed and improved for industrial applications. However, due to the shortcomings of asymmetric RO membranes, TFC membranes were optimized for the materials and manufacturing methods regarding fabrication of selective thin layer with microporous support. Today, several TFC membrane materials are in play to dominate the RO membrane market. Polyamide TFC RO membranes are broadly utilized in the heavy metal removal from wastewater streams because of their high rejection of salt and organic matter, excellent hydrolytic resistance, high durability, and wide operating pH range [30].

Although TFC membranes are largely successful in RO, they face foremost constraint, i.e. degradation due to chlorine, a common disinfectant used in wastewater treatment. A need of the day is to develop RO membranes that are more chlorine resistant. It will eradicate the need for prior rechlorination of the RO permeate and dechlorination of RO feed, thus pulling down the overall system cost. Current RO membranes have shown a remarkable performance in terms of permeability and selectivity in commercial-scale seawater desalination plants. However, the future RO membrane development should target key issues like higher fouling and chlorine resistance and removal of trace contaminants like pesticides, persistent organic compounds, endocrine-disrupting chemicals, and pharmaceutically active compounds. Nanoparticles have also been exploited in RO membrane development. The two major issues that need to be addressed in these membranes are cost and scale-up. For drinking water desalination, health and safety aspects also need to be carefully estimated in the case of added nanoparticles [36, 102].

6.14 Membrane Materials for Forward Osmosis (FO)

Either forward osmosis (FO) membranes can be developed from existing commercial NF/RO membranes, or new membranes can be synthesized that have a precise design for FO applications. The first case is simple, efficient, and more cost-effective. FO membranes' performance criteria are the same as those for NF/RO membranes, i.e. high water flux and salt retention. However, the difference lies in the balance between low internal concentration polarization (ICP) of the support layer and high-pressure tolerance. FO process enjoys the benefit of low-pressure tolerance requirements. On the other hand, low ICP is essential to keep low salt leakage and high water flux. So, low ICP is achieved through the FO membrane's somewhat low mechanical strength, increasing porosity, and reducing tortuosity. The ideal FO membranes should have a strong support layer that is sufficiently hydrophilic and porous to control ICP effects. Moreover, these membranes should have an ultrathin and highly selective layer to get high salt rejection and water flux and then minimum reverse solute diffusion. The development strategy of FO membranes is similar to that of RO membranes in the material field.

Similarly, the FO also focuses on the control of membrane morphologies. Cellulose triacetate (CTA) FO membranes are currently leading the global market. TFC polyamide membrane has also been given special attention due to their remarkable properties like higher water fluxes, higher solute rejection rates, better pH stability, and resistance to hydrolysis and biological degradation. Membrane fouling, scaling, concentration polarization, and cake-enhanced osmotic pressure are some of the problems that stand in the way of better FO membranes' performance. Proper chemical cleaning protocols are crucial in achieving sustainable operation of the real and long-term FO process [33–38].

6.15 Challenges in Membrane Materials to Prevent Fouling

As discussed earlier, membrane fouling is still the biggest challenge in all membrane processes. Higher membrane fouling leads to lower water flux, increased cleaning requirements, and reduced membrane life. Fouling can be classified according to the type of foulants. It may be organic, scaling, or biofouling. Table 6.4 shows an overview of these foulants. Fouling is more rigorous in a pressure-driven membrane process like NF or RO where high hydraulic pressure is applied. Membrane fouling can be combated to some extent with feed pretreatment, operation optimization, and intermittent membrane cleaning. Another important strategy for fouling prevention is membrane modification. It includes modification in the chemical structure and morphology of the membrane. The chemical structure includes a charge, and functional groups, while morphology includes surface pattern, pore size of the membrane. Lower fouling tendency is generally favored by low surface roughness and high hydrophilicity. The downside of membrane modification is that it negatively affects flux and tends to be more costly. Membrane modification may effectively delay the early stage of fouling but cannot hinder it for extended periods of time. Therefore, new membrane materials must be explored that have the antifouling capacity and higher permeability. Similarly, advanced membrane modification techniques should be developed that are more cost-effective and simpler [33–40].

Nanocomposite membranes are a new class of membranes that are synthesized by combining polymer and nanomaterials. The four types of these membranes are (i)

Table 6.4 Fouling categories.

Fouling category	Fouling mechanism	Common foulants
Colloidal	Cake layer formed due to the particles building up on the membrane surface and inside the pores	Silicate, ferric oxide, iron oxide, aluminum oxide
Inorganic	Bulk and surface crystallization occur due to the dumping of precipitates on the membrane surface	Calcium carbonate, calcium sulfate, calcium phosphate, sodium chloride
Organic	Gel formation takes place because of the adsorption of natural organic compounds	Fulvic acid, protein, polysaccharides, polyacrylic polymer
Biofouling	Formation of biofilm on the membrane surface	Fungi, algae, yeast, bacteria

Source: Drioli et al. [40].

conventional nanocomposite, (ii) thin film nanocomposite, (iii) thin film composite with nanocomposite, and (iv) surface modified with nanocomposite. This classification is based on the location of the nanomaterials and the membrane structure. These membranes' structure and characteristics can target the specific water treatment applications by introducing unique functionalities such as catalytic, antibacterial, or adsorptive capabilities [41].

6.16 Conclusions and Perspectives

Notwithstanding the fact that much success has been achieved in the resource extraction from wastewater using membrane contactor, the membrane contactor that combined extraction and stripping processes is a highly efficient extraction process because of its less footprint and small amount of organic extractant. However, there is still more work to be done. The fabrication of a solvent stable membrane is urgently needed to overcome the swelling induced by the organic extractant. The development of mild extractants is another important research direction for membrane contactor to prolong the lifetime. IL is a thermally stable green solvent, which has been validated as efficient extractants for various heavy metals. The major challenge is more expensive in comparison with traditional extractants. The loss of ILs is still an issue as used in the conventional membrane contactor. This is a promising direction of research for achieving the high permeable and solvent resistant membrane [97].

Another significant drawback of membrane extraction contactor is high mass transfer resistance of a membrane. The resistance can be reduced by decreasing membrane roughness and optimizing the membrane module. Electrospun nanofibrous membrane has been reported as an attractive strategy for improving diffusion resistance of mass transfer in the membrane distillation. Therefore, it is an attractive strategy for decreasing the resistance of membrane extraction contactor. In sum, the present selective membrane contactor system could serve as a promising and highly efficient platform for the recovery of resources from wastewater.

The modern developments in manufacturing technology have allowed the membrane manufacturers to synthesize ultrathin high-surface-area membranes into compact modules. As a result, the overall cost of the membrane system has declined significantly over the years. The initial anisotropic membranes had an effective thickness of 0.2–0.4 μm, which can now further be reduced to a thickness of less than 0.05 μm through membrane surface modification and advanced composite membrane preparation techniques. Similarly, the permeability and selectivity have improved two to three times in the current period. All these factors have led to a rise of 5–10 folds in flux as compared with the membranes that were available 30 years back.

The major challenges involved in the large scale applications of nanocomposite membranes include precise understanding of the influence of various nanomaterials on membrane structure and membrane performance, homogeneous dispersion of nanomaterials in the polymer matrices, compatibility of nanofillers with the polymers, environmental toxicity of leached nanomaterials, and cost-effective manufacturing of large-scale membranes for industrial applications. The long-term stability of membranes under practical application conditions also need to be monitored.

Other challenges in membrane performance are concentration polarization and membrane fouling. Both of these factors equally affect all membrane processes. Further research is required to design and fabricate membrane modules that offer improved flow characteristics to control concentration polarization and easy module cleaning in membrane fouling.

References

1 Agrahari, G.K., Verma, N., and Bhattacharya, P.K. (2014). Removal of benzoic acid from water by reactive extraction using hollow fiber membrane contactor: experiment and modeling. *Clean: Soil, Air, Water* 42 (7): 901–908.

2 Sheikh, M. et al. (2020). Application of ZnO nanostructures in ceramic and polymeric membranes for water and wastewater technologies: a review. *Chemical Engineering Journal* 391: 123475.

3 Bade, R. and Lee, S.H. (2011). A review of studies on micellar enhanced ultrafiltration for heavy metals removal from wastewater. *Journal of Water Sustainability* 1: 85–102.

4 Balasubramanian, A. and Venkatesan, S. (2012). Removal of phenolic compounds from aqueous solutions by emulsion liquid membrane containing ionic liquid $[BMIM]^+[PF6]^-$ in tributyl phosphate. *Desalination* 289: 27–34.

5 Hajilary, N., Rezakazemi, M., and Shirazian, S. (2019). Biofuel types and membrane separation. *Environmental Chemistry Letters* 17 (1): 1–18.

6 Ayaz, M. et al. (2019). Enhanced water flux by fabrication of polysulfone/alumina nanocomposite membrane for copper(II) removal. *Macromolecular Research* 27 (6): 565–571.

7 Rezakazemi, M., Khajeh, A., and Mesbah, M. (2018). Membrane filtration of wastewater from gas and oil production. *Environmental Chemistry Letters* 16 (2): 367–388.

8 Azimi, A. et al. (2017). Removal of heavy metals from industrial wastewaters: a review. *ChemBioEng Reviews* 4 (1): 37–59.

9 Rezakazemi, M. (2018). CFD simulation of seawater purification using direct contact membrane desalination (DCMD) system. *Desalination* 443: 323–332.

10 Rezakazemi, M., Sadrzadeh, M., and Matsuura, T. (2018). Thermally stable polymers for advanced high-performance gas separation membranes. *Progress in Energy and Combustion Science* 66: 1–41.

11 Bartolomeu, M. et al. (2018). Wastewater chemical contaminants: remediation by advanced oxidation processes. *Photochemical & Photobiological Sciences* 17 (11): 1573–1598.

12 Rezakazemi, M. et al. (2018). Fouling-resistant membranes for water reuse. *Environmental Chemistry Letters* 16 (3): 715–763.

13 Baudot, A., Floury, J., and Smorenburg, H.E. (2001). Liquid–liquid extraction of aroma compounds with hollow fiber contactor. *AIChE Journal* 47 (8): 1780–1793.

14 Tavangar, T. et al. (2020). Textile waste, dyes/inorganic salts separation of cerium oxide-loaded loose nanofiltration polyethersulfone membranes. *Chemical Engineering Journal* 385: 123787.

15 Ahmadijokani, F. et al. (2020). Superior chemical stability of UiO-66 metal-organic frameworks (MOFs) for selective dye adsorption. *Chemical Engineering Journal* 399: 125346.

16 Mengting, Z. et al. (2019). Applicability of $BaTiO_3$/graphene oxide (GO) composite for enhanced photodegradation of methylene blue (MB) in synthetic wastewater under UV–vis irradiation. *Environmental Pollution* 255: 113182.

17 Mustafa, K. et al. *Nanotechnology and the Generation of Sustainable Hydrogen.*

18 Hussain, A., Farrukh, S., and Minhas, F.T. (2015). Two-stage membrane system for post-combustion CO_2 capture application. *Energy & Fuels* 29 (10): 6664–6669.

19 Baylan, N. and Cehreli, S. (2018). Ionic liquids as bulk liquid membranes on levulinic acid removal: a design study. *Journal of Molecular Liquids* 266: 299–308.

20 Berrios, J., Pyle, D.L., and Aroca, G. (2010). Gibberellic acid extraction from aqueous solutions and fermentation broths by using emulsion liquid membranes. *Journal of Membrane Science* 348 (1–2): 91–98.

21 Bessadok, A. et al. (2007). Effect of chemical treatments of Alfa (*Stipa tenacissima*) fibres on water-sorption properties. *Composites Science and Technology* 67 (3–4): 685–697.

22 Boller, M. (1997). Small wastewater treatment plants—a challenge to wastewater engineers. *Water Science and Technology* 35 (6): 1–12.

23 Bottino, A. (2009). Membrane technologies for water treatment and agroindustrial sectors. *Comptes Rendus Chimie* 12 (8).

24 Brahmia, N. et al. (2014). Recovery of 4-chlorophenol from an aqueous solution by ELM: stability of the membrane, modeling, and optimization of the extraction using experimental designs. *Desalination and Water Treatment* 52 (1–3): 375–383.

25 Busca, G. et al. (2008). Technologies for the removal of phenol from fluid streams: a short review of recent developments. *Journal of Hazardous Materials* 160 (2–3): 265–288.

26 Calabrò, V., Drioli, E., and Matera, F. (1991). Membrane distillation in the textile wastewater treatment. *Desalination* 83 (1–3): 209–224.

27 Rehman, A. et al. (2019). Adsorption of CO_2 on amine-functionalized green metal-organic framework: an interaction between amine and CO_2 molecules. *Environmental Science and Pollution Research* 26 (36): 36214–36225.

28 Rehman, A. et al. (2020). Synthesis and effect of metal–organic frame works on CO_2 adsorption capacity at various pressures: a contemplating review. *Energy & Environment* 31 (3): 367–388.

29 Campos, J.L. et al. (2016). Greenhouse gases emissions from wastewater treatment plants: minimization, treatment, and prevention. *Journal of Chemistry* 2016.

30 Chaouchi, S. and Hamdaoui, O. (2014). A cetaminophen extraction by emulsion liquid membrane using Aliquat 336 as extractant. *Separation and Purification Technology* 129: 32–40.

31 Darwish, M. (2014). Reclaimed wastewater for agriculture irrigation in Qatar. *Global Journal of Agricultural Research and Reviews* 3 (1): 106–120.

32 Das, C. et al. (2008). Removal of dyes and their mixtures from aqueous solution using liquid emulsion membrane. *Journal of Hazardous Materials* 159 (2–3): 365–371.

33 Diban, N. et al. (2008). Ethanol and aroma compounds transfer study for partial dealcoholization of wine using membrane contactor. *Journal of Membrane Science* 311 (1–2): 136–146.

34 Djenouhat, M. et al. (2008). Ultrasonication-assisted preparation of water-in-oil emulsions and application to the removal of cationic dyes from water by emulsion liquid membrane - Part 1: Membrane stability. *Separation and Purification Technology* 62 (3): 636–641.

35 Dlamini, D.S., Mishra, A.K., and Mamba, B.B. (2012). Kinetic and equilibrium studies of the removal of Pb^{2+} from aqueous solutions using Na_2SO_4-EVA/Cloisite®20A composite. *Materials Chemistry and Physics* 133: 369–375.

36 Devold, H. (2013). Oil and gas production handbook: an introduction to oil and gas production, transport, refining and petrochemical industry. In: *ABB Oil and Gas*.

37 Dhote, J., Ingole, S., and Chavhan, A. (2012). Review on wastewater treatment technologies. *International Journal of Engineering Research & Technology* 1: 1–10.

38 Dlamini, D.S., Mishra, A.K., and Mamba, B.B. (2012). Structural, transport and adsorptive properties of Lantana camara-reinforced ethylene vinyl acetate composites. *Water, Air, and Soil Pollution* 223 (7): 3831–3843.

39 Dlamini, D.S., Mishra, A.K., and Mamba, B.B. (2012). Morphological, transport, and adsorption properties of ethylene vinyl acetate/polyurethane/bentonite clay composites. *Journal of Applied Polymer Science* 124 (6): 4978–4985.

40 Drioli, E., Criscuoli, A., and Curcio, E. (2011). *Membrane Contactors: Fundamentals, Applications and Potentialities*. Elsevier.

41 Edokpayi, J.N. et al. (2017). Impact of wastewater on surface water quality in developing countries: a case study of South Africa. *Water Quality*: 401–416.

42 El-Shahawi, M.S. et al. (2010). An overview on the accumulation, distribution, transformations, toxicity and analytical methods for the monitoring of persistent organic pollutants. *Talanta* 80 (5): 1587–1597.

43 Rezakazemi, M., Shirazian, S., and Ashrafizadeh, S.N. (2012). Simulation of ammonia removal from industrial wastewater streams by means of a hollow-fiber membrane contactor. *Desalination* 285: 383–392.

44 Muhammad, A., Younas, M., and Rezakazemi, M. (2017). Quasi-dynamic modeling of dispersion-free extraction of aroma compounds using hollow fiber membrane contactor. *Chemical Engineering Research and Design* 127: 52–61.

45 Muhammad, A., Younas, M., and Rezakazemi, M. (2018). CFD simulation of copper(II) extraction with TFA in non-dispersive hollow fiber membrane contactors. *Environmental Science and Pollution Research* 25 (12): 12053–12063.

46 Sohaib, Q. et al. (2021). Rigorous non-isothermal modeling approach for mass and energy transport during CO_2 absorption into aqueous solution of amino acid ionic liquids in hollow fiber membrane contactors. *Separation and Purification Technology* 254: 117644.

47 Younas, M. et al. (2020). Post-combustion CO_2 capture with sweep gas in thin film composite (TFC) hollow fiber membrane (HFM) contactor. *Journal of CO_2 Utilization* 40: 101266.

48 Sohaib, Q. et al. (2020). Modeling pre-combustion CO_2 capture with tubular membrane contactor using ionic liquids at elevated temperatures. *Separation and Purification Technology* 241: 116677.

49 Ahmad, S. et al. (2020). Mass transfer modelling of hollow fiber membrane contactor for apple juice concentration using osmotic membrane distillation. *Separation and Purification Technology* 250: 117209.

50 Gabelman, A. and Hwang, S.-T. (1999). Hollow fiber membrane contactors. *Journal of Membrane Science* 159 (1–2): 61–106.

51 Rezakazemi, M., Maghami, M., and Mohammadi, T. (2018). High loaded synthetic hazardous wastewater treatment using lab-scale submerged ceramic membrane bioreactor. *Periodica Polytechnica, Chemical Engineering* 62 (3): 299–304.

52 Rezakazemi, M., Maghami, M., and Mohammadi, T. (2018). Wastewaters treatment containing phenol and ammonium using aerobic submerged membrane bioreactor. *Chemistry Central Journal* 12 (1): 79.

53 Förstner, U., Mader, P., and Salomons, W. (1995). *Heavy Metals: Problems and Solutions*. Springer.

54 González-Muñoz, M.J. et al. (2003). Recovery of phenol from aqueous solutions using hollow fibre contactors. *Journal of Membrane Science* 213 (1–2): 181–193.

55 Giles, D.E. et al. (2011). Iron and aluminium based adsorption strategies for removing arsenic from water. *Journal of Environmental Management* 92 (12): 3011–3022.

56 Pierre, F., Souchon, I., and Marin, M. (2001). Recovery of sulfur aroma compounds using membrane-based solvent extraction. *Journal of Membrane Science* 187 (1–2): 239–253.

57 Rodriguez, M. et al. (1997). Removal of valeric acid from wastewaters by membrane contactors. *Journal of Membrane Science* 137 (1–2): 45–53.

58 Mandal, D., Guha, A., and Sirkar, K. (1998). Isomer separation by a hollow fiber contained liquid membrane permeator. *Journal of Membrane Science* 144 (1–2): 13–24.

59 Mota, F.J. et al. (2003). Optimisation of extraction procedures for analysis of benzoic and sorbic acids in foodstuffs. *Food Chemistry* 82 (3): 469–473.

60 Zhang, X. et al. (2006). Degradability of five aromatic compounds in a pilot wastewater treatment system. *International Biodeterioration & Biodegradation* 58 (2): 94–98.

61 Hannuksela, M. and Haahtela, T. (1987). Hypersensitivity reactions to food additives. *Allergy* 42 (8): 561–575.

62 Iasimone, F. et al. (2017). Experimental study for the reduction of CO_2 emissions in wastewater treatment plant using microalgal cultivation. *Journal of CO_2 Utilization* 22: 1–8.

63 Jin, T. et al. (2017). The role of the surfactant sodium dodecyl sulfate todynamically reduce mass transfer resistance of SPEEK coatedmembrane for oil-in-water emulsion treatment. *Journal of Membrane Science* 541: 9–18.

64 Gryta, M., Karakulski, K., and Morawski, A. (2001). Purification of oily wastewater by hybrid UF/MD. *Water Research* 35 (15): 3665–3669.

65 Juang, R.S. and Huang, H.L. (2003). Mechanistic analysis of solvent extraction of heavy metals in membrane contactors. *Journal of Membrane Science* 213 (1–2): 125–135.

66 Khalid, S. et al. (2018). A review of environmental contamination and health risk assessment of wastewater use for crop irrigation with a focus on low and high-income countries. *International Journal of Environmental Research and Public Health* 15 (5).

67 Kouki, N., Tayeb, R., and Dhahbi, M. (2014). Recovery of acetaminophen fromaqueous solutions using a supported liquid membrane based on a quaternary ammonium salt as ionophore. *Chemistry Papers* 68: 457–464.

68 Kumar, A., Thakur, A., and Panesar, P.S. (2018). Statistical optimization of lactic acid extraction using green emulsion ionic liquid membrane (GEILM). *Journal of Environmental Chemical Engineering* 6 (2): 1855–1864.

69 Kumar, A., Thakur, A., and Panesar, P.S. (2019). A review on emulsion liquid membrane (ELM) for the treatment of various industrial effluent streams. *Reviews in Environmental Science and Bio-Technology* 18 (1): 153–182.

70 Lazo-Cannata, J.C. et al. (2011). Adsorption of phenol and nitrophenols by carbon nanospheres: effect of pH and ionic strength. *Separation and Purification Technology* 80 (2): 217–224.

71 Lee, S.C. (2011). Extraction of succinic acid from simulated media by emulsion liquid membranes. *Journal of Membrane Science* 381 (1–2): 237–243.

72 Licon, E. et al. (2015). Ammonium removal by liquid–liquid membrane contactors in water purification process for hydrogen production. *Desalination and Water Treatment* 56: 3607–3616.

73 Liguori, L. et al. (2013). Effect of process parameters on partial Dealcoholization of wine by osmotic distillation. *Food and Bioprocess Technology* 6 (9): 2514–2524.

74 Mahajan, S. (1985). *Pollution Control in Process industries.* Tata McGraw-Hill Education.

75 Mahmud, H. et al. (2002). Mass transport in the membrane air-stripping process using microporous polypropylene hollow fibers: effect of toluene in aqueous feed. *Journal of Membrane Science* 209 (1): 207–219.

76 Messikh, N., Bousba, S., and Bougdah, N. (2017). The use of a multilayer perceptron (MLP) for modelling the phenol removal by emulsion liquid membrane. *Journal of Environmental Chemical Engineering* 5 (4): 3483–3489.

77 Manzak, A. and Melek, S. (2010). Extraction of acetic acid from aqueous solutions by emulsion type of liquid membranes using alamine300 as carrier. *Indian Journal of Chemical Technology* 17: 41–445.

78 Masindi, V. and Muedi, K.L. (2018). Environmental contamination by heavy metals. *Heavy Metals* 10: 115–132.

79 Minier-Matar, J. et al. (2017). Application of membrane contactors to remove hydrogen sulfide from sour water. *Journal of Membrane Science* 541: 378–385.

80 Motsa, M. et al. (2011). Preparation, characterization, and application of polypropylene-clinoptilolite composites for the selective adsorption of lead from aqueous media. *Journal of Colloid and Interface Science* 359 (1): 210–219.

81 Ng, Y.S., Jayakumar, N.S., and Hashim, M.A. (2011). Behavior of hydrophobic ionic liquids as liquid membranes on phenol removal: experimental study and optimization. *Desalination* 278 (1–3): 250–258.

82 Panday, K., Prasad, G., and Singh, V. (1985). Copper(II) removal from aqueous solutions by fly ash. *Water Research* 19 (7): 869–873.

83 Panigrahi, A., Pilli, S.R., and Mohanty, K. (2013). Selective separation of Bisphenol a from aqueous solution using supported ionic liquid membrane. *Separation and Purification Technology* 107: 70–78.

84 Rajan, R.J., J. Sudarsan, and S. Nithiyanantham 2019 Microbial population in constructed wetlands–review of recent advancements for water treatment.

85 Schwarze, M. et al. (2015). Micellar enhanced ultrafiltration (MEUF) of metal cations with oleylethoxycarboxylate. *Journal of Membrane Science* 478: 140–147.

86 Wasana, H.M.S. et al. (2017). WHO water quality standards Vs synergic effect(s) of fluoride, heavy metals and hardness in drinking water on kidney tissues. *Scientific Reports* 7.

87 Trtic-Petrovic, T.M. et al. (2002). Analysis of concentration boundary layer in thallium (III) extraction with butyl acetate using membrane modules of different length. *Desalination* 148 (1–3): 241–246.

88 Younas, M., Druon-Bocquet, S., and Sanchez, J. (2010). Kinetic and dynamic study of liquid–liquid extraction of copper in a HFMC: experimentation, modeling, and simulation. *AIChE Journal* 56 (6): 1469–1480.

89 Sims, R.C. and Overcash, M. (1983). Fate of polynuclear aromatic compounds (PNAs) in soil-plant systems. In: *Residue Reviews*, 1–68. Editor: Francis A. Gunther Springer.

90 Zwane, S. et al. (2019). Polyethersulfone/chromolaena odorata (PES/CO) adsorptive membranes for removal of Congo red from water. *Journal of Water Process Engineering* 30: 100498.

91 Taghipour, M. and Jalali, M. (2015). Effect of clay minerals and nanoparticles on chromium fractionation in soil contaminated with leather factory waste. *Journal of Hazardous Materials* 297: 127–133.

92 Vilakati, G.D. et al. (2012). Metal ion adsorption behavior of lignocellulosic fiber-ethylene vinyl acetate composites. *Polymer Engineering and Science* 52 (4): 760–767.

93 Wiesler, F. and Sodaro, R. (1996). Degasification of water using novel membrane technology. *Ultrapure Water* 13 (6): 53–56.

94 Yang, X. et al. (2011). Effect of pH, ionic strength and temperature on sorption of Pb(II) on NKF-6 zeolite studied by batch technique. *Chemical Engineering Journal* 168 (1): 86–93.

95 Yan, C.N. and Othman, O.Z.Y.N. (2014). Prediction of Kraft lignin extraction performance using emulsion liquid membrane carrier-diffusion model. *Journal of Engineering and Technological Sciences* 67: 17–21.

96 Yordanov, B. and Boyadzhiev, L. (2004). Pertraction of citric acid by means of emulsion liquid membranes. *Journal of Membrane Science* 238 (1–2): 191–197.

97 Zakrzewska-Trznadel, G., Harasimowicz, M., and Chmielewski, A.G. (1999). Concentration of radioactive components in liquid low-level radioactive waste by membrane distillation. *Journal of Membrane Science* 163 (2): 257–264.

98 Zarebska, A. et al. (2017). Ammonia recovery from pig slurry using a membrane contactor-influence of slurry pretreatment. *Water, Air, and Soil Pollution* 228 (4):150.

99 Zheng, C. (2013). Treatment technologies for organic wastewater. *Water Treatment* 11: 250–286.

100 Rezakazemi, M. and Shirazian, S. (2018). Separation performance of nanostructured ceramic membranes: analytical model development. *Journal of Non-Equilibrium Thermodynamics* 43: 245.

101 Fasihi, M. et al. (2012). Computational fluid dynamics simulation of transport phenomena in ceramic membranes for SO2 separation. *Mathematical and Computer Modelling* 56 (11): 278–286.

102 Esfandiari, A. et al. (2019). Computational fluid dynamic modeling of water desalination using low-energy continuous direct contact membrane distillation process. *Applied Thermal Engineering* 163: 114391.

7

Applications of Membrane Contactors in Food Industry

Waheed Ur Rehman[1], Bazla Sarwar[2], Sidra Saqib[2], Ahmad Mukhtar[3], Mohammad Younas[4], and Mashallah Rezakazemi[5]

[1] *PCSIR Laboratories Complex, Jamrud Road, Peshawar 25120, Pakistan*
[2] *COMSATS University Islamabad (CUI), Department of Chemical Engineering, Lahore Campus, Punjab 54000, Pakistan*
[3] *NFC Institute of Engineering and Fertilizer Research, Department of Chemical Engineering, Faisalabad, Punjab 38000, Pakistan*
[4] *University of Engineering and Technology, Department of Chemical Engineering, Peshawar 25120, Pakistan*
[5] *Shahrood University of Technology, Faculty of Chemical and Materials Engineering, Shahrood, Iran*

7.1 Introduction

In **Chapter 5**, different operating modes of membrane contactors (MCs), their basic operating concepts, mass and heat transfer relationships, types of membranes used in these modes, and the advantages and drawbacks of each type of operating mode are addressed. In this chapter food industry applications of membrane contactors will be discussed. MC also has wide uses in agricultural and medicinal sectors and requires the concentration of liquid foods [1] and dealdolization of beverages [2]. Osmotic membrane distillation (OMD) and membrane distillation (MD) have been extensively tested and implemented in food-related operations. In traditional membrane activities, instead of pressure gradient, temperature gradient or concentration is the separation stimulator. Maintenance of fluid/fluid interface at membrane pore juncture may only be accomplished by lowering slight pressure through the membrane. Moreover, membrane separation is not selective [3]. The guiding factor in MD and OMD is a disparity in vapor pressure around the membrane. Either a temperature gradient for MD or a variation in water behavior for OMD produces this strain. The emphasis will be on a brief analysis of the application of these two modes in liquid food concentration, particularly fruit juices. The standard MD and OMD concentration profile is shown in Figure 7.1.

7.2 Membrane Distillation (MD) Applications in Food Industry

The membrane contactor enables contact of gas–liquid or liquid–liquid phases by serving as a barricade between them and having a well-defined contact interface.

Membrane Contactor Technology: Water Treatment, Food Processing, Gas Separation, and Carbon Capture, First Edition. Edited by Mohammad Younas and Mashallah Rezakazemi.

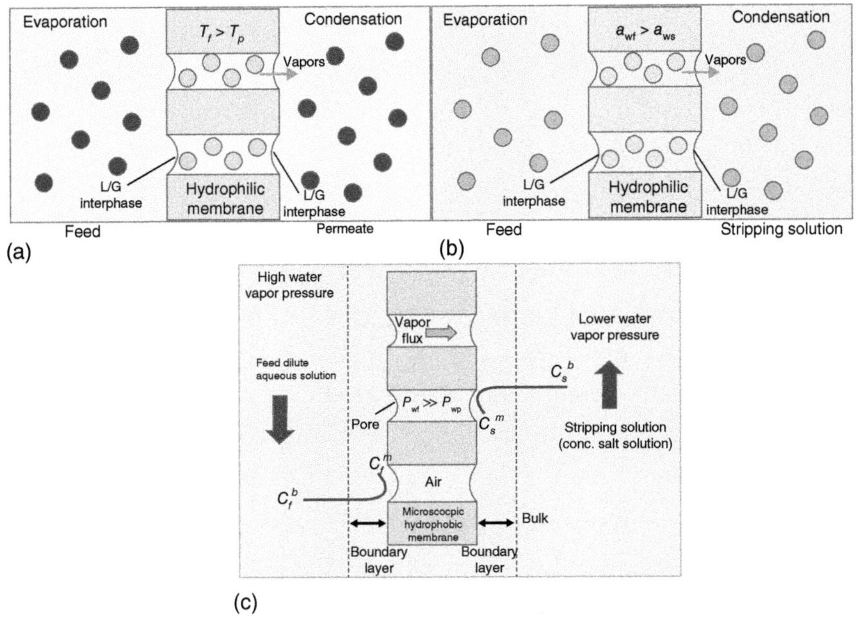

Figure 7.1 (a) MD, (b) OMD, and (c) concentration profile.

Since a membrane divides phases, no mixing or dispersion process happens, and mass transfer exists only by diffusion [4, 5]. The membrane may be symmetrical and microporous, water repelling, or water related [6, 7]. In almost all operations, a membrane contactor's key objective is to obtain optimum mass transfer, reduce fouling, operate with low pressure rises, and promote reliable output over the entire module duration.

The concept of MD was first patented by Bodell [8]. MD was introduced for first time as the membrane contactors (MCs) for water treatment and desalination [9]. The well-established membrane-based processes that use MD as their successor are microfiltration (MF), ultrafiltration (UF), nanofiltration (NF), electrodialysis (ED), reverse osmosis (RO), and pervaporation (PV). Low pressures or theoretical variations are the advantage of utilizing MD in operating processes, instead on concentration gradients or temperature for water transfer into pores of membrane. MD has other significant uses in combination with other membrane-based technologies to improve performance or actually introduce itself [3, 10]. It was found that gas absorption using MC has high efficiency relative to traditional columns with optimized conditions [11–13].

MD is a method of separating the electrical membrane where the vapor travels through the porous hydrophobic membrane and wraps around the full side [14, 15]. The guiding factor of MD is the pulse pressure gradient caused by the temperature difference between the feeder and the side of the full membrane [16, 17]. Step-by-step evaporation occurs as follows: (i) the transfer of heat from the solution of the mass to the surface of the membrane, (ii) only evaporation in hot surface membrane surface interface, (iii) the transfer of vapor to all porous membranes due to temperature variations, and (iv) cooling or cooling plate summary [18–20].

MD exhibited promise in many applications. Its key point is creating a highly filtered permeate stream. Contaminants are often easily isolated from liquid feeds. The primary applications of MD processes in the food industry include fruit juice concentration, taste recovery and fragrance compounds, milk processing, and elimination of volatile components from fermentation broth [21].

7.2.1 MD in the Concentration of Apple Juice

An et al. [22] investigated the combined FO-MD process for extracting apple juice. Potassium sorbate has been used as a novel solution. Polyamide TFC membrane was used to concentrate apple juice through the FO process. Figure 7.2 shows the detailed morphology and the structural parameter of the membrane used. The diluted solution was then concentrated in the MD process using poly(vinylidene fluoride-*co*-hexafluoropropylene) (PVDF-HFP) for nanofibrous membranes. The long-term efficacy of the FO-MD process was assessed by water flow. At a temperature of $25 \pm 1\,°C$, stable water availability of ~5 l/m^2 h was obtained during operation of 240 hours. The quality of the juice was well maintained, and a constant stimulation of the extract solution was obtained in the MD process. The integrated FO-MD process has shown a cost-saving advantage over conventional heating or cooling methods. Julian et al. [23] studied the process of direct immersion membrane

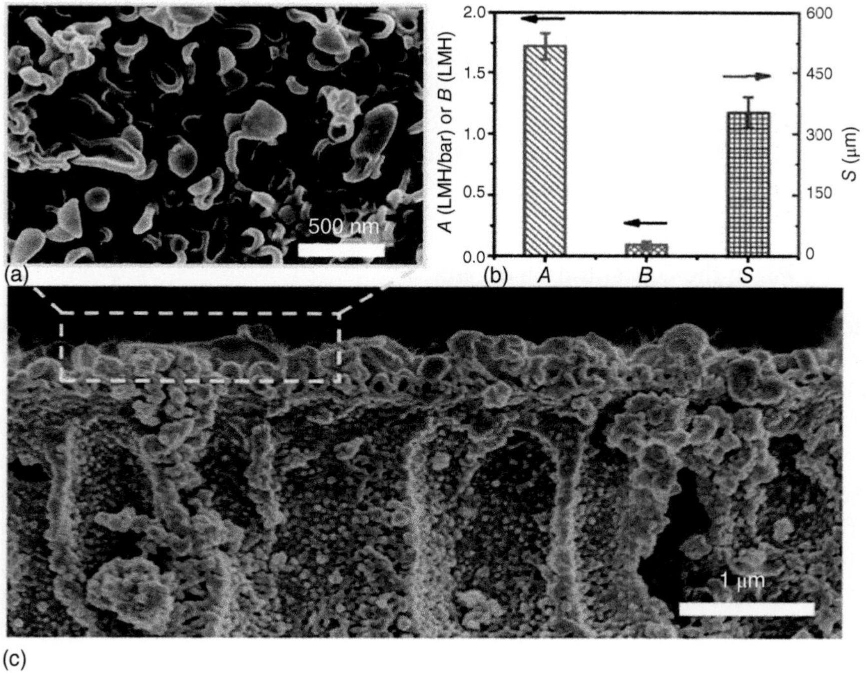

Figure 7.2 Surface (a) and cross-sectional (c) morphologies of TFC polyamide membrane and its intrinsic transport parameter including water permeability coefficient *A*, NaCl permeability coefficient *B*, and structural parameter *S* (b). Source: Reproduced with permission from Elsevier [22].

immersion (DCMD) to concentrate apple juice using potassium pyrophosphate ($K_4P_2O_7$) as a dye solution. Process parameters such as feed temperature (30–50 °C), concentration (545–630 g/l), and flow rate (2–4 ml/s) of the dilution solution varied to assess their effect on water flow. In addition, compounds of phenolic, vitamin C, acid, solid foods, and the color of the juice were analyzed to obtain an opinion on product quality. It was observed that the flow of water increased as the rate of feed and filtration and the flow rate of the suction solution increased. However, phenolic and vitamin C compounds were severely restricted as the temperature was kept very high. However, the chemical degradation of phenolic and vitamin C was still relatively small compared with highly concentrated apple juice. The ideal conditions for obtaining high-quality focus are feed temperature of 30 °C, concentration solution of 630 g/l, and flow rate solution of 3 ml/s. In these cases, the apple juice was squeezed twice.

7.2.2 MD in the Concentration of Orange Juice

Quist-Jensen et al. [24] evaluated the probability of DCMD in orange juice concentration pre-clarified by ultrafiltration. Two polypropylene (PP) hollow fiber membrane contactor (HFMC) modules were used in the DCMD setup, each with 0.2 μm pore size and 0.1 m^2 effective membrane surface. The juice concentration increased from an initial 9.5 °Bx to a final 24 °Bx in the ultrafiltration stage. DCMD stage further raised juice concentration to 65 °Bx. The flux decay was due to the increased exponential juice viscosity with rising juice concentration. Total suspended solids (TSS), total antioxidant activity (TAA), and phenolic compounds were well preserved after the DCMD method. Juice flavonoids and polyphenols are well maintained during concentration measures. During the concentration phase, crystals were created. This would contribute to their future application in the dairy, nutraceutical, or medicinal industries.

7.2.3 MD in the Concentration of Milk

Moejes et al. [25] assessed AGMD process for the concentration of milk. They evaluated the potential of different process configurations, including RO and MD. It was found that RO treatment resulted in a maximum possible milk concentration of 18% solids, which was further enhanced to 50% solids by AGMD process. AGMD process enjoys the advantage of internal heat recovery and is, therefore, a preferred choice over DCMD. However, the AGMD process required a good deal of recirculation of products to achieve a sufficient cross flow over the membrane surface, which ultimately led to a higher energy cost. Moreover, the deposition of fouling layer over the membrane surface also reduced the water flux and heat transfer. They found that the optimal AGMD configuration was a single-stage unit that operated at low water flux and high feed concentration. Furthermore, they suggested that the AGMD system could further improve by using maximum possible feed and permeate-side temperatures, using spacers to allow lower cross-flow velocities and lower circulation rates, and finally, using the available waste heat. Hausmann

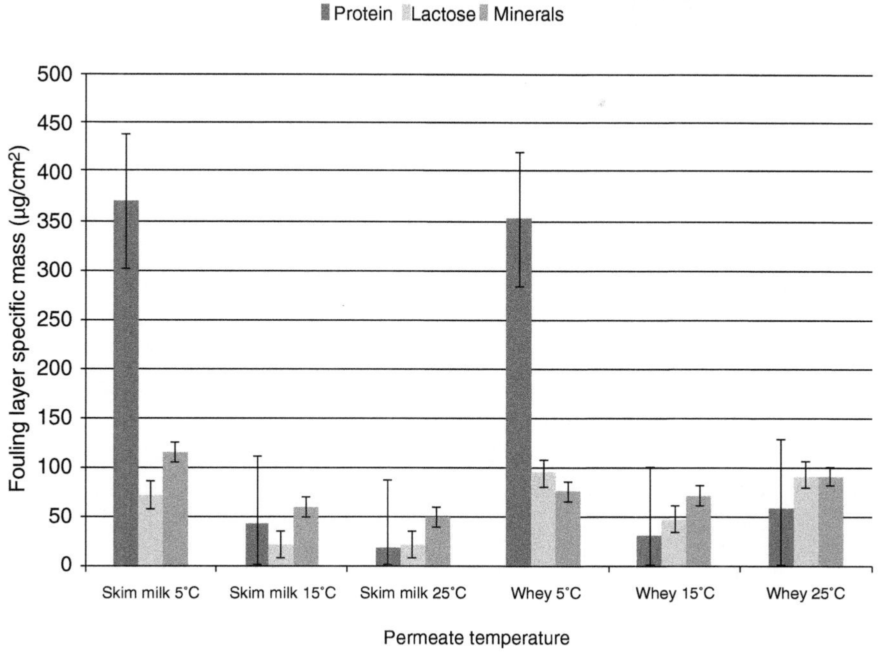

Figure 7.3 Effect of feed temperature on fouling layer deposition. Source: Hausmann et al. [26].

et al. [26] carried out the performance assessment of DCMD for skim milk and whey processing using a flat-sheet PTFE membrane. A lower feed temperature reduced the fouling propensity over the membrane surface in both cases as shown in Figure 7.3. Moreover, higher cross-flow velocities lead to higher water fluxes. During 18 hours of DCMD operation, a maximum water flux of 12 and 20 kg/m^2 h was achieved for skim milk and whey feed, respectively.

7.2.4 MD in the Treatment of Saline Dairy Waste Water

Direct contact with the MD may also apply to the dairy industry for the concentration of milk protein [27]. Temperature is an important distraction because proteins can be expressed warmly, given a temperature of 70 °C, which reflects at least some whey protein, reduces the nutrient content. Therefore, recommended operating temperature is 55 °C, reducing the cost reduction with a small driving cost. High flow velocity and low viscosity are important for increasing flow. Hausmann et al. [28] used a flat-sheet MC for the concentration of DCMD milk proteins, at 54 °C feed temperature and 5 °C higher temperatures. Fat-free silk milk and whey solutions have been used, with a protein content of 20 wt%, to keep fatty acids free and phospholipids in water. Within 18 hours of operation, a decrease in 80% flux was observed in all whey samples, which could be combined with membrane fouling. Lactose plays an important role in protein in fouling phenomena as it is found in polluted parts. Abdelkader et al. [29] used the DCMD process for the treatment of saline fluid-producing fluids. The raw feed was initially microfiltration using a 30 μm polyester filter cartridge and then a tubular polyether sulfone (PES) module for the ultrafiltration membrane. After previous treatment steps, a feed solution was obtained from the DCMD system containing PTFE membranes under different operating parameters. Premade feed solution leads to better quality in the DCMD process compared with non-treatable ones. This was because an untreated feed solution leads to wetting and placing a contaminant layer over the surface of the membrane. In a PTFE membrane with a pore size of 0.45 μm, the water flow remained stable at 17 kg/m^2 h but decreased slightly after 7 hours to 15 kg/m^2 h during 10 hours' operation. A full concentration of 60 μS/cm and a TOC of less than 2 mg/l were achieved. With a low-temperature drop between feeder and filling, membrane hydrophobicity is improved, although water flow is reduced, protectively cleanses the membrane, and reduces membrane hydrophobicity. Figure 7.4 shows that as the temperature difference ($\Delta T = 30$ and 15 °C) between feed and saturation solution increased, water flow was also improved.

7.2.5 MD in the Concentration of Muscadine Grape Pomace

Anari et al. [30] isolated and condensed anthocyanins from muscadine grape pomace by OMD, DCMD, and combination of OMD and DCMD. PP and poly(ethylene chlorotrifluoroethylene) (ECTFE) membranes were examined in each activity mode. OMD alone kept the feed and stripper tanks at 22 °C. Owing to the thermal vulnerability of anthocyanins, feed and permeate were held at 40

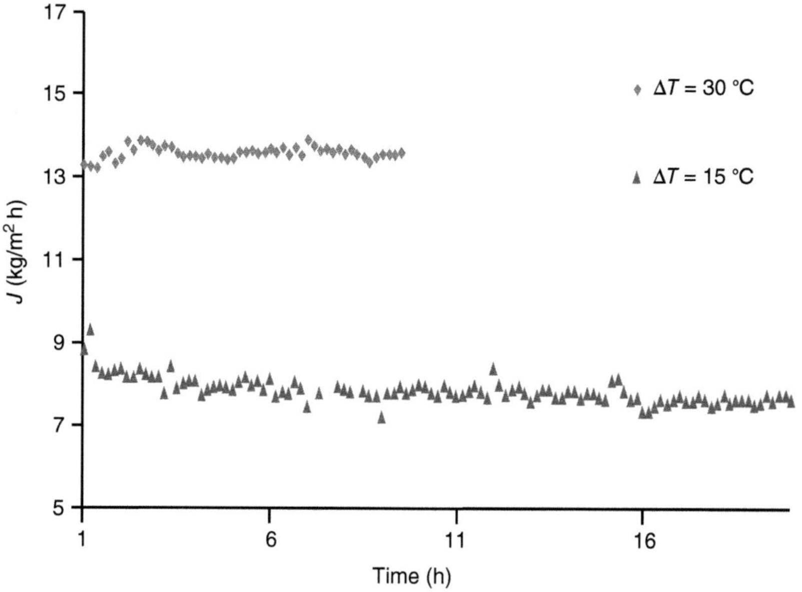

Figure 7.4 Water flux behavior during DCMD treatment of saline dairy wastewater at two different temperature differences ($\Delta T = 30$ and 15 °C). Source: Abdelkader et al. [29].

and 10 °C, respectively, resulting in a maximum temperature discrepancy of 30 °C. Likewise, feed and permeate were held at 40 and 10 °C, respectively, in combined OMD-DCMD service. The combined OMD-DCMD mode provided the maximum water flux of 16.7 l/m^2 h, utilizing the PP membrane, from three operating modes. Moreover, after OMD, DCMD, and combined OMD-DCMD activity, the maximum concentration factor of total anthocyanins was 1.07, 1.6, and 2.78. The OMD-DCMD mode was therefore the best choice for water flux and anthocyanin concentration.

7.2.6 MD in the Recovery of Phenols from Olive Mill Wastewater

Carnevale et al. [31] used DCMD and VMD methods to produce a filtered stream coupled with a phenol-rich wastewater concentrate. For this reason, the PP membrane was used with 0.2 μm pore size and 30 cm^2 effective membrane area. An eight-layer tissue prefiltered the feed. Rejections and water flux as high as 99.9% and 6.5 kg/m^2 h, respectively, were achieved at 50 °C at DCMD mode. With VMD mode, rejections and water flux of 99.6% and 19 kg/m^2 h, respectively, were achieved at 50 °C. Thus, the VMD process demonstrated superior efficiency over the DCMD process, and the expense and complexity of commercial plant was minimized.

7.2.7 MD in the Concentration of Sucrose Solution

Chen et al. [32] used the hydrophobic alumina HFMC module to focus on sucrose solution in the VMD system. The effect of feed flow rate, concentration, temperature, and vacuum pressure on permeate flux was analyzed in detail. The sucrose solution was concentrated from the initial 10 °Bx to the final concentration of 50 °Bx with a full flow of 19 kg/m^2 h when the feed temperature, flow rate, and vacuum pressure were 70 °C, 5 l/min, and −100 kPa, respectively. As sucrose concentrations increase, membrane contamination and temperature fluctuations are increased, which can be reduced by increasing feed flow rate. As shown in Figure 7.5, the viscosity increased significantly due to the increase in sucrose solution, which caused decay in the flow of water. Their results demonstrate the possible reliability of the VMD process for the filtering of sucrose solution. In a similar study by Hirota et al. [33], hydrophobic alumina membranes have been used to concentrate sucrose solution through the VMD process. Water filled with a flow of more than 7 kg/m^2 h obtained at a feed temperature of 60 °C and a resting pressure of 0.5 kPa. They have developed a mathematical model based on the Knudsen and Knudsen–viscous models. It has been found that the water content of alumina asymmetric membranes depends on the size of the substrate layer.

7.2.8 Effect of Operating Parameters on MD Flux

With feed temperature increases, MD evaporation fluxes boost. This is because a gradual increase in evaporation pressure from the feed liquid at room temperature

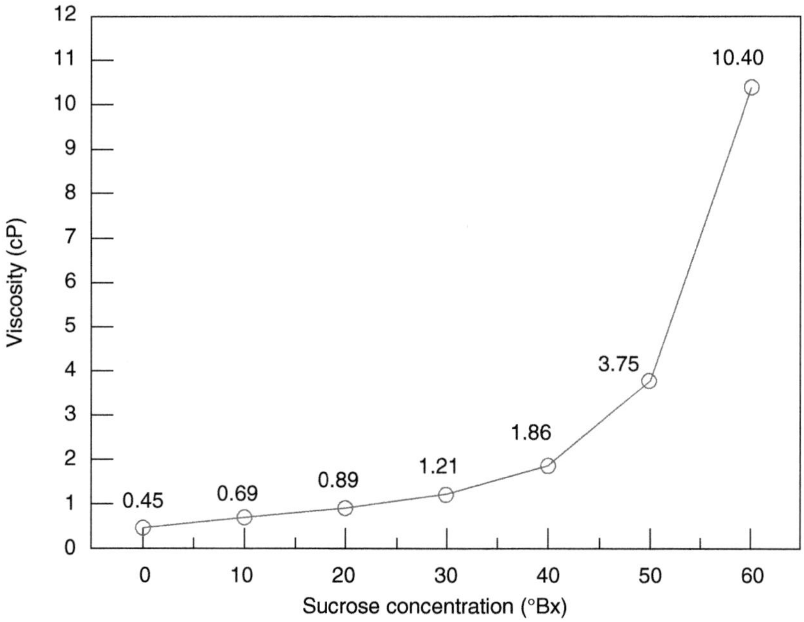

Figure 7.5 Exponential increase in sucrose solution viscosity with an increase in concentration. Source: Chen et al. [32].

enhances the vapor pressure difference in the membrane that is why the driving force works. In DCMD systems, the effect is enhanced evaporation flow by reducing the permeation temperature [9]. Growing the flow rate includes lowering Reynolds quantity and thereby rising transport coefficients. Permeate flow increases linearly with temperature differential, if the mean temperature is kept stable. On the other hand, the flow rate increases significantly while the normal temperature remains unchanged [34]. Considering non-flexible solutions, the flexibility of all MD settings is reduced by increasing feed intake: due to the reduced vapor pressure from the feed solution and increased urine viscosity of the feed solution, the effect can be linked to reduced driving capacity. MD fluxes are higher with greater concentration than other pressurized membrane processes [35]. Permeate fluxes typically increase when reactive elements in aqueous solutions, such as alcohols, are removed, expanding their concentration of solute. This may be understood since an increased concentration of the volatile compound on the feed side is linked to an improvement in its partial transmembrane strain. Improved permeate flow speed leads to improved permeate pressure in the MD process. Rising flow velocities in the membrane's permeated side increases heat transfer, reducing temperature and concentration phenomenon, while the membrane surface reaches the permeated bulk layer temperature, maximizing driving force and permeate surge. In the above cases of MD, a constant increase in MD flux and a variation of transmembrane vapor pressure are often observed.

7.3 Application of Osmotic Membrane Distillation (OMD) in Food Industry

OMD is a process of differentiating membranes called transmembrane distillation, thermopervaporation, isothermal fluid, osmotic membrane accumulation, and membrane and osmotic evaporation [36]. Unlike MD, OMD was operated at atmospheric temperatures and atmospheric pressure with minimal disruption of the thermal and mechanical solute. These substances allow OMD, a strong process, to concentrate temperature-sensitive liquids such as fruit juices and herbal ingredients. OMD has a low risk of overdose and temperature fluctuations [37, 38]. Boór et al. [39] proposed a cascade method of torture of wild juice of OMD juice. The modules contain a blank fiber membrane with a tube around 38 °C and a shell-side $CaCl_2$ solution brine at 18 °C for countercurrent touch action. Setting up two stages allows for at least 15% of the time required to achieve the target juice concentration, up to 21% with additional steps.

In the OMD phase, water vapors are evaporated due to the higher water activity of the feed fluid, which is then transferred through the pores and diluted on the stripper side due to the lower water activity of the solution [40]. A hydrophobic microporous membrane secures two liquid states with two separate solvent quantities with a condensed solution and a hypertonic solution at opposite ends. Hydrophobic membrane has the benefit of producing air openings in the membrane and avoiding water solution absorption in pores. The vapor pressure gradient is formed by the membrane due to the differential filtering of the visible moisture-liquid connector. This pressure gradient reduces evaporation movement through the pores from high pressure to low pressure [36, 38, 41, 42]. Steps involved in the major evaporation of water vapor are (i) evaporation of water vapor at the pore door on the feed side, (ii) the transport of rainwater through an air-filled pore, and (iii) access to water vapor at the escape of the pore on the side of the arrow [1, 43]. Under atmospheric conditions, the difference between the feeder and the temperature of a small liquid is very small, and the process may be herbal. In addition to the equilibrium heat dissipation of the solutions around the membrane, temperature differences are due to the reduction and evaporation of water vapor. In addition, heat transfer should be considered as it reduces the driving force of water transport [43, 44].

Reconcentration of diluted stripping solvent from the feed stream may be achieved by evaporation and reused in OMD. This should also have low expense, temperature tolerance, and nontoxicity and non-corrosive design. NaCl, $CaCl_2$, and ideally K_2HPO_4 are several more water-soluble salts that display strong solubility with elevated temperature so they can be condensed without crystallization. Potassium salts of phosphoric acids often have fewer weight, strictly positive temperature solubility, and protection in the food and medicine industries [45]. The value of normal transmembrane pressure across OMD is 140 kPa.

As ambient temperature and air pressure may be regulated OMD, it was with limited solutes thermal and mechanical damage. Concentrations of fluid foods, i.e. fruits, milk, vegetable, coffee, medicinal goods, and tea, were effectively implemented. The low temperature used inhibits chemical or enzymatic heat treatment responses. The low pressures result in decreased energy usage and capital expenditure, fewer fogging, and the likelihood of utilizing membranes that are less reactive than in pressurized membrane processes [46]. This results in better energy production and decreased expenditures in capital. OMD is especially ideal for the concentration of fruit juice, providing good quality and structure concentrated juices that are similar to fresh juices and are higher in thermal evaporation than traditional products [47, 48]. Base and frame systems have been built for concentrations of whole juice with a large pulp content with a net-shaped spacer on the side of the extract and smooth juice sides [49].

Table 7.1 summarizes the results of the research work of various researchers who used OMD process for the concentration of various fruit juices. It can be observed that several fruit juices have been tested for OMD concentration. This depicts the increasing interest in this particular process for the application in fruit juice industry. Mostly, PP membrane has been used due to its cheaper cost and ease of availability. However, satisfactory results have also been achieved with PVDF and PTFE membranes [70].

As it is clear from Table 7.1, the OMD process has been tested to concentrate several fruit juices. A few of these studies will be discussed briefly to highlight the importance and potential of this membrane separation technique in the food industry.

7.3.1 Effect of Operating Conditions on OMD Water Flux

The efficiency of water transport through evaporation of OMD is largely determined by operating conditions. Food allergies directly affect the effectiveness of the OMD process. Courel et al. [71] reported sudden decrease in the process from 10.3 kg/m^2 h. This occurs when sugar solutions with a high sucrose content of 0–65% w/w are heated at 25 °C using a 45.5 w solution/w first $CaCl_2$.

Babu et al. [61] detected transmembrane flux degradation by increasing the concentration of feed in the juice of the sweet lemon of OMD and phycocyanin solution. Solute materials are subject to an exponential enhanced solution. In comparison, the distribution coefficient falls significantly compared with light water. The increased viscosity causes an increase in the impact of the concentration separation that reduces the driving force and thus the flux ratio [72].

With the orange and grape juice concentrate across the PTFE membrane having a 0.2 mm pore size and 100 mm thickness, Sheng et al. [37] studied the effects of OMD flux conditions. The osmotic pressure gap between the liquid streams reduces the outflow of transmembrane vapor by 33%, five times the fall.

The quality of the solution often influences OD flexibility. Courel et al. reported a 64% reduction in vapor discharge with a 30% reduction in brine solution for separation of water by using $CaCl_2$ as an extractor at 25 °C [71]. Babu et al. found a consistent pattern in the concentration of lemon juice and phycocyanin solution [61]. These findings suggest a heavy reliance on stripping the behavior of the solution to salt content. While evaporation is required as the salt content decreases as the brine density and viscosity begin to decrease, reducing the resistance of the mass transfer to the salt solution. Indeed, the impact of performance prevents this change.

Table 7.1 Summary of the OMD concentration of fruit juices.

	Membrane specifications			Flux		Feed concentration		
Feed juice/brine	Material	Effective area (m^2)	Pore diameter (μm)	Initial (kg/m^2 h)	Final (kg/m^2 h)	Initial (°Bx)	Final (°Bx)	References
Apple/$CaCl_2$	HF PP	10.2	0.2	1.4	1.0	11.0	57.0	[50]
Apple/$CaCl_2$	FS PP FS PTFE FS PVDF	0.0012	0.10, 0.20 0.45 0.45	2.6 (PTFE)	2.2	10.34 (PTFE)	12.91	[51]
Apple/$CaCl_2$	HF PP	0.54	0.04	0.264	0.089	13.2	53.0	[52]
Grape/$CaCl_2$	HF PP	10.2	0.2	1.4	0.7	15.0	66.0	[50]
Grape/$CaCl_2 \cdot 2H_2O$	HF PVDF	0.013	0.2	2.02	0.58	40.0	55.0	[43]
Grape/$CaCl_2$		1.0		2.75	0.50	23.0	72.0	[53]
Red grape/$CaCl_2$	FS PTFE	0.001 25	0.2 0.45 1.2	7.0 8.0 7.0	4.2 4.2 4.3	5	7.9 7.85 7.9	[54]
Pomegranate/$CaCl_2 \cdot 2H_2O$	HF PP	0.1	0.2	0.82	0.53	17.1	55.5	[55]
Pomegranate/$CaCl_2 \cdot 2H_2O$	HF PP	1.4	0.2	0.85	0.38	16.0	52.0	[56]
Pomegranate/$CaCl_2 \cdot 2H_2O$	HF PP	0.58	0.04	1.053	0.186	13.0	52.0	[16]
Pomegranate/$CaCl_2 \cdot 2H_2O$	FS PVDF	0.009 63	0.20	1.246	0.208	10.3	18.0	[57]
Pomegranate/$CaCl_2 \cdot 2H_2O$	FS PTFE	0.009 63	0.20	2.077	0.935	10.3	41.0	[57]

Pomegranate/$CaCl_2 \cdot 2H_2O$	FS PVDF-HF PET	0.009 63	0.78	8.621	0.312	14.3	51.3	[58]
Pineapple/$CaCl_2$	FS PTFE	0.005	0.2	7.5	4.6	12.6	56.7	[59]
Blood orange/$CaCl_2 \cdot 2H_2O$	HF PP	1.4	0.03	1.25	0.41	10.5	61.4	[60]
Sweet lime/$CaCl_2 \cdot 2H_2O$	FS PP	0.012	0.05, 0.2	2.23	1.37	5.0	53.0	[61]
Orange/$CaCl_2 \cdot 2H_2O$	HF PP	0.16	0.2	1.55	0.36	13.5	62.0	[62]
Blood orange/$CaCl_2 \cdot 2H_2O$	HF PP	1.4	0.03	1.40	0.48	12.6	60.6	[63]
Carrot/$CaCl_2 \cdot 2H_2O$	HF PP	1.4	0.03	1.40	0.10	13.6	63.0	[63]
Black mulberry/$CaCl_2 \cdot 2H_2O$	HF PP	0.1	0.2	1.40	0.48	14.5	65.0	[64]
Cranberry/$CaCl_2 \cdot 2H_2O$	HF PP	0.58		Max flux: 1.21		18.0	48.0	[1]
Tomato/$CaCl_2$	HF PP	0.1	0.2	1.07	0.88	5.5	40.0	[65]
Passion/$CaCl_2$	HF PP	10.2	0.2	0.62	0.50	12.6	63.0	[66]
Melon/$CaCl_2$	HF PP	10	0.2	0.70	0.57	7.0	55.0	[67]
Noni/$CaCl_2$	HF PP	0.58	0.3	Max flux: 0.413		8.0	32.0	[68]
Kiwi/$CaCl_2 \cdot 2H_2O$	HF PP	1.4	—	1.30	0.40	11.2	61.2	[69]

OMD flux often influences the type of osmotic agent. $CaCl_2$ reduces the higher flow of transmembrane in the concentration of OMD than the brine solution [44]. A major contributing factor behind this is the high osmotic behavior of $CaCl_2$, which leads to a significant increase in vapor pressure through the membrane, but $CaCl_2$ causes rapid corrosion. As an alternative to $CaCl_2$, aqueous solutions of glycerol, propylene glycol, and glycerol–salt mixtures were also tested to deal with the rust and measuring problem associated with brine use [45]. Propylene glycol and 70–75 wt% glycerol demonstrated less effectiveness than $CaCl_2$ and showed the same level of excretion. Glycol cannot be recommended as a juice extractor due to its negative penetration pressure on all membrane pores and nonnegligable instability propylene. Water–glycerol–NaCl is a ternary compound separated by lower viscosities than glycerol alone, which provides similar flexibility.

The OMD transmembrane increases flux with the flow volume of the osmotic agent. This may be associated with high shear pressure around the condensation side of the membrane. This will reduce the thickness of the hydrodynamic layer and the impact of the filter separation. Ravindra Babu et al. [73] reported an 8% increase in flux as the pineapple flow at 12 °Bx increased to 100 ml/min. This increase in flux can be explained by a decreasing the effect of the split feed side. The flux increase was almost 20% with an increase in osmotic agent velocity. This effect can be caused by lower separation on the feed side and not on the brine side. Bui and Nguyen found a small effect of feed velocity in reducing the OMD polarization problem [74] and focus on strong glucose solutions. According to them, heat transfer to OD is less or less involved; on the other hand, the feed flow rate is important to feed the thermal current through the MD membrane.

The evaporation flow in the OMD is also influenced by feed temperature. Courel et al. [71] also demonstrate temperature dependency under isothermal conditions for pure water and a sugar solution of 35–65% w/w of sucrose content at 20–35 °C. The first solution was 45% w/w calcium chloride used as a dehydration solution. The sugar content at 20 °C ranges from 0.5 to 12.8 kg/m^2 h of stagnant water at 35 °C. Increasing the mass transfer depends on the content of the solution. At a region of 20–35 °C, the evaporation flow increased by 120% to obtain a saccharose solution of 35 w/w and by only 32% of the 60 w/w solvent substance. This effect can be explained by considering the exponential form of the relationship between the differences in vapor pressure through the membrane and the temperature under Clapeyron's rule. In addition, an increase in temperature leads to a decrease in feed viscosities and increase in the coefficient of solvent diffusion. The higher the solute content, the greater the heat effect. Bui and Nguyen [72] recorded a 200% improvement in the evaporation of 20 °C supply temperatures at 40% w/w and 50 w/w in strong sugary solutions using PVDF-free fibers.

7.3.2 OMD in the Concentration of Apple Juice

Onsekizoglu et al. [75] carried out the concentration of clarified apple juice through OMD, MD, and combined OMD-MD operations and then compared them with conventional thermal evaporation process. All three membrane processes were able to

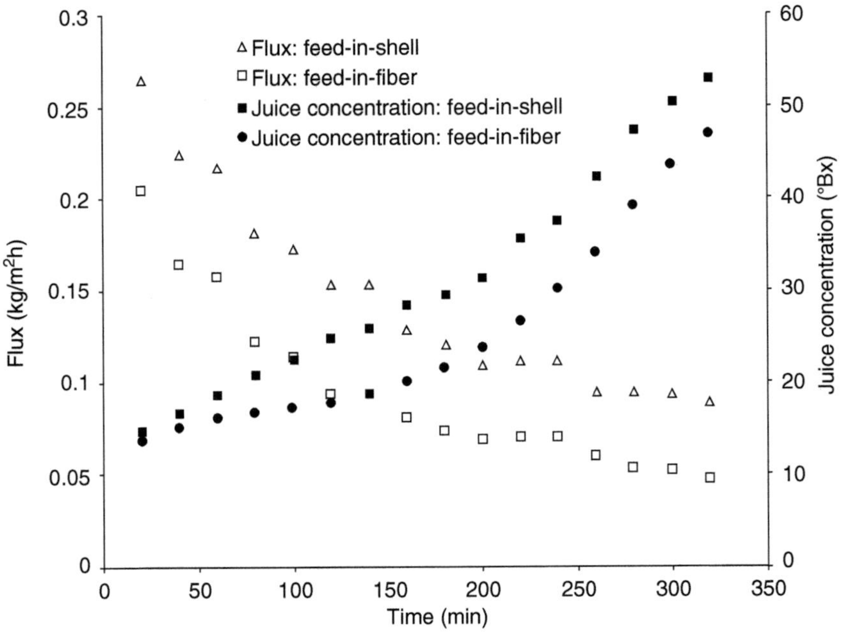

Figure 7.6 Water flux and juice concentration during OMD of apple juice. Source: Rehman et al. [52].

achieve the final juice concentration of 65 °Bx. The nutritional and sensorial quality parameters like color and aroma were well preserved in the membrane processes; otherwise there will be lost in the thermal process. Moreover, the 5-hydroxymethylfurfural (HMF), which is an indicator of Maillard reactions, was detected in the thermal process, while not found in the membrane processes. Other quality characteristics like in phenolic compounds, organic acids, and sugars were not altered in either thermal or membrane separation processes. The combined OMD-MD operation showed the best results among all the tested membrane techniques. Rehman et al. [52] concentrated apple juice through OMD process using PP HFMC. Figure 7.6 [52] indicates that water vapor flux in the feed-in-fiber flow configuration was 0.047 kg/m^2 h after 320 minutes of OMD service, while the feed-in-shell flow configuration was 0.089 kg/m^2 h. The normalized flux decline was 77% and 66% for feed-in-shell flow setups. On the other side, in feed-in-fiber flow configuration, final juice concentration was 47.0 °Bx, while in feed-in-shell flow configuration, it was 53.0 °Bx after 320 minutes of OMD operation. Therefore, feed-in-shell flow arrangement culminated in an approximately 90% improvement in water vapor flux than feed-in-fiber flow setup. In the later flow arrangement, the final juice concentration was 13% greater. Performance parameters such as total titratable acidity (TTA), ascorbic acid, and total phenols were well preserved after OMD.

7.3.3 OMD in the Concentration of Grape Juice

Kujawski et al. [54] applied the OMD process in the concentration of red grape juice using flat-sheet PTFE membrane. The juice was concentrated from an initial value of 5 °Bx to a final value of 20 °Bx during nine hours of operation. The quality parameters of the concentrated juice like total polyphenols and antioxidant activity were not affected by the OMD process. Cissé et al. [50] conducted grape juice OMD using a PP HFMC pilot plant with an effective surface area of 10.2 m^2 and an average pore diameter of 0.2 μm. From an initial value of 16 °Bx to a final value of 67 °Bx, the grape juice was condensed, with initial and final water permeate flux of 1.4 and 0.6 kg/m^2 h. Physicochemical, biochemical, and aromatic properties of the final grape juice concentrate were similar to the raw juice.

7.3.4 OMD in the Concentration of Pomegranate Juice

Onsekizoglu [55] concentrated pomegranate juice with OMD processes and mixed OMD and MD processes. The analysis used PP HFMC with an acceptable diameter of 0.10 m^2 and pore size of 0.20 μm. At 390 minutes, final concentrations of 0.53 kg/m^2 h and 55.5 °Bx juice concentration were obtained. Both membranes have been very effective in maintaining similar properties such as pH, color, TTA, total phenolic content (TPC), total monomeric anthocyanins (TMA), and TAA; phenolics are individual and are specific to organic acid. On the other hand, the concentration of pomegranate by thermal processing contributes to the production of HMF, which has not been observed in membrane concentration techniques. Bagci et al. [76] combined RO and OMD pomegranate filters. RO membrane modified with plasma is used for pre-specified juice purposes. Using a capillary membrane module, the final detection of 60 °Bx juice was obtained in 330 minutes with a final flow of 0.40 kg/m^2 h. Cassano et al. [56] indicate the occurrence of an membrane pathway in the clarity for pomegranate juice using PP HFMC with an appropriate region of 1.4 m^2 and a pore diameter of 0.20 μm. A final detection of 0.38 kg/m^2 h flux and 52 °Bx juice was obtained at 140 minutes' OMD service. Rehman et al. [16] used PP HFMC to concentrate pomegranate. Contamination status was studied in depth, and the presence of a contaminant layer on the surface of the membrane was found to have a significant impact on pore prevention during extended activity. Figure 7.7 shows the obvious contamination, exhibiting the SEM image of a single HFMC fiber. The process of cleaning the membrane can help reduce the tendency to get dirty. In the purified membrane, vapor flux values were higher than untreated, indicating that membrane cleaning has a significant effect on reducing the effects of contamination and maintaining water flow. In another study, Rehman et al. [57] also processed pomegranate juice extract with flat PVDF and PTFE membrane, suppressing membrane moisturizing. In the case of PVDF membrane, the water vapor decreased from 1.246 to 0.208 kg/m^2 h, a similar decrease in 83% equilibrium and a final filter of 18.5 °Bx juice after 24 hours of OMD. A high decrease in normal flux and only a slight increase in the final juice filter are caused by the liquid moisturizing during the OMD filter. In the case of the PTFE membrane, the water vapor decreased from 2.077 to 0.935 kg/m^2 h, a similar decrease of 55% similarity and a final filter of 41 °Bx juice after 24 hours of OMD. Unlike the PVDF membrane, the PTFE membrane was found to be more resistant to wetting. By conducting pore geometry studies, as shown in Figure 7.8, pore enlargement occurs in the PVDF membrane due to the membrane wetting. The pore correction also changed to a higher circle, mainly

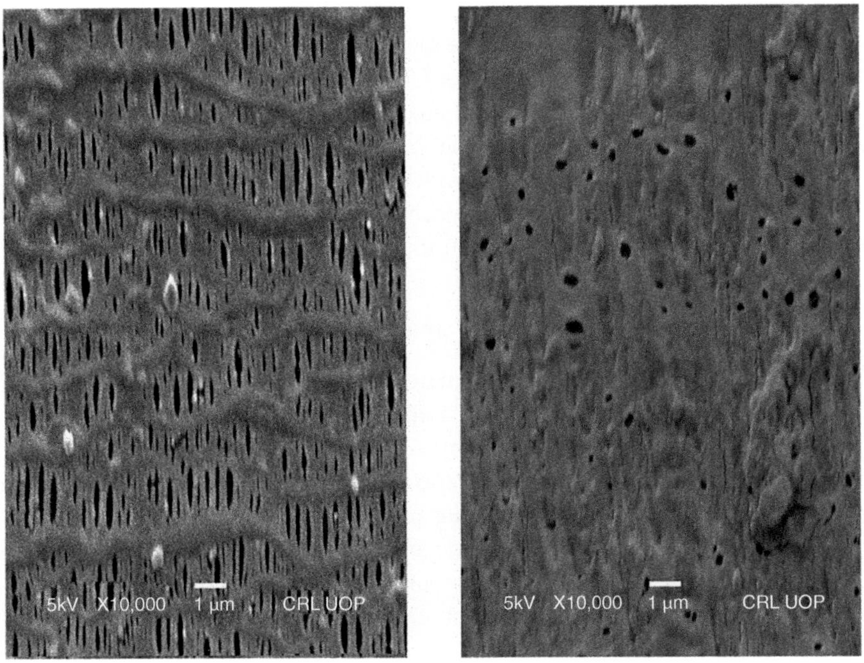

Figure 7.7 Surface of HFMC single fiber – fresh (a) and spent (b). Source: Reproduced with permission from Elsevier [16].

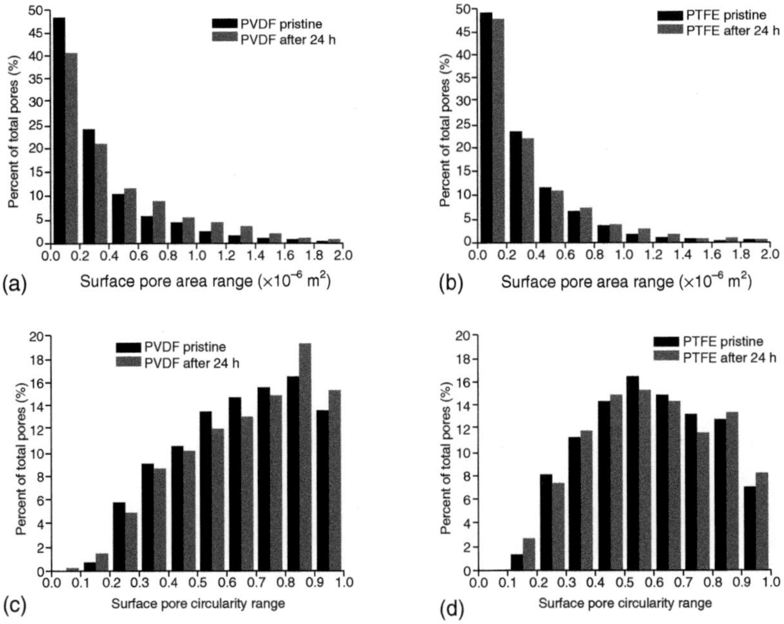

Figure 7.8 Pore geometry of the membrane surface. (a) PVDF (distribution of pore area). (b) PTFE (distribution of pore area). (c) PVDF (distribution of pore circularity). (d) PTFE (distribution of pore circularity). Source: Rehman et al. [57].

due to the pressure of the liquid entering the membrane pores that moved the pore geometry from the irregular form into a circle. In the case of the PTFE membrane, the surface area and pore rotation remain constant. These results showed that PTFE membranes showed higher yields than PVDF formation in terms of water evaporation, juice concentration, and resistance to membrane wetting. In another study, Rehman et al. [58] used PVDF-HFP and poly(ethylene terephthalate) membrane with double-layer electrospun hierarchical fibrous composite (HFC) membranes as effective and supportive layers for the extraction of pomegranate juice. The flow of water vapor decreased from 8.621 to 0.312 kg/m^2 h, an average decrease of 83% flux and a final filtering of 51.3 °Bx juice after 25.5 hours of OMD. Unlike the previous OMD studies on the simple PVDF membrane, hydrophobicity analysis reported that the dual-double HFC membrane showed greater tolerance to membrane water. After OMD treatment, the contact angle and fluid-induced pressure decreased to 120.1° ± 3° and 25.0 ± 1 kPa, respectively. Due to the rapid increase in juice viscosity and the temporary irregularities around the membrane, the decrease in the flow of water vapor is mainly due to the formation of gel caps. A favorable finding, however, is that such easily drained water washes away jelly and contaminated surfaces and retains a clear surface layer.

7.3.5 OMD in the Concentration of Orange Juice

Destani et al. [60] recovered and condensed by OMD in orange juice. Ultrafiltration pre-clarified the raw juice and is subjected to OMD, where juice concentration increased from 10.5 to 61.4 °Bx. Figure 7.9 demonstrates as time function water flux, juice concentration, stripper concentration, and juice viscosity. Phenolic compounds were found to be well preserved in OMD concentrate, as shown by a constant value of the ratio between OMD concentrate concentration of phenolic compounds and UF permeate stream (range of 5.54–6.39). Rehman et al. [52] attempted OMD orange juice concentration. The juice was distilled in complete recycled mode to a final value of 53.8 °Bx at 30 °C in 360 minutes. Water vapor flux ranged from 0.294 to 0.081 l/m^2 h during juice concentration. Juice consistency was observed after concentration of OMD. Complete acidity and ascorbic acid were determined to be 70.2 mg per 100 ml citric acid and 420 mg per 100 ml ascorbic acid. The pH value of orange juice concentrate was 4 ± 1.5, which is more ideal for handling.

7.3.6 OMD in the Concentration of Cranberry and Noni Juices

Zambra et al. [1] performed the OMD cranberry lemon extraction process using the PP HFMC module. The water flow is 0.25 and 1.21 l/m^2 h. The amount of 500 ml juice was concentrated from the initial value of 8.6 °Bx to the final value of 48 °Bx, with a total duration of 18 minutes. The total phenols in the concentrate of the juice are preserved. Valdés et al. [68] tested OMD in noni juice with PP HFMC. In isothermal conditions of 30 °C, the water vapor flow is 0.090–0.413 kg/m^2 h, while the concentration of 32 °Bx juice was reached in 60 minutes of OMD operation. As shown in Figure 7.10, the phenolic content of noni juice concentrate is well maintained after OMD treatment.

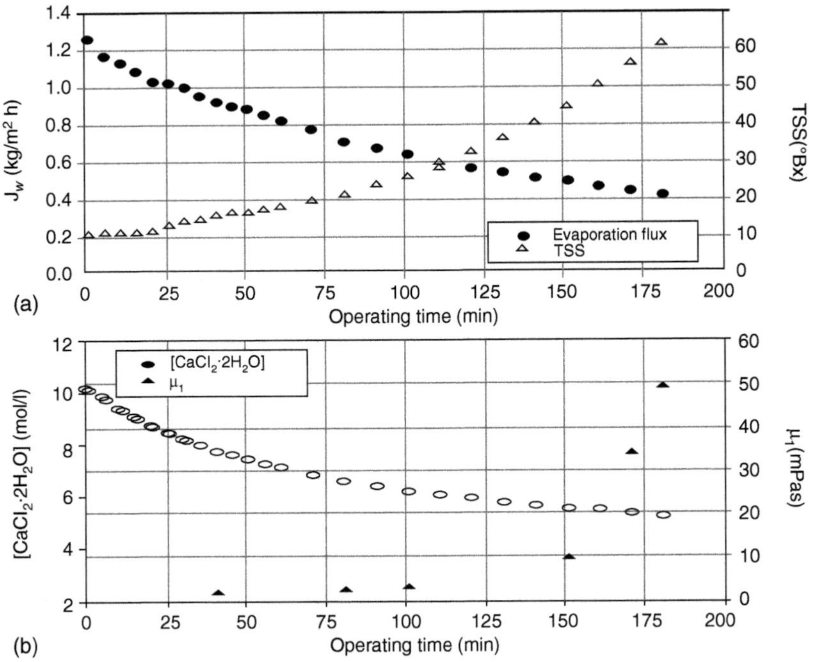

Figure 7.9 Time course of (a) evaporation flux and TSS concentration in the OMD of clarified red orange juice and (b) calcium chloride concentration and juice viscosity. Source: Destani et al. [60].

7.3.7 OMD in the Concentration of Kiwi and Pineapple Juices

Cassano and Drioli [47] worked on the OMD concentration of kiwi fruit juice with Liqui-Cel Extra-Flow 2.5–8 in. PP HFMC module. The specified juice is concentrated in the final TSS 66.6 °Bx, with an initial flux of 1.3 kg/m^2 h. OMD filtration of kiwi juice was compared with a hot concentration process. It has been found that the content of ascorbic acid and the TAA is maintained almost consistently in the OMD process while, in heat treatment, antioxidant activity is reduced by 50%. Hongvaleerat et al. [59] examined the concentration of pineapple juice in the OMD process. It has been found that maintaining a temperature difference of 15 °C between juice (35 °C) and brine (20 °C) on the side improved water vapor from 2 to 13 kg/m^2 h across the contactor membrane. No significant changes were observed in the quality characteristics of the pineapple juice.

7.3.8 OMD in the Concentration of Tea Extracts

Marques [77] focused on three different types of tea extracts through the OMD process. For this purpose, the PP HFMC module used has an operating area of 0.54 m^2. It was noted that, using a maximum area/feed volume of 774 m^2/m^3, tea density can reach 40% (w/w) in five hours, with constant evaporation of steam. In many teas, the phenolic content and antioxidant activity were not lost.

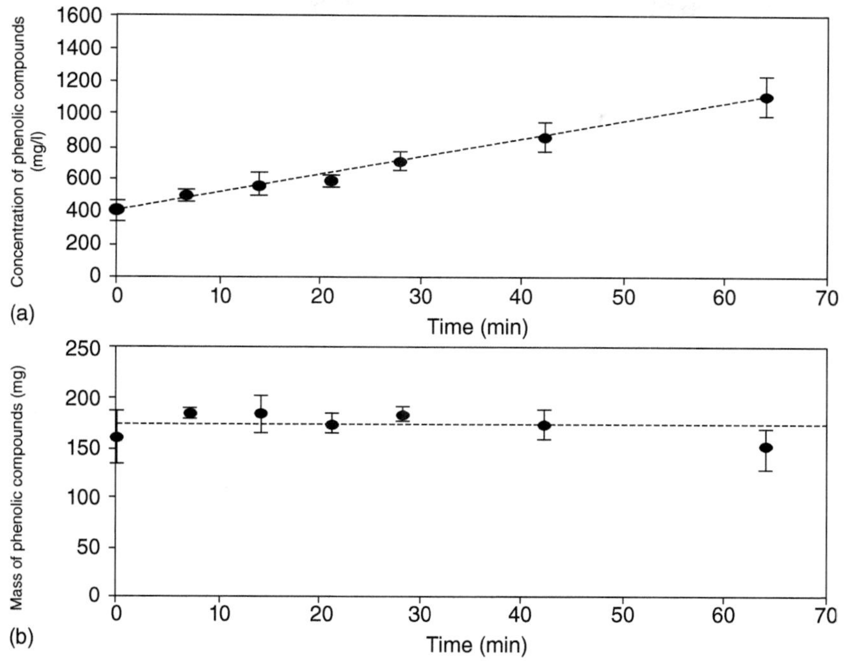

Figure 7.10 Phenolic contents in the noni juice. (a) Concentration of phenolic compounds as a function of the time during the OMD concentration of noni juice. (b) Overall mass of phenolic compounds as a function of the time during the OMD concentration of noni juice. Source: Valdés et al. [68].

7.3.9 Dealcoholization of Beer and Wine

OMD should be used in beer and wine dealcoholization, extracting ethanol instead of sugar. It was noted that using water as an osmotic agent when working inconsistently with the feed around the side of the tube can lead to the removal of alcohol by more than 50% regardless of the initial alcohol abuse. Therefore, alcohol can be completely eliminated using several cycles [2]. The main limitation of the process is simultaneous diffusion of water and ethanol vapor in osmotic distillation, and ethanol from the feed distributes the differential pressure of vapor pressure and water separation due to differences in water performance [78]. Fifty percent glycerol and 40% $CaCl_2$ aqueous solutions have been proposed as alternatives. Water was obtained with high ethanol flux dealcoholization containing 1.01 kg/m^2 h ethanol solution and 0.7 ms^{-1} stripping solution related to 0.98 and 0.97 kg/m^2 h $CaCl_2$ and glycerol, respectively. The effects of glycerol and $CaCl_2$ are used in the dispersion of water vapor from feed to dissolving solution due to their low water-related activity, with 0.70 and 0.36 kg/m^2 h, respectively, with a velocity of 0.7 ms^{-1} strip. With empty water, the water flow is less than 0.21 kg/m^2 h per dispersal from the digestive fluid compared with $CaCl_2$ and glycerol. Pure water is the best option as an effective alcohol solution.

Because the chemicals and physiological components of alcohol may contribute to the formation of final wine (e.g. flavonols, organic acids, computer content, and irregular formations) and their properties (e.g. vapor pressure, boiling point, water melting, hydrophobicity, color, and rainfall indicator), testing is required [79–81]. With a triangular test, the treated samples tasted a bit of their untreatable proportions. It has been shown that the process drastically changes the taste of wine. Motta et al. [82] compared the properties of blended wine using two therapeutic methods, namely, MC OMD and vacuum distillation. It was found that the concentration gradient of vacuum-treated wine was six times greater than that of MC. This is due to the high loss of water during the drinking process. OMD will help reduce water leaks. In contrast, vacuum-based wine was stronger on ethyl esters and isoamyl acetate than MC-treated wine.

OMD was also suggested for the sale of alcohol [83]. Alcohol reduction is done in four cycles. The process used an aqueous ethanol solution as a deterrent solution from previous experiments on pure water purification. The loop ends with a concentration gap of less than 0.5 vol between the beer and the dilute solution. The solvent was then reduced by 0.41%, 0.17%, and 0.08% more volume for the next three cycles in a lower ethanol concentration than the previous cycle. Experimental ethanol flux of approximately 12 g/m^2 s was collected for the first step and reduced to approximately 4 g/m^2 s in the fourth cycle, reducing alcohol concentration from 4.95 to 0.46 vol%. Various factors such as color, gravity, acidity, and polyphenol content were not affected during treatment, but significant losses of various chemicals were reported, including fusel alcohol (high alcohol content leading to beer flavor), 99% esters, and 93% aldehyde.

7.4 Coupled Operation of Osmotic Distillation and Membrane Distillation

Due to the relatively low water flow in OMD, the surface area used for the flow can be expanded (typically 0.07–7.2 kg/m^2 h). The thermal impact on OMD reduces the ability of water to move over the membrane, as mentioned above. These processes can be used to create an integrated approach in which a mixture of salt water and feed is controlled by temperature at different temperatures: an osmotic mixture on the cold-side and hot-side solutions. The combined performance of MD and OMD will facilitate the flow of water through the membrane [84]. Although the solution is slightly heated during service, it always works under reasonable conditions as the temperature difference is less than 15 °C [85]. With the influence of various mixed experiments, Godino et al. [86] examined the dynamic pressure, mean temperature, and mean temperature differences in the flow of water used in the mixed MD/OMD phase. In the study of Koroknai et al. [87, 88], distilled fruit juices (apple, red and black currante, sour cherry, and raspberry) were collected at 60 °Bx during operation for between 15 and 20 hours, maintaining a temperature difference of 15 °C and using $CaCl_2$6 M as an osmotic solution. There is also a reduction in energy consumption and a faster phase where the temperature of the liquid will be reduced to keep the osmotic fluid at room temperature. At the same time, the integrated approach is more effective than MD or MD alone. To reduce heat loss, it was suggested to use short membrane modules in cascade sequence for thermal exchanges between them.

7.5 Conclusions

During the past two decades, the academia and research community have shown a tremendous interest in the application of membrane contactors through MD and OMD techniques in food processing industry. The main reason for such keenness is the operation of these techniques at atmospheric pressures. The OMD process has a further advantage of carrying out the food concentration at ambient temperature. This added advantage helps in maintaining the genuine nutritional and aroma properties of the fruit juices, which are otherwise lost in conventional thermal treatment processes. This fact has been demonstrated in many research studies on the focus of various fruit juices. Moreover, the HFMCs offer a large surface area in a confined volume, which thus reduces the plant footprint and eliminates the need of large evaporators in traditional plants. Following rigorous membrane cleaning protocols increases the membrane service life and brings down the membrane replacement cost. Application of novel superhydrophobic membrane contactors can further enhance the efficiency of MD and OMD techniques.

7.6 Future Perspectives

- New membrane cleaning protocols should be developed that can help in further curtailing the fouling propensity of the membrane contactors.
- New cost-effective stripping solutions for OMD, having lower water activity, should be explored and tested for achieving higher water flux.
- Integrated strategies for the stripping solution recovery like MD and membrane crystallization should be evaluated.
- The existing commercial PP, PVDF, and PTFE membranes should be physically or chemically modified to achieve superhydrophobic membrane surfaces.
- Composite superhydrophobic membranes should be synthesized and tested for MD and OMD techniques.
- The surface-modified commercial membranes or newly synthesized superhydrophobic membranes should be tested, and their effect on fruit juice quality should be evaluated.
- Realistic mathematical models should be developed and simulated to facilitate the commercial scale OMD concentration of fruit juices.
- Computational fluid dynamics simulation of OMD concentration of fruit juices should be carried out to get a detailed insight into the hydrodynamics of the process.

References

1 Zambra, C., Romero, J., Pino, L. et al. (2015). Concentration of cranberry juice by osmotic distillation process. *Journal of Food Engineering* 144: 58–65.

2 Liguori, L., Russo, P., Albanese, D. et al. (2013). Evolution of quality parameters during red wine dealcoholization by osmotic distillation. *Food Chemistry* 140 (1–2): 68–75.

3 Gabelman, A. and Hwang, S.-T. (1999). Hollow fiber membrane contactors. *Journal of Membrane Science* 159 (1–2): 61–106.

4 González-Muñoz, M.J., Luque, S., Álvarez, J.R. et al. (2005). A predictive model for the extraction and simultaneous stripping of phenylglycine from alkaline solutions using membrane contactors. *Journal of Membrane Science* 255 (1–2): 133–140.

5 Sciubba, L., Di Gioia, D., Fava, F. et al. (2009). Membrane-based solvent extraction of vanillin in hollow fiber contactors. *Desalination* 241 (1–3): 357–364.

6 Crespo, J.G., Coelhoso, I.M., and Viegas, R. (2000). *Membrane Contactors: Membrane Separations*. Elsevier.

7 Drioli, E., Criscuoli, A., and Curcio, E. (2011). *Membrane Contactors: Fundamentals, Applications and Potentialities*, vol. 11. Elsevier.

8 Bodell, B.R. (1968). Distillation of saline water using silicone rubber membrane. US Patent 3,361,645, filed 9 August 1966 and issued 2 January 1968.

9 Lawson, K.W. and Lloyd, D.R. (1997). Membrane distillation. *Journal of Membrane Science* 124 (1): 1–25.

10 Kostyanaya, M., Bazhenov, S., Borisov, I. et al. (2019). Surface modified polysulfone hollow fiber membranes for ethane/ethylene separation using gas–liquid membrane contactors with ionic liquid-based absorbent. *Fibers* 7 (1): 4.

11 Demontigny, D., Tontiwachwuthikul, P., and Chakma, A. (2005). Comparing the absorption performance of packed columns and membrane contactors. *Industrial and Engineering Chemistry Research* 44 (15): 5726–5732.

12 Iliuta, I. and Iliuta, M.C. (2015). CO_2 removal in packed-bed columns and hollow-fiber membrane reactors. Investigation of reactor performance. *Industrial and Engineering Chemistry Research* 54 (49): 12455–12465.

13 Iliuta, I., Bougie, F., and Iliuta, M.C. (2015). CO_2 removal by single and mixed amines in a hollow-fiber membrane module – investigation of contactor performance. *AIChE Journal* 61 (3): 955–971.

14 Shokrollahi, M., Rezakazemi, M., and Younas, M. (2020). Producing water from saline streams using membrane distillation: modeling and optimization using CFD and design expert. *International Journal of Energy Research* 44 (11): 8841–8853.

15 Ahmad, S., Marson, G.V., Zeb, W. et al. (2020). Mass transfer modelling of hollow fiber membrane contactor for apple juice concentration using osmotic membrane distillation. *Separation and Purification Technology* 250: 117209.

16 Rehman, W.U., Muhammad, A., Khan, Q.A. et al. (2019). Pomegranate juice concentration using osmotic

distillation with membrane contactor. *Separation and Purification Technology* 224: 481–489.

17 Esfandiari, A., Monjezi, A.H., Rezakazemi, M. et al. (2019). Computational fluid dynamic modeling of water desalination using low-energy continuous direct contact membrane distillation process. *Applied Thermal Engineering* 163: 114391.

18 Alsebaeai, M.K. and Ahmad, A.L. (2020). Membrane distillation: progress in the improvement of dedicated membranes for enhanced hydrophobicity and desalination performance. *Journal of Industrial and Engineering Chemistry* 86: 13–34.

19 Yao, M., Tijing, L.D., Naidu, G. et al. (2020). A review of membrane wettability for the treatment of saline water deploying membrane distillation. *Desalination* 479: 114312.

20 Shirazi, M.M.A. and Kargari, A. (2015). A review on applications of membrane distillation (MD) process for wastewater treatment. *Journal of Membrane Science and Research* 1: 101–112.

21 Macedonio, F. and Drioli, E. (2019). Membrane distillation development. In: *Sustainable Water and Wastewater Processing* (eds. C.M. Galanakis and E. Agrafioti), 133–159. Elsevier.

22 An, X., Hu, Y., Wang, N. et al. (2019). Continuous juice concentration by integrating forward osmosis with membrane distillation using potassium sorbate preservative as a draw solute. *Journal of Membrane Science* 573: 192–199.

23 Julian, H., Yaohanny, F., Devina, A. et al. (2020). Apple juice concentration using submerged direct contact membrane distillation (SDCMD). *Journal of Food Engineering* 272: 109807.

24 Quist-Jensen, C.A., Macedonio, F., Conidi, et al. (2016). Direct contact membrane distillation for the
concentration of clarified orange juice. *Journal of Food Engineering* 187: 37–43.

25 Moejes, S.N., Van Wonderen, G.J., Bitter, J.H. et al. (2020). Assessment of air gap membrane distillation for milk concentration. *Journal of Membrane Science* 594: 117403.

26 Hausmann, A., Sanciolo, P., Vasiljevic, T. et al. (2014). Performance assessment of membrane distillation for skim milk and whey processing. *Journal of Dairy Science* 97 (1): 56–71.

27 Christensen, K., Andresen, R., Tandskov, I. et al. (2006). Using direct contact membrane distillation for whey protein concentration. *Desalination* 200 (1): 523–525.

28 Hausmann, A., Sanciolo, P., Vasiljevic, T. et al. (2013). Fouling of dairy components on hydrophobic polytetrafluoroethylene (PTFE) membranes for membrane distillation. *Journal of Membrane Science* 442: 149–159.

29 Abdelkader, S., Gross, F., Winter, D. et al. (2019). Application of direct contact membrane distillation for saline dairy effluent treatment: performance and fouling analysis. *Environmental Science and Pollution Research* 26 (19): 18979–18992.

30 Anari, Z., Mai, C., Sengupta, A. et al. (2019). Combined osmotic and membrane distillation for concentration of anthocyanin from muscadine pomace. *Journal of Food Science* 84 (8): 2199–2208.

31 Carnevale, M.C., Gnisci, E., Hilal, J. et al. (2016). Direct contact and vacuum membrane distillation application for the olive mill wastewater treatment. *Separation and Purification Technology* 169: 121–127.

32 Chen, Y.R., Chen, L.H., Chen, C.H. et al. (2018). Hydrophobic alumina hollow fiber membranes for sucrose concentration by vacuum membrane distillation. *Journal of Membrane Science* 555: 250–257.

33 Hirota, Y., Kayama, M., Kamiya, T. et al. (2016). Hydrophobic Al_2O_3 membrane for sucrose concentration via vacuum membrane distillation system. *Journal of Chemical Engineering of Japan* 49 (10): 915–919.

34 Mengual, J. and Peña, L. (1997). Membrane distillation. *Colloid Interface Science* 1: 17–29.

35 Cath, T.Y., Adams, V.D., and Childress, A.E. (2004). Experimental study of desalination using direct contact membrane distillation: a new approach to flux enhancement. *Journal of Membrane Science* 228 (1): 5–16.

36 Lefebvre, M.S. (1988). Method of performing osmotic distillation. US Patent 4,781,837, filed 21 November 1985 and issued 1 November 1988.

37 Sheng, J., Johnson, R., and Lefebvre, M. (1991). Mass and heat transfer mechanisms in the osmotic distillation process. *Desalination* 80 (2–3): 113–121.

38 Hogan, P.A., Canning, R.P., Peterson, P.A. et al. (1998). A new option: osmotic distillation. *Chemical Engineering Progress* 94 (7): 49–61.

39 Boór, A., Bélafi-Bakó, K., and Nemestóthy, N. (2016). Concentration of colourful wild berry fruit juices by membrane osmotic distillation via cascade model systems. *Journal of Membrane Science and Research* 2 (4): 201–206.

40 Alves, V.D., Koroknai, B., Bélafi-Bakó, K. et al. (2004). Using membrane contactors for fruit juice concentration. *Desalination* 162: 263–270.

41 Alves, V. and Coelhoso, I. (2002). Mass transfer in osmotic evaporation: effect of process parameters. *Journal of Membrane Science* 208 (1–2): 171–179.

42 Mengual, J.I., Pen, L., and Vela, A. (1993). Osmotic distillation through porous hydrophobic membranes. *Journal of Membrane Science* 82 (1–2): 129–140.

43 Thanedgunbaworn, R., Jiraratananon, R., and Nguyen, M. (2007). Mass and heat transfer analysis in fructose concentration by osmotic distillation process using hollow fibre module. *Journal of Food Engineering* 78 (1): 126–135.

44 Nagaraj, N., Patil, G., Babu, B.R. et al. (2006). Mass transfer in osmotic membrane distillation. *Journal of Membrane Science* 268 (1): 48–56.

45 Celere, M. and Gostoli, C. (2002). The heat and mass transfer phenomena in osmotic membrane distillation. *Desalination* 147 (1–3): 133–138.

46 Courel, M., Tronel-Peyroz, E., Rios, G.M. et al. (2001). The problem of membrane characterization for the process of osmotic distillation. *Desalination* 140 (1): 15–25.

47 Cassano, A. and Drioli, E. (2007). Concentration of clarified kiwifruit juice by osmotic distillation. *Journal of Food Engineering* 79 (4): 1397–1404.

48 Jiao, B., Cassano, A., and Drioli, E. (2004). Recent advances on membrane processes for the concentration of fruit juices: a review. *Journal of Food Engineering* 63 (3): 303–324.

49 Gostoli, C. (1998). Gas membrane extraction: a new technique for the production of high quality juices. *Fruit Processing* 8: 417–421.

50 Cissé, M., Vaillant, F., Bouquet, S. et al. (2011). Athermal concentration by osmotic evaporation of roselle extract, apple and grape juices and impact on quality. *Innovative Food Science & Emerging Technologies* 12 (3): 352–360.

51 Kujawa, J., Guillen-Burrieza, E., Arafat, H.A. et al. (2015). Raw juice concentration by osmotic membrane distillation process with hydrophobic polymeric membranes. *Food and Bioprocess Technology* 8 (10): 2146–2158.

52 Rehman, W.U., Zeb, W., Muhammad, A. et al. (2016). Osmotic distillation and quality evaluation of sucrose, apple and orange juices in hollow fiber membrane contactor. *Chemical Industry and Chemical Engineering Quarterly* 23: 35.

53 Bailey, A.F.G., Barbe, A.M., Hogan, P.A. et al. (2000). The effect of ultrafiltration on the subsequent concentration of grape juice by osmotic distillation. *Journal of Membrane Science* 164 (1–2): 195–204.

54 Kujawski, W., Sobolewska, A., Jarzynka, K. et al. (2013). Application of osmotic membrane distillation process in red grape juice concentration. *Journal of Food Engineering* 116 (4): 801–808.

55 Onsekizoglu, P. (2013). Production of high quality clarified pomegranate juice concentrate by membrane processes. *Journal of Membrane Science* 442: 264–271.

56 Cassano, A., Conidi, C., and Drioli, E. (2011). Clarification and concentration of pomegranate juice (*Punica granatum* L.) using membrane processes. *Journal of Food Engineering* 107 (3–4): 366–373.

57 Rehman, W.U., Muhammad, A., Younas, M. et al. (2019). Effect of membrane wetting on the performance of PVDF and PTFE membranes in the concentration of pomegranate juice through osmotic distillation. *Journal of Membrane Science* 584: 66–78.

58 Rehman, W.U., Khan, A., Mushtaq, N. et al. (2020). Electrospun hierarchical fibrous composite membrane for pomegranate juice concentration using osmotic membrane distillation. *Journal of Environmental Chemical Engineering* 8: 104475.

59 Hongvaleerat, C., Cabral, L.M., Dornier, M. et al. (2008). Concentration of pineapple juice by osmotic evaporation. *Journal of Food Engineering* 88 (4): 548–552.

60 Destani, F., Cassano, A., Fazio, A. et al. (2013). Recovery and concentration of phenolic compounds in blood orange juice by membrane operations. *Journal of Food Engineering* 117 (3): 263–271.

61 Ravindra Babu, B., Rastogi, N., and Raghavarao, K. (2006). Mass transfer in osmotic membrane distillation of phycocyanin colorant and sweet-lime juice. *Journal of Membrane Science* 272 (1–2): 58–69.

62 Alves, V. and Coelhoso, I. (2006). Orange juice concentration by osmotic evaporation and membrane distillation: a comparative study. *Journal of Food Engineering* 74 (1): 125–133.

63 Cassano, A., Drioli, E., Galaverna, G. et al. (2003). Clarification and concentration of citrus and carrot juices by integrated membrane processes. *Journal of Food Engineering* 57 (2): 153–163.

64 Dincer, C., Tontul, I., and Topuz, A. (2016). A comparative study of black mulberry juice concentrates by thermal evaporation and osmotic distillation as influenced by storage. *Innovative Food Science & Emerging Technologies* 38: 57–64.

65 Bahçeci, K.S., Akıllıoğlu, H.G., and Gökmen, V. (2015). Osmotic and membrane distillation for the concentration of tomato juice: effects on quality and safety characteristics. *Innovative Food Science & Emerging Technologies* 31: 131–138.

66 Vaillant, F., Jeanton, E., Dornier, M. et al. (2001). Concentration of passion fruit juice on an industrial pilot scale using osmotic evaporation. *Journal of Food Engineering* 47 (3): 195–202.

67 Vaillant, F., Cisse, M., Chaverri, M. et al. (2005). Clarification and concentration of melon juice using membrane processes. *Innovative Food Science & Emerging Technologies* 6 (2): 213–220.

68 Valdés, H., Romero, J., Saavedra, A. et al. (2009). Concentration of noni juice by means of osmotic distillation. *Journal of Membrane Science* 330 (1–2): 205–213.

69 Cassano, A., Marchio, M., and Drioli, E. (2007). Clarification of blood orange juice by ultrafiltration: analyses of operating parameters, membrane fouling and juice quality. *Desalination* 212 (1–3): 15–27.

70 Alves, V. and Coelhoso, I. (2007). Study of mass and heat transfer in the osmotic evaporation process using hollow fibre membrane contactors. *Journal of Membrane Science* 289 (1–2): 249–257.

71 Courel, M., Dornier, M., Herry, J.M. et al. (2000). Effect of operating conditions on water transport during the concentration of sucrose solutions by osmotic distillation. *Journal of Membrane Science* 170 (2): 281–289.

72 Bui, A., Nguyen, H., and Joachim, M. (2005). Characterisation of the polarisations in osmotic distillation of glucose solutions in hollow fibre module. *Journal of Food Engineering* 68 (3): 391–402.

73 Ravindra Babu, B., Rastogi, N., and Raghavarao, K. (2008). Concentration and temperature polarization effects during osmotic membrane distillation. *Journal of Membrane Science* 322 (1): 146–153.

74 Bui, V. and Nguyen, M.H. (2006). The role of operating conditions in osmotic distillation and direct contact membrane distillation: a comparative study. *International Journal of Food Engineering* 2 (5): 1–12.

75 Onsekizoglu, P., Bahceci, K.S., and Acar, M.J. (2010). Clarification and the concentration of apple juice using membrane processes: a comparative quality assessment. *Journal of Membrane Science* 352 (1–2): 160–165.

76 Bagci, P.O., Akbas, M., Gulec, H.A. et al. (2019). Coupling reverse osmosis and osmotic distillation for clarified pomegranate juice concentration: use of plasma modified reverse osmosis membranes for improved performance. *Innovative Food Science & Emerging Technologies* 52: 213–220.

77 Marques, C.M.P. (2014). *Concentration of Beverages by Osmotic Evaporation: Effect on Organoleptic Properties and Antioxidant Activity*. Universidade Nova de Lisboa.

78 Varavuth, S., Jiraratananon, R., and Atchariyawut, S. (2009). Experimental study on dealcoholization of wine by osmotic distillation process. *Separation and Purification Technology* 66 (2): 313–321.

79 Diban, N., Athes, V., Bes, M. et al. (2008). Ethanol and aroma compounds transfer study for partial dealcoholization of wine using membrane contactor. *Journal of Membrane Science* 311 (1–2): 136–146.

80 Gambuti, A., Rinaldi, A., Lisanti, M.T. et al. (2011). Partial dealcoholisation of red wines by membrane contactor technique: influence on colour, phenolic compounds and saliva precipitation index. *European Food Research and Technology* 233 (4): 647.

81 Lisanti, M.T., Gambuti, A., Genovese, A. et al. (2013). Partial dealcoholization of red wines by membrane contactor technique: effect on sensory characteristics and volatile composition. *Food and Bioprocess Technology* 6 (9): 2289–2305.

82 Motta, S., Guaita, M., Petrozziello, M. et al. (2017). Comparison of the physicochemical and volatile composition of wine fractions obtained by two different dealcoholization techniques. *Food Chemistry* 221: 1–10.

83 Liguori, L., De Francesco, G., Russo, P. et al. (2015). Production and characterization of alcohol-free beer by membrane process. *Food and Bioproducts Processing* 94: 158–168.

84 Wang, Z., Zheng, F., Wu, Y. et al. (2001). Membrane osmotic distillation and its mathematical simulation. *Desalination* 139 (1–3): 423–428.

85 Bélafi-Bakó, K. and Koroknai, B. (2006). Enhanced water flux in fruit juice concentration: coupled operation of osmotic evaporation and membrane distillation. *Journal of Membrane Science* 269 (1–2): 187–193.

86 Godino, M.P., Peña, L., Ortiz de Zárate, J.M. et al. (1995). Coupled phenomena membrane distillation and osmotic distillation through a porous hydrophobic membrane. *Separation Science and Technology* 30 (6): 993–1011.

87 Koroknai, B., Gubicza, L., and Bélafi-Bakó, K. (2006). Coupled membrane process applied to fruit juice concentration. *Chemical Papers* 60 (6): 399–403.

88 Koroknai, B., Kiss, K., Gubicza, L. et al. (2006). Coupled operation of membrane distillation and osmotic evaporation in fruit juice concentration. *Desalination (Amsterdam)* 200 (1–3): 526–527.

8

Applications of Membrane Contactor Technology for Pre-combustion Carbon Dioxide (CO_2) Capture

Zarrar Salahuddin[1], *Sarah Farrukh*[1], *Mohammad Younas*[2], *and Mashallah Rezakazemi*[3]

[1] *National University of Sciences and Technology, School of Chemical and Materials Engineering, Islamabad, Pakistan*
[2] *University of Engineering and Technology, Department of Chemical engineering, Peshawar, Pakistan*
[3] *Shahrood University of Technology, Faculty of Chemical and Materials Engineering, Shahrood, Iran*

8.1 Introduction

Energy is the essential requirement of the economic development of modern society. As the world economies continue to develop rapidly, the energy demand is continuously increasing. According to a report published by the International Energy Agency, the global energy demand touched a staggering figure of about 550 exajoules in 2011 [1]. This figure is further projected to increase by up to 36% by 2035 [2]. This increase in energy demand, in turn, means a total annual increase in the consumption of natural gas of about 2% to reach a value of about 5000 billion cubic meters by 2035 [3]. Until now, much of the energy requirements are still being fulfilled by using fossil fuels despite the work being done on the development of alternate energy resources, and it is projected to stay this way in the foreseeable future. However, the overall use of fossil fuels, which includes a substantial share of natural gas, has led to CO_2 emissions increasing dramatically over time. The global CO_2 emissions were registered at a value of 33.4 billion tons in 2011. Suppose the CO_2 emissions were left to increase without applying any carbon capture strategies; then according to the Intergovernmental Panel on Climate Change (IPCC), the atmosphere's CO_2 concentration will be in the range of 600–1550 ppm by 2030. Moreover, if CO_2 concentration increment is left unattended, the mean global surface temperature has been predicted to increase by 2.5–7.8 °C by 2100 [4]. Many strategies and policies have been put in place by IPCC to reduce CO_2 emissions into the atmosphere. One is the use of low-carbon energy-intensive resources, such as hydrogen fuel [5] and biomass [6]. Another way to reduce CO_2 emissions to the atmosphere is to use carbon capture and storage (CCS) techniques [7].

CO_2 capture and storage are a set of technologies that capture, transport, and store CO_2 in geological formations deep underground instead of being released into the atmosphere [8]. The main technologies used to capture CO_2 from different

Membrane Contactor Technology: Water Treatment, Food Processing, Gas Separation, and Carbon Capture, First Edition. Edited by Mohammad Younas and Mashallah Rezakazemi.

gas streams include pre-combustion carbon capture, post-combustion carbon capture, and oxy-fuel combustion carbon capture technologies [9, 10]. Of these technologies, oxy-fuel combustion uses pure oxygen instead of air to produce fossil fuel energy. This may reduce the gas separation difficulty as the flue gas stream only contains CO_2 and water, where water vapor can be easily separated by condensation. However, a large amount of energy is required to separate oxygen from nitrogen in the air to form a pure oxygen stream, which is used for the oxy-fuel combustion oxidant stream [11]. Furthermore, post-combustion carbon capture is used as an "end-of-pipe" approach, such that CO_2 is captured once it has undergone the process of combustion and all the energy has been acquired from the heat thus produced from the combustion. However, post-combustion carbon capture has a significant drawback. The gas stream from which CO_2 has to be separated contains very low concentrations or partial pressures of the hazardous gas. This makes the overall process uneconomical and very highly energy intensive [12].

In other attempts to reduce CO_2 emissions to the atmosphere, hydrogen fuel is being increasingly used because of its myriad of advantages. An ideal fuel is supposed to have certain properties, including inexhaustibility, cleanliness, convenience, and independence from foreign control. Hydrogen fuel has all these properties, so it has been promoted and marketed as an environmentally friendly alternative to fossil fuels [13]. Hydrogen fuel has been traditionally produced using steam reforming of methane or biomass gasification. Both these methods produce considerable amounts of CO_2 that, if not removed from the fuel produced, can pose serious environmental problems. However, their high energy and exergy efficiency makes them highly economically viable for hydrogen gas production as fuel [14]. However, the problem still remains of the high amount of CO_2 that normally accompanies hydrogen fuel production from these sources. The hydrogen gas produced by gasification or steam reforming of methane is most often used in fuel cells, within automobiles or spacecrafts. However, the hydrogen produced from natural gas or biomass cannot be directly used in fuel cells, mainly because the CO_2 formed during the processing of these two feedstocks causes poisoning of the electrocatalyst in fuel cell [15]. Therefore, the CO_2 must be removed from hydrogen in the fuel stream before it can be used for any energy-related applications. This CO_2 separation from hydrogen is achieved by a carbon capture technology called pre-combustion carbon capture.

Pre-combustion carbon capture is a technology used to separate CO_2 from fossil fuel or biomass-derived fuel stream before combustion process. Therefore, the separation of CO_2 from any gas stream, in which the CO_2 is produced by any method other than combustion, can be termed as pre-combustion carbon capture [16]. As pre-combustion carbon capture is used mainly in gasification processes, it is imperative to understand the overall gasification process and the utility of pre-combustion carbon capture. In a typical gasification process, biomass or coal is first gasified to form synthesis gas (or syngas). This syngas is then sent to cyclone separator where unwanted macroscopic impurities, such as dust, are removed. After that, carbon monoxide (CO) within the syngas mixture is converted into CO_2, through successive water–gas shift (WGS) reactions with steam. The product gas stream is then forwarded to desulfurization and CO_2 removal to produce highly pure hydrogen fuel, to be used for various power generation applications.

In a pre-combustion carbon capture process, the CO_2 to be removed comes from two different reaction pathways, i.e. the steam methane reforming pathway and the partial oxidation pathway. The general chemical equations, (Eqs. (8.1) and (8.2)), of both the reactions are summarized as follows [17]:

$$\text{Steam reforming}: \quad C_xH_y + xH_2O \leftrightarrow xCO + \left(x + \frac{y}{2}\right) H_2 \tag{8.1}$$

$$\text{Partial oxidation}: \quad C_xH_y + x/2O_2 \leftrightarrow xCO + y/2H_2 \tag{8.2}$$

Furthermore, the gas stream produced from both these reactions is then subjected to WGS reaction, as stated earlier. The chemical equation for the WGS reaction is shown in Eq. (8.3) as follows:

$$\text{Water – gas shift reaction}: \quad CO + H_2 \leftrightarrow CO_2 + H_2 \tag{8.3}$$

The pre-combustion carbon capture process is actually utilized at the end of hydrogen production through gasification, where sulfur and CO_2 are removed from the product gas stream [18, 19]. The schematic diagram of a general pre-combustion carbon capture system is shown in Figure 8.1 [20], and the schematic diagram of WGS reaction is shown in Figure 8.2 [21].

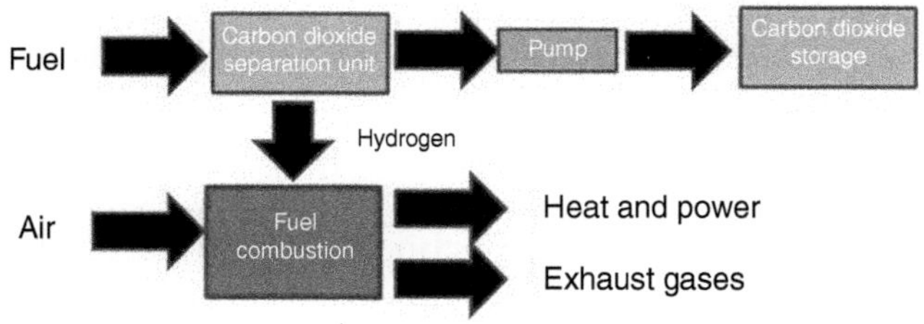

Figure 8.1 Schematic diagram of a pre-combustion carbon capture system. Source: Adapted from Theo et al. [20].

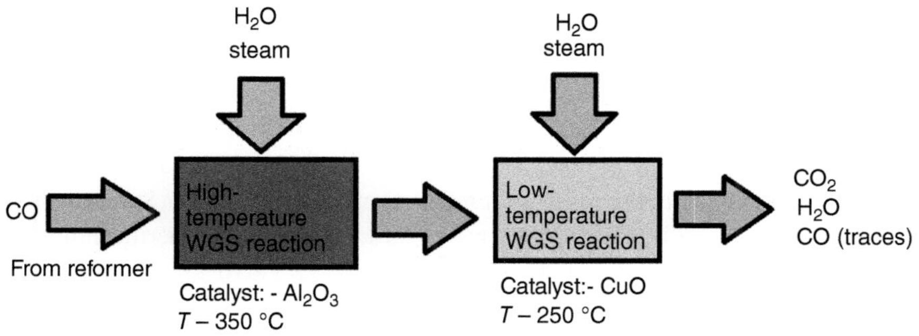

Figure 8.2 Schematic diagram of water–gas shift (WGS) reaction. Source: Adapted from Alfano and Cavazza [21].

8.2 Why Pre-combustion Carbon Capture?

This question would definitely arise in one's mind regarding the need for CCS technologies. Why would we need carbon capture technologies when we could just apply renewable energy systems and move away from fossil fuels altogether? The answer lies in the low energy efficiency of renewable energy resources, which makes them uneconomical. As an example, a normal combined cycle gas turbine (CCGT) power station produces 500 MW of electricity, whereas an offshore wind turbine produces only about 7 MW and onshore wind turbines have a power generation capacity of only 3.6 MW. This means that to match the performance and generation capacity of one CCGT power station, we would need roughly 150 onshore wind turbines [22]. Therefore, this example shows that fossil fuel-based power generation still dominates the energy sector, as can be seen from the fact that oil, natural gas, and coal account for about 80% of the world energy mix as of 2017 according to the International Energy Agency. In 2017, the amount of energy produced from natural gas was up to 3 106 799 tons of oil equivalent (ktoe), which was about roughly 21% of the total world energy production in that year [23]. This shows that as such a large amount of energy is still being produced from nonrenewable resources, it is more feasible to focus on carbon capture technologies. Moreover, the question arises, why pre-combustion carbon capture is needed? The need for pre-combustion carbon capture, especially and exclusively in gasification systems, is because the CO_2 produced from the overall gasification system would otherwise pose a threat to the environment [24]. The same is the case in a fuel cell; unless the CO_2 is removed from the hydrogen before it is used as fuel in the fuel cell, the technology will not be considered environmentally friendly [25].

8.3 Membranes for Pre-combustion Carbon Capture

Membranes used in pre-combustion carbon capture can be divided into broad categories, based on the gas that permeates more through them. These two classes of membranes are the H_2-selective membranes and the CO_2-selective membranes.

8.3.1 Hydrogen (H_2)-Selective Membranes

The most commonly used H_2-selective membranes are metallic membranes. The most often used metal for synthesizing these metallic membranes is palladium [26]; however, there have been reports of other metals being used for this purpose. For example, in one particular research work carried out by Steward [27], platinum-based metallic membranes were used to check the permeability of hydrogen. The results showed H_2 permeability of up to 8.1×10^{-7} mol/s-m-Pa$^{1/2}$ at a pressure of up to 100 kPa and temperature in the range of 705–827 K. In this same review, many other metals were also shown to be used for hydrogen-permeable membranes, for example, copper, molybdenum, niobium, nickel, and others as well [27]. Alloys have also been used for making membranes for hydrogen permeability testing. Some examples include using a nickel–palladium alloy, as detailed in work carried out by Zhang et al. [28]. Their research showed that the resulting membranes had an H_2 permeability of 2×10^{-12} mol/s-m-Pa$^{1/2}$ at a temperature of 500 K and a pressure of 20 kPa. In another research work carried out by Cheng et al. [29], iron–aluminum (Fe_3Al) alloy was used as the hydrogen-permeable membrane. It gave a hydrogen permeability of 1×10^{-10} mol/s-m-Pa$^{1/2}$. The working temperature was kept at 300 K and a working pressure at 101 kPa. Other research works using alloys as the main matrix material for H_2-selective membranes include using a niobium–titanium–nickel alloy ($Nb_{21}Ti_{50}Ni_{29}$) as used by Hashi et al. [30]. Their fabricated membrane samples gave a maximum H_2 permeability of 1×10^{-8} mol/s-m-Pa$^{1/2}$. In yet another research work carried out by Tong et al. [31], palladium–silver (Pd–Ag) alloy was used to form hydrogen-permeable membranes, and it gave hydrogen permeability of 5×10^{-5} mol/s-m-Pa$^{1/2}$ at a temperature of 723 K and pressure of 20–75 kPa. Whichever metal is used to fabricate membrane for H_2-selective separation from H_2/CO_2, the mechanism for hydrogen permeation through metals is one of dissociative chemisorption on the metal surface [32]. Under this mechanism, hydrogen molecules are first to split into atomic hydrogen on the surface. They diffuse through the membrane because of the partial pressure drop between the two ends of the membrane and finally emerge and recombine into hydrogen molecules on the membrane's permeate side. Equation (8.4) describes hydrogen permeability through a metallic membrane and is stated as follows [33]:

$$J_{H_2} = \frac{P_{H_2}}{t}\left(\sqrt{f_{\text{feed}}} - \sqrt{f_{\text{permeate}}}\right) \tag{8.4}$$

where f is the fugacity of the gas streams, J_{H_2} is the flux of hydrogen gas through the membrane, and P_{H_2} is the permeability of hydrogen through the membrane samples. Despite the fact that metallic membranes are completely impermeable to CO_2 and give infinite ideal selectivity, some problems need to be addressed as far as their utilization in commercial applications is concerned. As palladium is still the most widely used metal for metallic membranes for hydrogen separation, so we shall look into the different limitations that have thus far impeded its use on the commercial level. One major limitation is the fact that Pd undergoes degradation at lower temperatures, and this is attributed to hydrogen embrittlement occurring at lower temperatures (temperatures less than 300 °C) [34]. Another disadvantage of Pd membranes is that they are susceptible to inhibition of their hydrogen dissociation and recombination reactions due to impurities, mainly sulfur, carbon monoxide, chlorine, and ammonia, all of which are present in the synthesis gas produced from the gasification process [33, 35]. Even though much of the problems and limitations related to the use of metallic membranes, especially Pd membranes, have been addressed, the biggest issue still remains, which is the short lifespan of Pd membranes, of the order of a few months. This means that there is frequent replacement required [36]. This ultimately adds to the operating cost, as Pd is a precious metal.

Seeing the problems associated with metallic membranes, another class of membrane material has been investigated as the matrix for H_2-selective membranes. This class of membranes is inorganic membranes. Over the past years, many different

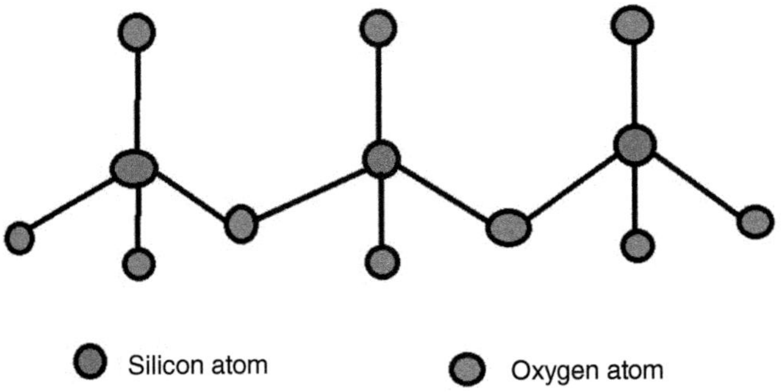

Figure 8.3 Structure of silica. Source: Adapted from Hamza et al. [38].

inorganic materials have been used for membrane fabrication in H_2 permeation and separation applications. In one such work carried out by Giessler et al. [37], their research used hydrophilic silica membranes, which gave an H_2/CO_2 selectivity of 11 with H_2 permeance of 20 GPU (gas permeation unit). The temperature and pressure conditions were fixed at 200 °C and 101 kPa, respectively. The structure of silica is shown in Figure 8.3 [38].

In another research work carried out by Gu et al. [39], MFI zeolite was used to carry out H_2/CO_2 separation, and this membrane sample gave a maximum selectivity of 18 at hydrogen permeance value of 560 GPU. In this experiment, the temperature and pressure were kept at 450 °C and 101 kPa. As an example of using another zeolite membrane, Guan et al. [40] carried out H_2/CO_2 separation using MFI zeolite membrane. The membrane samples gave a maximum H_2/CO_2 selectivity of 24, with a maximum H_2 permeance of 1000 GPU. The authors quoted the temperature conditions to be fixed at 35 °C. Extensive research has been done on the use of silica membranes for H_2/CO_2 separation, and it has been seen that silica membranes give excellent separation coefficients with high permeability [41]. However, they suffer from a major drawback that they lose their permeability (as much as 50%) within a short period on being exposed to moisture [42]. An example of the instability of silica membranes, when treated with moisture, can be seen by a work carried out by Wu et al. [43]. In their research, they synthesized silica membranes using chemical vapor deposition (CVD). The fabricated membranes were tested with He and N_2 gas streams. It was observed that there was 62% and 70% reduction of He and N_2 permeabilities, respectively, when the membranes were treated with gas stream containing up to 20 mol% water vapor, at a temperature of 600 °C. Permeability reduction on the exposure of silica membranes to water vapor is a big problem because H_2 production process almost always uses steam as one of its feedstocks, and so steam is often also present in the product stream [44]. Therefore, it is a big aim of recent studies on silica membranes to improve and enhance the stability of silica membranes in the presence of moisture. To achieve this, many researchers in recent works have focused on the incorporation of metal oxides, such as TiO_2 [45], ZrO_2 [46], Co_3O_4 [47], and NiO [48]. In one particular work carried out by Yang and Chen, Ag-doped alumina was used as the H_2-selective membrane. It was found that with silver doping, the permeability of H_2 increased from 3.62×10^{-6} to 8.99×10^{-6} mol/m^2-s-Pa, whereas the H_2/CO_2 selectivity increased from 9.26 for pure silica membrane to 10.22 for Ag-doped silica membranes. Moreover, the permeability was not greatly affected by the moisture content. The experiments were conducted at a temperature of 200 °C and a pressure drop across the membrane of 0.35 MPa [49]. In a research endeavor carried out by Smart et al. [50], cobalt oxide-doped alumina membranes were synthesized on alumina supports. Then, the synthesized membranes were then tested for single and mixed gas permeation of both H_2 and CO_2 at temperatures of up to 600 °C and pressures of up to 600 kPa. The hydrogen permeance of the membranes was found to be 1.9×10^{-7} mol/m^2-s-Pa with an H_2/CO_2 selectivity of up to 1500. Moreover, it was seen that cobalt oxide-doped silica membranes did not undergo thermally induced densification, even at temperatures as high as 600 °C. This showed that doping with metal ions proved to be very helpful in improving hydrothermal stability of silica membranes. More recently, inorganic membranes have been synthesized by using metal–organic framework (MOF) membranes. MOF membranes are promising in this application, as is detailed by a research work carried out by Huang et al. [51]. In that particular work, Zn(benzene dicarboxylate [BDC])(TED)$_{0.5}$ membranes were synthesized and tested for H_2 separation at a temperature of 180 °C and a pressure of 1 bar. The hydrogen permeance for these membranes comes out to be 2.65×10^{-6} mol/s-m^2-Pa with H_2/CO_2 permselectivity of 12.1. The structure of a typical MOF is shown in Figure 8.4 [52].

In another more recent work performed by Yang et al. [53], in situ interfacial assembly was used to prepare layered $Zn_2(bim)_4$ was synthesized and then tested for H_2/CO_2 separation. The membrane samples showed H_2 permeance in the range of 6300–10 680 GPU with H_2/CO_2 selectivity of 86.5. The results obtained from this research showed that $Zn_2(bim)_4$ can easily surpass the Robeson upper bound curve for H_2/CO_2 separation, and it is also highly hydrophobic, making it very resistant to hydrothermally induced densification and subsequent permeability reduction. Another similar kind of material is also being used for H_2/CO_2 separation, and it is called a covalent organic framework (COF). COFs are very difficult to fabricate into membranes, as shown in the work carried out by Tang et al. [54]. They used TpPA as the COF matrix and dispersed graphene oxide (GO) in the structure to give the structure of the COF dimensional stability. The resulting membrane had H_2/CO_2 selectivity of about 25.57 with H_2 permeance of up to 1.067×10^{-6} mol/m^2-s-Pa. Another class of inorganic membranes is zeolite membranes. In a research work

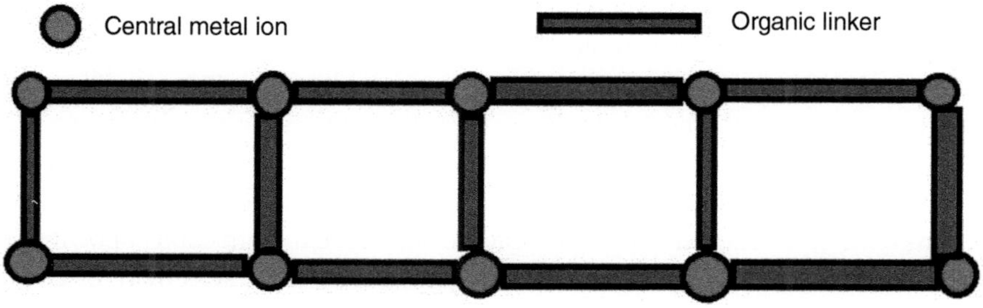

Figure 8.4 Structure of a typical MOF. Source: Adapted from Bell [52].

carried out by Roy and Das [55], graphite amine fiber-incorporated NaX zeolite membranes were used for H_2/CO_2 separation. It was found that pure NaX zeolite membrane gave H_2 permeance of about 5.1×10^{-7} mol/m^2-s-kPa and H_2/CO_2 selectivity of 6.5. This value increased to H_2 permeance of 36.5×10^{-7} mol/m^2-s-kPa and H_2/CO_2 selectivity of 31. These tests were carried out under hydrothermal conditions, so the increase in both permeance and selectivity is a proof to the zeolite membranes' hydrothermal stability. A newly emerging and interesting class of inorganic membranes that can further be worked on is zeolitic imidazolate framework (ZIF) membranes. In one very recent work carried out by Li et al. [56], a special kind of ZIF membrane was designed that had suitable aperture size and superhydrophobic properties. The resulting membrane showed H_2/CO_2 selectivity of 26.3 and H_2 permeability of about 9.75×10^{-7} mol/m^2-s-Pa. This new type of ZIF membrane is called JUC-160, and the testing of its separation performance was carried out at 200 °C. Moreover, the JUC-160 membranes did not show much decrease in separation performance, even in the presence of steam. There are many other examples of the use of inorganic membranes for H_2/CO_2 separation; however, the limitations mentioned here are still being worked on.

To work on the disadvantages and limitations of inorganic and metallic membranes, a class of membranes that would give low cost and durable structure needed to be developed. This led to the use of polymeric membranes for H_2/CO_2 separation. In one such work carried out by Chung et al. [57], 6-Food and Drug Administration (FDA) durene was modified using propane-1,3-diamine, and the resulting membranes gave an ideal selectivity for pure gases of about 101, whereas ideal mixed gas selectivity was found to be 42. The permeability of H_2 was found to be 120 barrer, and the tests were conducted at 35 °C and 3.5 bar. It was also seen that these membranes are modified by the addition of nanomaterials. In a particular work conducted by Choi et al. [58], polybenzimidazole (PBI) membranes were prepared, and then nanoporous silica was added as a filler into the matrix to form a mixed matrix membrane (MMM), where the loading of nanoporous silica was up to 14 wt%. The ideal H_2/CO_2 selectivity was found to be 45, with a hydrogen permeability of about 1 barrer. The permeation and gas separation tests were conducted at a temperature of 35 °C at varying pressure values. Another modification in the simple polymeric membrane to improve separation performance is to use blending and cross-linking as a dual technique to produce a membrane that has enhanced separation characteristics for H_2/CO_2 gas mixtures. In one such research endeavor carried out by Hosseini et al. [59], Matrimid®-based membranes were cross-linked with *p*-xylene diamine and then blended with PBI to give membrane samples with maximum H_2/CO_2 selectivity of 9.43 and H_2 permeability of 5.47 barrer. The tests were conducted at 35 °C and 3.5 bar. In another attempt to study the effect of modification with cross-linker on the separation characteristics of a membrane, Shao et al. [60] synthesized a polyimide-based membrane, which was modified using ethylene diamine (EDA). It was seen that the samples in which the modification of polyimide membranes was carried out for 10 minutes gave a single gas selectivity of about 102 and mixed gas selectivity of 16.6. The experiments were worked out at a CO_2 partial pressure of 3.5 atm. Polymeric membranes have indeed been modified in several ways to produce membranes with enhanced permeabilities and selectivities. In an attempt to enhance the permeability of H_2 through polymeric membranes, one research work carried out by Choi et al. [61] used the copolymerization of polybenzoxazole (PBO) with polypropylene polybenzyl isocyanate (PPI). The resulting PBO-*co*-PPI membranes gave a maximum H_2 permeability of 1680 barrer, which is a very high value. However, the membranes gave a modest H_2/CO_2 selectivity of 3.2. Therefore, work needed to be done to improve both selectivity and permeability to an appreciable extent. In a very recent work carried out by Abdul Nasir et al. [62], butylene diamine (BuDA) was used to modify the structure of polysulfone (PSf) membranes. It was found that the synthesized membranes had a higher permeability for both gases, and it also showed a maximum ideal H_2/CO_2 selectivity of 4.97. Heat treatment and chemical modification is another way of improving the performance of polymeric membranes, which has been extensively studied. In further work carried out by Omidvar et al. [63], carbonized PBI was prepared by using carbon molecular sieves (CMS) in PBI matrix, which were then subjected to harsh heat treatment at 900 °C. The final membrane samples gave a maximum H_2 permeability of 36 barrer and a maximum H_2/CO_2 selectivity of 53, when the testing was carried out at 150 °C. Table 8.1 summarizes the different H_2-selective membranes being studied and their corresponding performance.

In addition, the summary of gas separation performance of H_2-selective membranes is presented in Table 8.1.

8.3.2 CO_2-Selective Membranes

As compared to the H_2-selective membrane, CO_2-selective membranes are much more difficult to synthesize. Many different types of CO_2-selective membranes have been tried and tested, including carbon membranes and other inorganic membranes, as well as polymeric membranes [64–68]. However, substantial work has only been done in the use of polymeric membranes for CO_2/H_2 separation as far as CO_2-selective membranes are concerned [69–72]. There are many examples of polymeric materials being used for CO_2 separation applications [73–75].

In their work, they used semicrystalline polyethylene oxide (PEO) to synthesize the gas separation membranes. The resultant membranes gave CO_2/H_2 selectivity of 9.9 and CO_2 permeability of 17 barrer. The experiments were carried out at 35 °C and 14.6 atm. It was also seen from this research that the ether linkage in PEO led to better CO_2/H_2 separation as compared with a polymer without such linkage, for example, polyethylene (PE). This showed that we could exploit the advantages of different functional groups on the overall separation performance of a membrane. In a more recent research work carried out by the same authors [76], cross-linking was used to improve both permeability and selectivity. In this regard, they used cross-linked polyethylene glycol (PEG) acrylate. The resulting membranes gave CO_2/H_2 selectivity of 12, which is a significant improvement from the previous results obtained for PEO without cross-linking. Moreover, the cross-linked PEG acrylate gave a remarkably high CO_2 permeability of 570 barrer. Therefore, it can be seen that cross-linking caused both selectivity and permeability to increase.

Table 8.1 Summary of gas separation performance for H_2-selective membranes.

Membrane material	Temperature	Pressure	H_2 permeability	Selectivity (H_2/CO_2)	References
Pt	705–827 K	100 kPa	8.1×10^{-7} mol/s-m-Pa$^{1/2}$	—	[25]
Ni–Pd alloy	500 K	20 kPa	2×10^{-12} mol/s-m-Pa$^{1/2}$	—	[26]
Fe_3Al	300 K	101 kPa	1×10^{-10} mol/s-m-Pa$^{1/2}$	—	[27]
$Nb_{21}Ti_{50}Ni_{29}$	—	—	1×10^{-8} mol/s-m-Pa$^{1/2}$	—	[28]
Pd–Ag	723 K	20–75 kPa	5×10^{-5} mol/s-m-Pa$^{1/2}$	—	[29]
Hydrophilic silica	200 °C	101 kPa	20 GPU	11	[35]
MFI zeolite	450 °C	101 kPa	560 GPU	18	[36]
MFI zeolite	35 °C	—	1000 GPU	24	[37]
Ag-doped alumina	200 °C	0.35 MPa	8.99×10^{-6} mol/m^2-s-Pa	10.22	[46]
Co_3O_4-doped alumina	600 °C	600 kPa	1.9×10^{-7} mol/m^2-s-Pa	1500	[47]
Zn(BDC)(TED)$_{0.5}$	180 °C	1 bar	2.65×10^{-6} mol/s-m^2-Pa	12.1	[48]
$Zn_2(bim)_4$	—	—	6300–10 680 GPU	86.5	[49]
GO-TpPA	—	—	1.067×10^{-6} mol/m^2-s-Pa	25.57	[50]
Graphite amine fibers/NaX			36.5×10^{-7} mol/m^2-s-kPa	31	[51]
JUC-160			9.75×10^{-7} mol/m^2-s-Pa	26.3	[52]
NH_2-modified 6FDA durene	35 °C	3.5 bars	120 barrer	101 (pure) 42 (mixed)	[53]
PBI/SiO_2	35 °C	1–5 bars	1 barrer	45	[54]
p-XDA/Matrimid	35 °C	3.5 bars	5.47 barrer	9.43	[55]
PI/EDA	—	3.5 atm (CO_2 partial)		102 (pure) 16.6 (mixed)	[56]
PBO/PPI	—	—	1680 barrer	3.2	[57]
BuDA/PSf	—	—	—	4.97	[58]
CMS/PBI	150 °C	—		53	[59]

It is a well-known fact that CO_2 causes plasticization in polymeric materials, which reduces the polymer matrix's glass transition temperature, as high-pressure gas is dissolved into the polymer matrix [77]. Plasticization would normally reduce selectivity and separation performance for different gas mixtures, such as CO_2/CH_4 and CO_2/N_2. However, in the case of CO_2/H_2, the separation is enhanced if diffusivity selectivity is reduced. This can only be achieved if the polymer chains are made more flexible, and this can be done in several ways. In one such research work carried out by Lin et al. [78], high-molecular-weight and highly branched and cross-linked copolymers were produced by the combination of polyethylene glycol diacrylate (PEGDA) and polyethylene glycol methyl ether acrylate (PEGMEA). In this way, the plasticization that occurs in polymers due to the CO_2 was used to enhance the membranes' selectivity and permeability when compared with the neat polymeric phase. The copolymeric membrane thus formed had CO_2/H_2 selectivity of 9.4 with an exceptional CO_2 permeability of 440 barrer. These results were obtained at temperatures of about 35 °C and a pressure as high as 17 atm.

Another method of improving the plasticization of the copolymeric membranes and hence the CO_2/H_2 selectivity is done by adding plasticizers, such as PEG. It was found in a research work carried out by Car et al. [79] that when low-molecular-weight PEG was added to a copolymeric membrane made of PEO and polybutylene terephthalate (PBT), the resulting blend produced CO_2/H_2 selectivity of about 13 with CO_2 permeability of about 190 barrer at 35 °C and at varying pressures. In another research endeavor carried out by Qiao et al. [80], the advantages of cross-linking and structural modification on CO_2/H_2 separation performance were shown. In their work, polyvinyl amide (PVAm) was modified using different small molecular amines (SMAs). It was found that monoethanol amine (MEA)-modified MEA-PVAm when coated onto the surface of PSf porous membrane support provided CO_2/H_2 selectivity of about 62 with H_2 permeance of about 0.015 μmol/m^2-s-Pa. This exceptional result for selectivity and permeability shows that modification of any pure polymeric membrane is necessary, if the separation performance is enhanced.

As we have formulated and established that changing the structure or the overall makeup of a polymeric membrane can improve its separation performance, so now we will be studying one very interesting and very commonly used method of modification of the structure of a polymeric membrane, which the use of certain filler particles, called *facilitators*. The formulated membrane samples are called facilitated transport membranes [81]. The question here is, how facilitated transport membranes work? Facilitated transport membranes were initially developed by Mahajan and Koros [82]. They work by a mechanism according to which the permeance of gas, like CO_2, is improved by a reversible reaction of the gas with certain functional groups within the polymer matrix, such as amines, carbonates, and carboxylates [83]. Thus, the use of facilitated transport membranes causes improved transport properties by complementing solution–diffusion mechanism [84]. One of the most commonly used configurations of facilitated transport membranes is the supported liquid membranes (SLMs) [85]. The working of a facilitated transport membrane is shown in Figure 8.5 [86].

In one such research work carried out by Myers et al. [87], the amine-modified ionic liquid was incorporated onto an ionic liquid-supported membrane to give the final facilitated transport membrane. This membrane exhibited CO_2/H_2 selectivity up to 10 and CO_2 permeability of up to 1000 barrer. The experiments were carried out at a temperature of 100 °C and pressure of about 1 atm.

In another work carried out by Bai and Ho [88], sulfonated polybenzimidazole (SPBI) was modified using polyethylene imine (PEI) to form the final facilitated transport membrane. The biggest advantage of this type of membrane was that it contained both fixed-site carriers and facilitators, so there was a coordinated

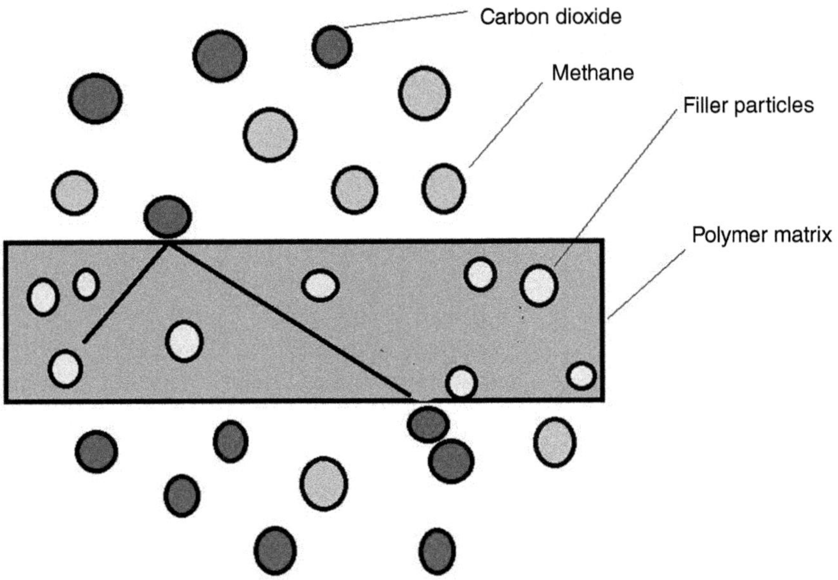

Figure 8.5 Working of a facilitated transport membrane. Source: Adapted from Xing and Ho [86].

effect of both to give much more enhanced CO_2/H_2 separation. These membrane samples gave maximum CO_2/H_2 selectivity of up to 50 with permeability values reaching up to 1262 barrer. The experiments were carried out at 100 °C and 2 atm. These exceptional permeation results showed that facilitated transport membranes have great potential for gas separation. In yet another research effort carried out by El-Azzami and Grulke [89], chitosan-facilitated transport membranes were formed by incorporating 40 wt% sodium alginate in the chitosan matrix. These membranes provided CO_2/H_2 selectivity of about 144, with CO_2 permeability of almost 1500 barrer. The temperature and pressure for these experiments were kept at 110 °C and 1.5 atm, respectively. These results further consolidated the usefulness of facilitators in improving the separation performance of gas separation membranes.

There are many recent examples of using polymer blends for the synthesis of gas separation membranes. One such work carried out by Reijerkerk et al. [90] used a specially designed polyether block amide (Pebax® 1657) with PEG-based polymeric material. The resulting membranes were tested at a temperature of 35 °C and pressure of about 3.95 atm. These blend membranes showed a maximum CO_2 permeability of 98 barrer, with CO_2/H_2 selectivity coming out to be 9.5. The use of polymer blends has often yielded exceptional CO_2/H_2 separation results, as was detailed in another work carried out by Chen et al. [91]. In their work, they used pyromellitic dianhydride (PMDA), 4,4-oxydianiline (ODA), and PEO to form polymer blends with varying compositions. The membranes were then tested for their gas separation performance at 35 °C and up to 2 atm. The experiments yielded CO_2/H_2 selectivity of 22.7 with CO_2 permeability of 179 barrer.

Furthermore, there has been extensive research on the use of facilitated transport membranes in more recent research works as well. In a research work carried out by Mehmood et al. [92], thermoplastic polyurethane (TPU)-based MMMs were prepared using γ-cyclodextrin metal–organic framework (γ-CD MOF) as the facilitator. The composition of γ-CD MOF was varied from 0.2 to 1.0 wt%. The resulting membranes were tested at ambient temperature and a maximum pressure of 5 bar. The maximum selectivity of CO_2/CH_4 was found to be 28.56 for filler concentration of 0.2 wt%.

Another research work was done by A. Hussain et al. [93], using cellulose acetate (CA)-based MMMs with multiwalled carbon nanotubes (MWCNTs) as the facilitating filler and PEG as the polymer blend component. Maximum selectivity and optimized permeability were obtained for membrane samples containing 10 wt% PEG and 10 wt% MWCNTs. The value of optimized CO_2 permeability and maximum selectivity came out to be 13.17 barrer and 38.5, respectively.

In one research work carried out by Zhao and Ho [94], polyallyl amine was blended with cross-linked polyvinyl alcohol (PVA) to form a cross-linked facilitated transport, fixed-site carrier membrane. The cross-linking of the PVA was carried out by glutaraldehyde as cross-linker and potassium hydroxide as the cross-linking catalyst. The final synthesized membrane was tested at 110 °C and 2 atm for gas separation performance. The feed gas composition was kept at 20% CO_2, 40% H_2, and 40% N_2. The membrane samples thus formed gave CO_2/H_2 selectivity of 45 and CO_2 permeability of about 300 barrer. The effectiveness of a membrane for gas separation can be increased using a special kind of technology called *membrane contactors*. These are a type of phase contacting device used in both gas separation and liquid–liquid extraction [95–98]. One such type of membrane contactors is SLM. More specifically, ionic liquids have been used as the supporting medium in such membranes. In one particular work carried out by Chen et al. [99], a blend membrane was fabricated by using polyvinylidene fluoride (PVDF) as the polymer matrix and incorporating room-temperature ionic liquid (RTIL) into the matrix. For this research work, 1-ethyl-3-methylimidazolium tetracyanoborate ([emim][$B(CN)_4$]) is used. In these membranes, the ionic liquid provides enhanced CO_2 solubility, and PVDF provides the mechanical strength as the membrane's main polymer matrix. For these membranes, the pure gas permeation testing was done at 35 °C and 2 atm. The tests yielded CO_2/H_2 selectivity of 12.9 with a maximum CO_2 permeability of 1778 barrer. This exceptionally high permeability can be attributed to the improved solubility of CO_2 in the blend membranes. Blend membranes, when combined with a certain nanofiller, give a blend MMM, which has superior gas separation properties, far more than either blend membrane or a simple MMM sample. In a research work carried out by Ansaloni et al. [100], a three-polymer blend membrane was formed by the mixing together of cross-linked PVA, polysiloxanes (POS), and polyallyl amine (PAAm). Afterward, amine-functionalized multiwalled carbon nanotubes (AF-MWCNTs) were incorporated within these membranes. The resulting mixed matrix blend membranes gave CO_2/H_2 selectivity of around 16 with CO_2 permeability of about 224 barrer. These experiments were carried out using gas mixtures containing 20% CO_2, 40% H_2, and 40% N_2. These results showed that by using a facilitator such as MWCNT, the separation performance is improved by a significant degree. Another work using membrane contactor technology used SLMs once again for the enhancement of gas separation. In one particular work of this type carried out by Klepić et al. [101], PVA was used as the polymer matrix, whereas 1-ethyl-3-methylimidazolium dicyanamide ([emim][DCA]) was used as the ionic liquid. In their research, these authors found that incorporation of small amount of ionic liquid produces membranes in which the gas transport is diffusion-controlled. Therefore, the membranes with no or very little ionic liquid incorporated within them were actually H_2 selective. This H_2 selectivity of the resultant membranes increased up to a point, after which the transport properties of gas through the membrane became solubility controlled. As CO_2 is more soluble in the ionic liquid and the PVA matrix, the facilitated transport membranes with higher ionic liquid content gave reasonably good CO_2/H_2 selectivity. As of this research work, the membrane samples containing 53 wt% ionic liquid produced CO_2 permeability of about 66.9 barrer, which is about 2400 times improvement as compared with pure PVA matrix. In addition to that, the ionic liquid-supported membranes also gave CO_2/H_2 selectivity of about 7.73. This value of selectivity is slightly above the Robeson 2008 plot. Table 8.2 shows the performance of CO_2-selective membranes that have been researched in recent years. The results of these experimental works are summarized in Table 8.2.

Table 8.2 Summary of gas separation performance of CO_2-selective membranes.

Membrane material	Temperature	Pressure	CO_2 permeability	Selectivity (CO_2/H_2)	References
PEO	35 °C	14.6 atm	17 barrer	9.9	[71]
PEG acrylate	35 °C	14.6 atm	570 barrer	12	[72]
PEGDA-PEGMEA	35 °C	17 atm	440 barrer	9.4	[74]
PEO-PEG-PBT	35 °C	1–5 atm	190 barrer	13	[75]
MEA-PVAm/PSf	—	—	0.015 μmol/m²-s-Pa	62	[76]
NH_2-IL/IL-SLM	100 °C	1 atm	1000 barrer	10	[82]
PEI-SPBI	100 °C	2 atm	1262 barrer	50	[83]
Na alginate/chitosan	110 °C	1.5 atm	144 barrer	1500	[84]
Pebax® 1657/PEG	35 °C	3.95 atm	98 barrer	9.5	[85]
PMDA-ODA-PEO	35 °C	2 atm	179 barrer	22.7	[86]
TPU/γ-CD-MOF	25 °C	5 bars	220 barrer	28.56	[87]
CA/PEG/MWCNTs	25 °C	5 bars	13.17 barrer	38.5	[88]
PAAm/PVA	110 °C	2 atm	300 barrer	45	[89]
[emim][B(CN)$_4$]/PVDF	35 °C	2 atm	1778 barrer	12.9	[94]
PVA-POS-PAAm/AF-MWCNT	—	—	224 barrer	16	[95]
[emim][DCA]/PVA	—	—	66.9 barrer	7.73	[96]

Although there is exceptional separation performance obtained from the different membrane gas separation systems described earlier, there are also several limitations that these systems suffer from, and these can be overcome by using a special kind of technology known as *membrane contactors*. One advantage that membrane contactors provide over conventional processes is reducing equipment size by about 20 times as far as gas separation is concerned. Other advantages of membrane contactors include no flooding or loading limitations, ability to accommodate similar density fluids, ease of scale-up, and reduction in back-mixing [102]. There are three main types of membrane contactor technologies, namely, dispersion-free membrane extraction [102], emulsion liquid membranes (ELMs) [103], and supported liquid membranes [104]. Still, only SLMs and ELMs have been extensively reported to have been used for gas separation applications. An example of the use of ELMs for CO_2 separation from gas mixtures was reported in work carried out by Kamarudin et al. [105]. In this work, water-in-oil ELM was used for CO_2 separation using kerosene as the organic phase, Span 80 as the surfactant, and a mixture of NaOH solution and triethanolamine (TEA) as the aqueous phase. In this work, 6 vol% TEA was used in the ELM, and this resulted in 87% removal of CO_2 from the gas mixture. In another research work involving ELM carried out by Dolmat et al. [106], a mixture of methyl diethanol amine (MDEA) and piperazine (PZ) was used as extractants for CO_2 removal. In the research work, varying amounts of MDEA and PZ were used in the ELM, and the most superior CO_2 removal performance was shown in the samples using 6 ml PZ and 2 ml MDEA that gave 56.54% CO_1 removal.

The second type of membrane contactors that is frequently used for pre-combustion carbon capture is the SLM. Previously, we have quoted several research works that used this technology, and some more works will be quoted here. In one such work carried out by Ilyas et al. [107], SLMs were fabricated by using 3-(trimethoxysilyl)propan-1-aminium acetate as the ionic liquid phase, whereas P-84 polyimide was used as the polymer support matrix. The resulting membranes gave a maximum CO_2/CH_4 selectivity of 38.1 with CO_2 permeance of 21 GPU. In another work related to SLM carried out by Saeed et al. [108], in which natural deep eutectic solvents (NADESs) were prepared by using betaine, maleic acid, and tartaric acid. Afterward, NADES was incorporated into the pores of already-synthesized PVDF microporous support. NADESs were synthesized using a combination of either betaine and maleic acid or betaine and tartaric acid, and the maximum CO_2/CH_4 selectivity was obtained for the SLM containing the combination of betaine and tartaric acid. Also, it gave a value of pure gas selectivity of up to 56.1. In a more recent work carried out by Craveiro et al. [109], deep eutectic solvent (DES) was synthesized using choline chloride along with either urea, glycerol, ethylene glycol, or oxalic acid. The prepared DES was then incorporated into the pores of polytetrafluoroethylene (PTFE) microporous support membrane. Of all the DES SLMs, the SLM with choline chloride–urea incorporated into PTFE showed the highest CO_2/CH_4 selectivity, which gave a value of 98. Such results showed how useful membrane contactors, mainly SLM and ELM, are in the separation of CO_2 from pre-combustion gas mixtures.

8.4 Advantages and Limitations of Pre-combustion Carbon Capture Using Membrane Technology

One of the major advantages of pre-combustion carbon capture technology is that it is free from combustion-based pollutants, such as SO_x and NO_x, as this process occurs before combustion. This reduces the complexity of the process required for the separation unit to be installed [110]. Another advantage of pre-combustion carbon capture is that it operates with a higher CO_2 concentration than post-combustion carbon capture. Therefore, it requires less amount of energy for downstream CO_2 capture and compression [111]. Another notable advantage of pre-combustion carbon capture is that it results in a marked reduction in SO_2 and NO_2 emissions. This was seen by a study conducted by the US Department of Energy [112]. The report stated that on the installation of carbon capture system in the Shell Integrated Gasification Combined Cycle (IGCC) plant, the SO_2 emissions reduced by 15% and NO_2 emissions reduced by 17%. Moreover, reducing the overall amount of NO_x and SO_x results in much lower operating costs for any process taking place downstream [113]. Another advantage of pre-combustion carbon capture is that it handles much smaller gas volume, as the treatment occurs before syngas is diluted with air or nitrogen [110]. In addition to the inherent advantages obtained from pre-combustion carbon capture, there are also some advantages of using membranes for this process. These include the fact that membranes are generally low cost, require much lower energy, and are smaller in size than other conventional carbon capture technologies [110].

Despite its multiple advantages, some limitations are encountered in pre-combustion carbon capture processes using membrane technology. There is a major problem of cost-effectiveness when it comes to pre-combustion carbon capture membranes. The most efficient separation is achieved by metallic membranes such as palladium and its alloys, and they are extremely expensive to operate on a large scale. Pre-combustion carbon capture membranes are also brittle, such as carbon membranes and other inorganic membranes. Moreover, in the case of alumina membranes, pre-combustion carbon capture membranes also exhibit poor hydrothermal stability [114]. Another major drawback of pre-combustion carbon capture is that the system's overall capital cost is very high [19]. It requires an air separation unit (ASU) for producing oxygen to be fed into the gasifier so that a minimum amount of unwanted impurities are present in the gas exiting the gasification unit [115]. Another disadvantage or limitation of using membrane technology for pre-combustion carbon capture is its low selectivity as compared with conventional processes, so much more work needs to be done to bring the selectivity of these emerging processes at par with conventional ones so that they can replace the conventional processes [35]. Furthermore, it is difficult to retrofit an existing IGCC plant to accommodate higher hydrogen concentrations in the feed gas because the gas turbine will have to be redesigned among other equipment as well [116]. The advantages and disadvantages of pre-combustion carbon capture systems using membrane technology are summarized in Table 8.3.

Table 8.3 Advantages and limitations of pre-combustion carbon capture using membrane technology.

Advantages	Limitations
Less complex downstream operation	Membranes expensive for the process
Less energy requirements	Brittle, such as carbon membranes
Lower SO_2 and NO_2 emissions	Low hydrothermal stability, such as alumina
Lower cost of operation	High capital cost, due to air separation unit
Handling much smaller volume of gas	Low selectivity of membranes
Much more compact membranes	Difficulties in retrofitting to existing plant

8.5 Applications of Pre-combustion Carbon Capture

The most important application of pre-combustion carbon capture is in IGCC plants. For this purpose, coke or coal is often used as the feed to produce syngas [20]. Moreover, some plants also use pre-combustion carbon capture technology in plants that use pulverized coal or natural gas as feed for the combined cycle process [117]. Moreover, pre-combustion carbon capture can also be used as a key step in the producing valuable chemicals, such as methanol, which is being done by Prosint Produtos Sinteticos in Rio de Janeiro, Brazil [118]. This process is also used in the production of soda ash [119], for enhanced oil recovery [120], for urea production [121, 122], for beverage production [123], for freezing and chilling purposes [124], and for ammonia production [125], among many other applications [126–128].

8.6 Current Trends and Future Prospects

Various research, design, and development (RD&D) projects have been started all over the world for the pre-combustion capture of CO_2. Recent research efforts have focused on the engineering of membrane materials to improve H_2/CO_2 separation. It has been seen that there is an increased interest in the use of such specially engineered polymeric materials that they are capable of working under elevated temperatures, i.e. 100–300 °C. The advantage of using such materials and operating them at higher temperatures is that there is improved solubility, selectivity, and prevention of plasticization at higher temperature [129]. The effect of temperature on solubility of a gas in a polymeric phase is represented in Eq. (8.5) [130]:

$$S_A = S_{A0} \exp\left(-\frac{\Delta H_{SA}}{RT}\right) \tag{8.5}$$

This shows that the relationship between solubility S_A and temperature T is very much an Arrhenius-type equation. Moreover, it is seen that the term ΔH_{SA} is related to the factor ε/k shown in Eq. (8.6) [130]:

$$\Delta H_s = R\left(500 - 10\frac{\varepsilon}{k}\right) \tag{8.6}$$

We must keep these equations in mind and also the fact that solubility increases with temperature, and with that, we study different polymeric systems with a commercial viewpoint. From a commercial standpoint, Membrane Technology and Research (MTR) developed a special membrane given the brand name Proteus™ [131], which was then tested for its efficacy in H_2/CO_2 separations. It was found that these membranes gave selectivity of up to 32, with H_2 permeance of 300 GPU. The tests were carried out at 150 psig and 200 °C, and the feed gas contained about 150 ppmv H_2S. These membranes are synthesized by the proper engineering of the three layers that make up the complete membrane. As we know that it is imperative to increase the permeability of a membrane to make it commercially viable, so the Proteus membrane is made by tuning and engineering the thickness of the dense skin layer and also tuning the gutter layer and the porous support layer in terms of its thickness and its porosity [132]. Therefore, the development of this novel type of membrane by MTR initiated the process of engineered polymeric membranes that is being currently widely studied. After the initial development of the Proteus by MTR, there was a boost in developing novel membranes with different materials. One such research endeavor was taken up by the Los Alamos National Laboratory (LANL) [133]. These membranes showed excellent H_2 permeance values ranging from 200 to 400 GPU with H_2/CO_2 selectivity of around 20 at temperatures of up to 250 °C. In another research effort carried out by SRI International [134], these membranes showed H_2 permeance in the range of 100–200 GPU while giving H_2/CO_2 selectivity ranging from 20 to 40 in the temperature range of 150–250 °C. The selectivity and permeance values depend on the thickness of the selective layer in the membrane samples. Other than the individual research projects and programs running in the industry, there is a general trend toward using new polymeric materials to manufacture membranes targeting H_2/CO_2 separation. One such example is the use of functionalized PBI for this separation. In this particular research work, which was carried out by Li et al. [135], different functional groups were used to modify PBI to form different membranes with selective layers of varying thicknesses. The best results for gas separation were obtained by using perfluorocyclobutyl (PFCB) functional groups, which, when incorporated into the structure of pure PBI, gave superior gas separation performance in terms of both hydrogen permeance and H_2/CO_2 selectivity. It was seen that for a selective layer thickness in the range of 5–20 μm, the H_2 permeability values are 39 and 323 barrer at 35 and 200 °C, respectively. Additionally, those membranes provided H_2/CO_2 selectivity values of 2.4 at 35 °C and 6.6 at 200 °C [135].

Another factor that was noted in these research works was that the fractional free volume (FFV) in polymer structure had a significant effect on the permeability of gas through the polymeric membrane. The relationship between penetrant molecular size, FFV, and permeability value is represented in Eq. (8.7) [136]:

$$P_A = A_A \exp\left(-\frac{B_A}{\text{FFV}}\right) \quad (8.7)$$

where P_A is the permeability of the gas (in barrer), A_A is a pre-exponent factor, B_A is a constant whose value increases with the increasing size of penetrant molecules, and FFV is the fractional free volume in the polymer chains of the membrane matrix. In turn, FFV can be calculated using Eq. (8.8):

$$\mathrm{FFV} = \frac{V - V_0}{V} \tag{8.8}$$

where V is the specific volume and V_0 is the occupied volume as estimated by the van der Waals' volume [129]. Other than the functionalization of existing polymers to improve the membrane's permeability and separation properties, other techniques may also be used for the improvement of the separation performance of these membranes. One method of doing so is by cross-linking, and this is being extensively used in current research works. Cross-linking plays an important role in improving the H_2/CO_2 selectivity by reducing FFV, thus making it more size selective and, hence, more H_2 permeable as compared with CO_2 [137]. There are many methods to achieve cross-linking. Interfacial polymerization is one of those methods and was used to prepare cross-linked polymeric membranes in a research work carried out by Ali et al. [138]. In their work, they produced gas separation membranes by the cross-linking of *m*-phenylenediamine (MPD) and trimesoylchloride (TMC). The resultant membranes gave mixed gas H_2/CO_2 selectivity of 50 with H_2 permeance of 350 GPU at a temperature of about 140 °C. Another method of cross-linking of polymers is by using a layer-by-layer approach. In one particular research work carried out by Kim et al. [139], polyether imide (PEI) was cross-linked using polyacrylic acid to form uniform membranes for gas separation. These membranes showed remarkable H_2/CO_2 selectivity of up to 300 with H_2 permeability of 4 barrer, and these experiments were carried out at 23 °C.

Another method of improving the separation performance and mechanical strength of certain polymers that form unstable and defective free-standing films is by making composite membranes from these polymeric materials. In one such research work carried out by Illing et al. [140], polyaniline dense films were coated onto the surface of a PVDF porous support membrane to form the final composite membranes with varying thickness of the dense top layer. The synthesized membrane gave a maximum H_2/CO_2 selectivity of 23.1 with H_2 permeability of about 1.8 barrer. Polymer blends are another important method to improve the separation performance and often the membranes' permeability properties by blending polymers like PBI with other polymers that have slightly less selectivity than PBI but much higher H_2 permeability. PBI often forms immiscible blends with polymers, in which the two polymers clearly show phase separation. The permeability of gas through such a polymer blend or for mixed matrix is depicted by Maxwell model, as shown in Eq. (8.9) [141]:

$$P_b = P_c\left[\frac{P_d + 2P_c - 2\emptyset_d(P_c - P_d)}{P_d + 2P_c + \emptyset_d(P_c - P_d)}\right] \tag{8.9}$$

where P_b is the overall gas permeability of the polymer blend and P_c and P_d are the gas permeabilities of the continuous and dispersed phases, respectively. An example of such an immiscible blend being used for H_2/CO_2 separation is using a blend membrane made by the combination of PBI and Matrimid. The incorporation of *p*-xylene diamine further consolidated the blend to form the final blend membrane. When tested for gas separation performance, these membranes were shown to give H_2/CO_2 selectivity of 26 with H_2 permeability of about 3.6 barrer. On the other hand, pure PBI membrane gave H_2/CO_2 selectivity of only 9.4 and permeability of 5.5 barrer. Therefore, it can be said that the selectivity of the membrane is significantly increased while its permeability is not affected very adversely [59]. Other research works have also been carried out in this regard using a variety of polymeric dispersed phases with PBI, including 6FDA-DAM [142], sulfonated polyphenylsulfone (sPPSU) [143], and poly(melamine-*co*-formaldehyde) (PMF) [144].

Another special example of blend membrane is the synthesis of asymmetric blend membranes. In a particular research endeavor carried out by Sánchez-Laínez et al. [145], PBI was blended with polymer of intrinsic microporosity (PIM) to form a blend that gave H_2/CO_2 selectivity of 24 and H_2 permeance of about 58 GPU. Another recent trend to engineer materials to make their separation performance better is by thermal treatment. One such research work was carried out by Giel et al. [146], in which they synthesized PBI/polyaniline (PANI) in a ratio of 80 : 20 by weight and then subjected the fabricated membranes to thermal treatment. The final membranes had H_2/CO_2 selectivity of about 15 and H_2 permeability of about 5.6 barrer.

Another heat treatment process is the polymeric materials' carbonization to form CMS [147]. A research work carried out by Omidvar et al. [63] detailed the carbonization of PBI to form CMS. When carbonization was done at 700 °C, the resulting molecular sieves provided H_2 permeability about 2 orders of magnitude higher than that of pure PBI. Moreover, the H_2/CO_2 selectivity of the sieves was 14 compared with the selectivity of PBI, which was about 8.7. Furthermore, when PBI was carbonized at 900 °C, it yielded sieves that had remarkably high H_2/CO_2 selectivity of 80 with slightly lower permeability, as compared with PBI at 700 °C.

Another aspect of work currently being done is the fabrication of MMMs, comprising a polymeric phase as the matrix with nanofiller particles as the dispersed medium within the matrix [148–150]. Many filler particles have been used to fabricate MMMs and are therefore a topic of interest for current H_2/CO_2 separation research work. Examples of some filler particles thus used in research works for this purpose include ZIF-71 [151], ZIF-8 [152], and Christian-Albrechts-University (CAU)-1-NH_2 [153], among many others. An example of a research work in which MMMs were used for H_2/CO_2 separation includes a particular endeavor undertaken by Japip et al. [154]. In their work, 6-FDA durene was modified using amine functionality of tris-(2-aminoethyl)amine (TAEA) to form a cross-linked layer of nanometer-scale thickness. The amount of cross-linker added determines whether the membranes would be CO_2 selective or H_2 selective. After cross-linking, nanofiller particles were added, and the gas separation performance was measured. In this work, it was seen that the incorporation of about 20 wt% ZIF-71 led to an increase in H_2/CO_2 selectivity from 47 for the unloaded membranes to 252 for loaded membranes. Moreover, the H_2 permeability increased from 300 to 581 barrer on the incorporation of 20 wt% ZIF-71. The permeation and gas separation experiments were carried out at a temperature of 35 °C. Other research works have also been published in which PBI was used as a polymer matrix with different filler media to give diffusion selectivity improvement. In one particular research work carried out by Kang et al. [155], a novel COF named National University of Singapore (NUS)-2 was synthesized and exfoliated into nanosheets using sonication and stirring. When these nanosheets are incorporated into PBI polymer matrix up to a concentration of 20 wt%, the resulting MMMs provided H_2/CO_2 selectivity of up to 31.4. These tests were carried out at 35 °C.

Furthermore, it has been seen that temperature has a marked effect on the separation and permeation performance of MMMs. In one such research work undertaken by Yang et al. [156], ZIF-8 nanoparticles are incorporated into PBI matrix to form MMMs specifically designed for high separation performance at elevated temperatures. The membrane samples containing 30 wt% ZIF-8 were most superior in their separation performance and gave H_2/CO_2 selectivity of 12.3 with H_2 permeability of up to 105.4 barrer. Other than diffusivity selectivity, work is also being done to improve solubility selectivity. It is known that palladium dissolves hydrogen gas to produce palladium hydrides (PdH_x). It is also known that when Pd nanoparticles are dispersed in a polymer matrix, they give solubility values of about 500–1000 cm^3 (standard temperature and pressure [STP])/cm^3, whereas a lean polymer has much less H_2 solubility, i.e. about 0.1 cm^3 (STP)/cm^3 [157]. Many research publications have detailed different methods to use Pd nanoparticles with modifications to form MMMs with increased solubility selectivity for H_2-selective membrane-based gas separation. In a research work carried out by Zhu et al. [158], the main focus of the researchers was to improve the sorption properties of the MMMs. The result of incorporating up to 12 vol% Pd nanoparticles into the PBI matrix provides H_2 solubility of 10 cm^3 (STP)/cm^3-atm. Moreover, the resulting membranes also gave H_2/CO_2 solubility selectivity of up to 28.

Other than the engineering of novel membrane materials for the improvement of H_2/CO_2 separation performance, another strategy for the purpose could be to improve the design of the process, by incorporating several other strategies using membrane technology at the heart of the process. According to a study conducted by Kaldis et al. [159], the two most interesting and useful strategies to use membranes for the improvement of H_2/CO_2 separation performance are to use two-stage membrane configuration and also to use catalytic membrane CO shift reactor for simultaneous WGS reaction and CO_2 capture. Both these strategies proved to be most fruitful in terms of energy penalty and cost avoided due to the CO_2 capture.

Having talked about the current trends within the use of membrane technology for pre-combustion carbon capture, we will now be looking at the future prospects for the technology. First of all, it is necessary to note that membrane synthesis, characterization, and testing are not enough to implement the technology at the industrial scale. Once favorable results are obtained for a particular membrane system within the laboratory, it must be tested for longer durations at the pilot plant scale; therefore in the future, this aspect must be worked on more vigorously [159]. It has been seen that a combination of different carbon capture technologies need to be used to solve the problem of CO_2 emissions. An example of such a process is using a hybrid multistage membrane gas separation system, consisting of a two-stage CO_2-selective membrane and a single-stage H_2-selective membrane. This multistage membrane system then leads to a gas oil hydrogenation process to form the final stream with maximum purity of H_2 and CO_2. This process is detailed in a study carried out by Huang et al. [160]. In another research work, a membrane and pressure swing adsorption (PSA) hybrid system was used to produce high-purity hydrogen gas (>99.9%) with a high degree of recovery (>99.9%) as detailed in a work carried out by Li et al. [161]. Furthermore, to improve pre-combustion carbon capture using membranes, new types of materials need to be investigated. An emerging class of materials is the MOFs, which either can be integrated with hydrophobic polymers to form block copolymers or could be used as a filler material within membranes to improve gas separation performance [162]. Different materials are being researched for improving H_2/CO_2 separation, and new paths are opening for novel materials. Some examples of new materials being used include a hybrid material made from the combination of ZIF-93 and ZIF-11 [163], metal-doped organosilica membranes [50, 164], and MOF membranes [165], among many other materials.

Moreover, other than experimental methods to produce new and improved materials, work must be done to properly and accurately characterize new materials and using computational tools to guide the development of those structures of materials that are most promising for our application. Also, materials synthesis and characterization should be done in such a way that it allows molecular-level control [166]. Computational screening must also be done and studied for high-throughput materials like MOF adsorbents and membranes as is shown in a study conducted by Avci et al. [167]. In another study done by Tong et al. COFs were studied using computational methods to understand their CO_2/H_2 separation performance [168]. Computational methods can also be used to study newly produced and synthesized materials for their gas separation performance, for example, the new MOFs added to the MOF database. It must also be kept in mind that the process of CO_2 sequestration and storage follows a life cycle in itself. Therefore, life cycle assessment of the entire process must also be undertaken to understand the full extent of the problem and find ways to mitigate and solve it on a larger scale [169].

After discussing pre-combustion CO_2 capture using membrane technology, one can conclude that membrane technology has great potential for use in this field. However, for it to be at par or better than existing technologies, certain improvements must be made in the overall process. Membrane contactor technology can be better than existing technology if more research on the topic, extensive experimentation, testing, and characterization of the materials/process are done. In the near future, further explorations on the said topic play a more vital role in the capture of CO_2 prior to combustion processes.

As we have studied, the pre-combustion carbon capture process is mainly used for the purification of hydrogen produced from various sources, such as steam reforming or gasification, so the main aim of the pre-combustion carbon capture process is to improve the quality of hydrogen produced as a result. Therefore, any future pre-combustion carbon capture process must be designed, keeping in mind the requirement of cost reduction for hydrogen production [170].

8.7 Concluding and Future Directions

In this chapter, we focused on a very important application of membrane contactor technology, which is pre-combustion carbon capture. We looked through the working principle and why we need pre-combustion carbon capture. Later on, we also explored the different membrane materials for use in both H_2- and CO_2-selective membranes. The purpose of this was to see what developments can be and are being made in the existing technology, to improve carbon capture, and hence to enable the proper utilization of hydrogen fuel produced by steam reforming of natural gas or by coal or biomass gasification. Furthermore, we also studied the use of membrane contactors in the existing framework, and it was seen that membrane contactors, primarily SLMs, are very helpful in improving separation performance and hence have very high potential to be used in future membrane gas separation technology. Future aspects of this technology include its further integration into the existing system, both as a standalone alternative to conventional technologies and as a hybrid along with the other ordinarily used technologies. Whatever framework we use, one thing is certain that using membrane technologies to any extent can improve the performance of the system. The only thing that needs to be worked on is cost optimization and proper design of the system to avoid losses.

References

1 Levine, M. (2011). *Key China Energy Statistics 2011*. California Digital Library.

2 Tanaka and National Journal of International Energy Agency for Pacific Islands (2010). *World Energy Outlook 2010*. Organization of Economic Cooperation and Development (OECD).

3 Matsuo, Y., Yanagisawa, A., and Yamashita, Y. (2013). A global energy outlook to 2035 with strategic considerations for Asia and Middle East energy supply and demand interdependencies. 2: 79–91.

4 Pachauri, R.K., Allen, M.R., Barros, V.R. et al. (2014). *Climate Change 2014: Synthesis Report. Contribution of Working Groups I, II and III to the Fifth Assessment Report of the Intergovernmental Panel on Climate Change*. IPCC.

5 Metz, B., Davidson, O., Swart, R., and Pan, J. (2001). *Climate Change 2001: Mitigation: Contribution of Working Group III to the Third Assessment Report of the Intergovernmental Panel on Climate Change*. Cambridge University Press.

6 Christensen, J.M., Denton, F., Garg, A. et al. (2006). *Changing Climates. The Role of Renewable Energy in a Carbon-Constrained World*. Renewable Energy Policy Network.

7 Hajilary, N., Shahi, A., and Rezakazemi, M. (2018). Evaluation of socio-economic factors on CO_2 emissions in Iran: factorial design and multivariable methods. *Journal of Cleaner Production* 189: 108–115.

8 Rackley, S.A. (2017). *Carbon Capture and Storage*. Butterworth-Heinemann.

9 Ooi, R.E. and Foo, D.C. (2014). Carbon constrained energy planning (CCEP) with carbon capture and storage incorporating carbon credit exchange. In:

Computer Aided Chemical Engineering (eds. M.R. Eden, J.D. Siirola and G.P. Towler), 585–590. Elsevier.

10 Younas, M., Rezakazemi, M., Daud, M. et al. (2020). Recent progress and remaining challenges in post-combustion CO_2 capture using metal-organic frameworks (MOFs). *Progress in Energy and Combustion Science* 80: 100849.

11 Habib, M., Badr, H., Ahmed, S. et al. (2011). A review of recent developments in carbon capture utilizing oxy-fuel combustion in conventional and ion transport membrane systems. 35: 741–764.

12 Wang, Y., Zhao, L., Otto, A. et al. (2017). A review of post-combustion CO_2 capture technologies from coal-fired power plants. 114: 650–665.

13 Gupta, R.B. (2008). *Hydrogen Fuel: Production, Transport, and Storage*. CRC Press.

14 Dincer, I. and Acar, C. (2015). Review and evaluation of hydrogen production methods for better sustainability. 40: 11094–11111.

15 Acres, G., Frost, J., Hards, G. et al. (1997). Electrocatalysts for fuel cells. 38: 393–400.

16 Babu, P., Ong, H.W.N., and Linga, P. (2016). A systematic kinetic study to evaluate the effect of tetrahydrofuran on the clathrate process for pre-combustion capture of carbon dioxide. 94: 431–442.

17 Jansen, D., Gazzani, M., Manzolini, G. et al. (2015). Pre-combustion CO_2 capture. *International Journal of Greenhouse Gas Control* 40: 167–187.

18 In, I., Metz, B., Davidson, O. et al. (eds.) (2005). *Special Report on Carbon Dioxide Capture and Storage*. Intergovernmental Panel on Climate Change (IPCC).

19 Olajire, A.A. (2010). CO_2 capture and separation technologies for end-of-pipe applications–a review. 35: 2610–2628.

20 Theo, W.L., Lim, J.S., Hashim, H. et al. (2016). Review of pre-combustion capture and ionic liquid in carbon capture and storage. *Applied Energy* 183: 1633–1663.

21 Alfano, M. and Cavazza, C. (2018). The biologically mediated water–gas shift reaction: structure, function and biosynthesis of monofunctional [NiFe]-carbon monoxide dehydrogenases. *Sustainable Energy & Fuels* 2: 1653–1670.

22 Lima, M.S.R. (2019). Comparação entre o sistema ccs e o solar em um ciclo combinado: análise exergoeconômica e exergoambiental, in Brasil.

23 Crippa, M., Oreggioni, G., Guizzardi, D. et al. (2017). *Fossil CO_2 and GHG Emissions of all World Countries*. Joint Research Center Science for Policy.

24 Sifat, N.S. and Haseli, Y. (2019). A critical review of CO_2 capture technologies and prospects for clean power generation. 12: 4143.

25 Ahmadi, P., Torabi, S.H., Afsaneh, H. et al. (2020). The effects of driving patterns and PEM fuel cell degradation on the lifecycle assessment of hydrogen fuel cell vehicles. 45: 3595–3608.

26 Holleck, G.L. (1970). Diffusion and solubility of hydrogen in palladium and palladium–silver alloys. *The Journal of Physical Chemistry* 74: 503–511.

27 Steward, S. (1983). Review of hydrogen isotope permeability through materials. In: *Lawrence Livermore National Lab.(LLNL)*. Livermore, CA: United States Department of Energy.

28 Zhang, Y., Gwak, J., Murakoshi, Y. et al. (2006). Hydrogen permeation characteristics of thin Pd membrane prepared by microfabrication technology. 277: 203–209.

29 Cheng, X., Wu, Q., and Sun, Y. (2005). Hydrogen permeation behavior in a Fe_3Al-based alloy at high temperature. 389: 198–203.

30 Hashi, K., Ishikawa, K., Matsuda, T., and Aoki, K. (2005). Microstructures and hydrogen permeability of Nb-Ti-Ni alloys with high resistance to hydrogen embrittlement. 46: 1026–1031.

31 Tong, H., Gielens, F., Gardeniers, J.G. et al. (2004). Microfabricated palladium-silver alloy membranes and their application in hydrogen separation. 43: 4182–4187.

32 Phair, J.W. and Donelson, R. (2006). Developments and design of novel (non-palladium-based) metal membranes for hydrogen separation. 45: 5657–5674.

33 Paglieri, S.N. and Way, J.D. (2002). Innovations in palladium membrane research. 31: 1–169.

34 Adhikari, S. and Fernando, S. (2006). Hydrogen membrane separation techniques. 45: 875–881.

35 Scholes, C.A., Kentish, S.E., and Stevens, G.W. (2009). Effects of minor components in carbon dioxide capture using polymeric gas separation membranes. 38: 1–44.

36 Lu, G., da Costa, J.C.D., Duke, M. et al. (2007). Inorganic membranes for hydrogen production and purification: a critical review and perspective. 314: 589–603.

37 Giessler, S., Jordan, L., da Costa, J.C.D., and Lu, G.Q.(Max) (2003). Performance of hydrophobic and hydrophilic silica membrane reactors for the water gas shift reaction. 32: 255–264.

38 Hamza, T., Sherif, A., and Abdalla, E. (2017). A novel approach to reinforce provisional material using silica gel powder. *Stomatological Disease and Science* 1: 3–7.

39 Gu, X., Tang, Z., and Dong, J. (2008). On-stream modification of MFI zeolite membranes for enhancing hydrogen separation at high temperature. 111: 441–448.

40 Guan, G., Tanaka, T., Kusakabe, K. et al. (2003). Characterization of $AlPO_4$-type molecular sieving membranes formed on a porous α-alumina tube. 214: 191–198.

41 Lee, D., Zhang, L., Oyama, S. et al. (2004). Synthesis, characterization, and gas permeation properties of a hydrogen permeable silica membrane supported on porous alumina. 231: 117–126.

42 Sea, B.-K., Soewito, E., Watanabe, M. et al. (1998). Hydrogen recovery from a H_2–H_2O–HBr mixture utilizing silica-based membranes at elevated temperatures. 1. preparation of H_2O-and H_2-selective membranes. 37: 2502–2508.

43 Wu, J.C.S., Sabol, H., Smith, G.W. et al. (1994). Characterization of hydrogen-permselective microporous ceramic membranes. 96: 275–287.

44 Battersby, S., Smart, S., Ladewig, B. et al. (2009). Hydrothermal stability of cobalt silica membranes in a water gas shift membrane reactor. 66: 299–305.

45 Van Gestel, T., Sebold, D., Hauler, F. et al. (2010). Potentialities of microporous membranes for H_2/CO_2 separation in future fossil fuel power plants: evaluation of SiO_2, ZrO_2, Y_2O_3–ZrO_2 and TiO_2–ZrO_2 sol–gel membranes. 359: 64–79.

46 Yoshida, K., Hirano, Y., Fujii, H. et al. (2001). Hydrothermal stability and performance of silica-zirconia membranes for hydrogen separation in hydrothermal conditions. 34: 523–530.

47 Uhlmann, D., Liu, S., Ladewig, B.P., and da Costa, J.C.D. (2009). Cobalt-doped silica membranes for gas separation. 326: 316–321.

48 Kanezashi, M. and Asaeda, M. (2006). Hydrogen permeation characteristics and stability of Ni-doped silica membranes in steam at high temperature. 271: 86–93.

49 Yang, J. and Chen, J. (2013). Hydrophobic modification and silver doping of silica membranes for H_2/CO_2 separation. 3: 21–29.

50 Smart, S., Vente, J.F., and da Costa, J.C.D. (2012). High temperature H_2/CO_2 separation using cobalt oxide silica membranes. *International Journal of Hydrogen Energy* 37: 12700–12707.

51 Huang, A., Chen, Y., Liu, Q. et al. (2014). Synthesis of highly hydrophobic and permselective metal–organic framework $Zn(BDC)(TED)_{0.5}$ membranes for H_2/CO_2 separation. *Journal of Membrane Science* 454: 126–132.

52 Bell, D. (2018). *The Promise of Metal–Organic Frameworks for Use in Liquid Chromatography*, vol. 36. LCGC North America.

53 Yang, K., Ban, Y., Guo, A. et al. (2020). In-situ interfacial assembly of ultra-H_2-permeable metal-organic framework membranes for H_2/CO_2 separation. *Journal of Membrane Science* 611: 118419.

54 Tang, Y., Feng, S., Fan, L. et al. (2019). Covalent organic frameworks combined with graphene oxide to fabricate membranes for H_2/CO_2 separation. *Separation and Purification Technology* 223: 10–16.

55 Roy, P. and Das, N. (2019). Synthesis of NaX zeolite-graphite amine fiber composite membrane: role of graphite amine in membrane formation for H_2/CO_2 separation. *Applied Surface Science* 480: 934–944.

56 Li, Z., Yang, P., Yan, S. et al. (2019). A robust zeolitic imidazolate framework membrane with high H_2/CO_2 separation performance under hydrothermal conditions. 11: 15748–15755.

57 Chung, T.S., Shao, L., and Tin, P.S. (2006). Surface modification of polyimide membranes by diamines for H_2 and CO_2 separation. 27: 998–1003.

58 Choi, S., Coronas, J., Lai, Z. et al. (2008). Fabrication and gas separation properties of polybenzimidazole (PBI)/nanoporous silicates hybrid membranes. *Journal of Membrane Science* 316: 145–152.

59 Hosseini, S.S., Teoh, M.M., and Chung, T.S. (2008). Hydrogen separation and purification in membranes of miscible polymer blends with interpenetration networks. *Polymer* 49: 1594–1603.

60 Shao, L., Lau, C.-H., and Chung, T.-S. (2009). A novel strategy for surface modification of polyimide membranes by vapor-phase ethylenediamine (EDA) for hydrogen purification. *International Journal of Hydrogen Energy* 34: 8716–8722.

61 Choi, J.I., Jung, C.H., Han, S.H. et al. (2010). Thermally rearranged (TR) poly(benzoxazole-co-pyrrolone) membranes tuned for high gas permeability and selectivity. *Journal of Membrane Science* 349: 358–368.

62 Abdul Nasir, N., Junaidi, M.U.M., Hashim, N.A. et al. (2018). Diamine modified polymeric membranes for hydrogen/carbon dioxide separation. *International Conference on Sustainable Energy and Green Technology*: 75.

63 Omidvar, M., Nguyen, H., Liang, H. et al. (2019). Unexpectedly strong size-sieving ability in carbonized polybenzimidazole for membrane H_2/CO_2 separation. *ACS Applied Materials & Interfaces* 11: 47365–47372.

64 Sohaib, Q., Muhammad, A., Younas, M. et al. (2021). Rigorous non-isothermal modeling approach for mass and energy transport during CO_2 absorption into aqueous solution of amino acid ionic liquids in hollow fiber membrane contactors. *Separation and Purification Technology* 254: 117644.

65 Younas, M., Tahir, T., Wu, C. et al. (2020). Post-combustion CO_2 capture with sweep gas in thin film composite (TFC) hollow fiber membrane (HFM) contactor. *Journal of CO_2 Utilization* 40: 101266.

66 Swati, I.K., Sohaib, Q., Cao, S. et al. (2020). Protic/aprotic ionic liquids for effective CO_2 separation using supported ionic liquid membrane. *Chemosphere* 267: 128894.

67 Sohaib, Q., Muhammad, A., Younas, M., and Rezakazemi, M. (2020). Modeling pre-combustion CO_2 capture with tubular membrane contactor using ionic liquids at elevated temperatures. *Separation and Purification Technology* 241: 116677.

68 Tavangar, T., Karimi, M., Rezakazemi, M. et al. (2020). Textile waste, dyes/inorganic salts separation of cerium oxide-loaded loose nanofiltration polyethersulfone membranes. *Chemical Engineering Journal* 385: 123787.

69 Pishnamazi, M., Nakhjiri, A.T., Taleghani, A.S. et al. (2020). Molecular investigation into the effect of carbon nanotubes interaction with CO_2 in molecular separation using microporous polymeric membranes. *Scientific Reports* 10: 13285.

70 Dashti, A., Raji, M., Azarafza, A. et al. (2020). Computational simulation of CO_2 sorption in polymeric membranes using genetic programming. *Arabian Journal for Science and Engineering* 45: 7655–7666.

71 Sodeifian, G., Raji, M., Asghari, M. et al. (2019). Polyurethane-SAPO-34 mixed matrix membrane for CO_2/CH_4 and CO_2/N_2 separation. *Chinese Journal of Chemical Engineering* 27: 322–334.

72 Rezakazemi, M., Darabi, M., Soroush, E., and Mesbah, M. (2019). CO_2 absorption enhancement by water-based nanofluids of CNT and SiO_2 using hollow-fiber membrane contactor. *Separation and Purification Technology* 210: 920–926.

73 Rezakazemi, M., Sadrzadeh, M., and Matsuura, T. (2018). Thermally stable polymers for advanced high-performance gas separation membranes. *Progress in Energy and Combustion Science* 66: 1–41.

74 Rezakazemi, M., Ebadi Amooghin, A., Montazer-Rahmati, M.M. et al. (2014). State-of-the-art membrane based CO_2 separation using mixed matrix membranes (MMMs): an overview on current status and future directions. *Progress in Polymer Science* 39: 817–861.

75 Rezakazemi, M., Heydari, I., and Zhang, Z. (2017). Hybrid systems: combining membrane and absorption technologies leads to more efficient acid gases (CO_2 and H_2S) removal from natural gas. *Journal of CO_2 Utilization* 18: 362–369.

76 Lin, H. and Freeman, B.D. (2005). Materials selection guidelines for membranes that remove CO_2 from gas mixtures. *Journal of Molecular Structure* 739: 57–74.

77 Handa, Y.P., Lampron, S., and O'neill, M.L. (1994). On the plasticization of poly(2,6-dimethyl phenylene oxide) by CO_2. 32: 2549–2553.

78 Lin, H., Van Wagner, E., Freeman, B.D. et al. (2006). Plasticization-enhanced hydrogen purification using polymeric membranes. 311: 639–642.

79 Car, A., Stropnik, C., Yave, W., and Peinemann, K.-V. (2008). Tailor-made polymeric membranes based on segmented block copolymers for CO_2 separation. 18: 2815–2823.

80 Qiao, Z., Wang, Z., Yuan, S. et al. (2015). Preparation and characterization of small molecular amine modified PVAM membranes for CO_2/H_2 separation. 475: 290–302.

81 Quinn, R. and Laciak, D.V. (1997). Polyelectrolyte membranes for acid gas separations. 131: 49–60.

82 Mahajan, R. and Koros, W.J. (2002). Mixed matrix membrane materials with glassy polymers. Part 1. *Science* 42: 1420–1431.

83 Song, C., Wang, T., Wang, X. et al. (2008). Preparation and gas separation properties of poly(furfuryl alcohol)-based C/CMS composite membranes. 58: 412–418.

84 Ebner, A.D. and Ritter, J.A. (2009). State-of-the-art adsorption and membrane separation processes for carbon dioxide production from carbon dioxide emitting industries. 44: 1273–1421.

85 Llosa Tanco, M.A., Medrano, J.A., Gallucci, F., and Pacheco Tanaka, D.A. (2018). Membrane optimization and process condition investigation for enhancing the CO_2 separation from natural gas. In: *Current Trends and Future Developments on (Bio-) Membranes*, Chapter 17 (eds. A. Basile and E.P. Favvas), 469–509. Elsevier.

86 Xing, R. and Ho, W.S.W. (2009). Crosslinked polyvinylalcohol–polysiloxane/fumed silica composite membranes containing amines for CO_2/H_2 separation. *Journal of Membrane Science* 367: 91–102.

87 Myers, C., Pennline, H., Luebke, D. et al. (2008). High temperature separation of carbon dioxide/hydrogen mixtures using facilitated supported ionic liquid membranes. *Journal of Membrane Science* 322: 28–31.

88 Bai, H. and Ho, W.S.W. (2009). New carbon dioxide-selective membranes based on sulfonated polybenzimidazole (SPBI) copolymer matrix for fuel cell applications. *Industrial & Engineering Chemistry Research* 48: 2344–2354.

89 El-Azzami, L.A. and Grulke, E.A. (2009). Carbon dioxide separation from hydrogen and nitrogen: facilitated transport in arginine salt–chitosan membranes. *Journal of Membrane Science* 328: 15–22.

90 Reijerkerk, S.R., Wessling, M., and Nijmeijer, K. (2011). Pushing the limits of block copolymer membranes for CO_2 separation. 378: 479–484.

91 Chen, H., Xiao, Y., and Chung, T.-S. (2010). Synthesis and characterization of poly(ethylene oxide) containing copolyimides for hydrogen purification. 51: 4077–4086.

92 Mehmood, O., Farrukh, S., Hussain, A. et al. (2020). Optimization analysis of polyurethane based mixed matrix gas separation membranes by incorporation of gamma-cyclodextrin metal organic frame work. *Chemical Papers* 74: 3527–3543.

93 Hussain, A., Farrukh, S., Hussain, A., and Ayoub, M. (2019). Carbon capture from natural gas using multi-walled CNTs based mixed matrix membranes. *Environmental Technology* 40: 843–854.

94 Zhao, Y. and Ho, W.S.W. (2012). Steric hindrance effect on amine demonstrated in solid polymer membranes for CO_2 transport. 415: 132–138.

95 Ahmad, S., Marson, G.V., Zeb, W. et al. (2020). Mass transfer modelling of hollow fiber membrane contactor for apple juice concentration using osmotic membrane distillation. *Separation and Purification Technology* 250: 117209.

96 Rehman, W.U., Muhammad, A., Khan, Q.A. et al. (2019). Pomegranate juice concentration using osmotic distillation with membrane contactor. *Separation and Purification Technology* 224: 481–489.

97 Muhammad, A., Younas, M., and Rezakazemi, M. (2018). CFD simulation of copper(II) extraction with TFA in non-dispersive hollow fiber membrane contactors. *Environmental Science and Pollution Research* 25: 12053–12063.

98 Hajilary, N. and Rezakazemi, M. (2018). CFD modeling of CO_2 capture by water-based nanofluids using hollow fiber membrane contactor. *International Journal of Greenhouse Gas Control* 77: 88–95.

99 Chen, H.Z., Li, P., and Chung, T.-S. (2012). PVDF/ionic liquid polymer blends with superior separation performance for removing CO_2 from hydrogen and flue gas. 37: 11796–11804.

100 Ansaloni, L., Zhao, Y., Jung, B.T. et al. (2015). Facilitated transport membranes containing amino-functionalized multi-walled carbon nanotubes for high-pressure CO_2 separations. *Journal of Membrane Science* 490: 18–28.

101 Klepić, M., Setničková, K., Lanč, M. et al. (2020). Permeation and sorption properties of CO_2-selective blend membranes based on polyvinyl alcohol (PVA) and 1-ethyl-3-methylimidazolium dicyanamide ([EMIM][DCA]) ionic liquid for effective CO_2/H_2 separation. *Journal of Membrane Science* 597: 117623.

102 Reed, B.W., Semmens, M.J., and Cussler, E.L. (1995). Membrane contactors. In: *Membrane Science and Technology*, Chapter 10 (eds. R.D. Noble and S.A. Stern), 467–498. Elsevier.

103 Chakraborty, M., Bhattacharya, C., and Datta, S. (2010). Emulsion liquid membranes: definitions and classification, theories, module design, applications, new directions and perspectives. In: *Liquid Membranes*, Chapter 4 (ed. V.S. Kislik), 141–199. Amsterdam: Elsevier.

104 Jiang, B., Dou, H., Zhang, L. et al. (2017). Novel supported liquid membranes based on deep eutectic solvents for olefin-paraffin separation via facilitated transport. *Journal of Membrane Science* 536: 123–132.

105 Kamarudin, K., Bhatti, I., Suahadah, S. et al. (2010). Removal of carbon dioxide using water-in-oil emulsion liquid membrane containing triethanolamine. 6: 2251–2256.

106 Dolmat, N., Kamarudin, K., and Paiman, S.H. (2014). Removal of carbon dioxide using emulsion liquid membrane containing *N*-methyldiethanolamine and piperazine in rotating disc contactor. 14: 1477–1480.

107 Ilyas, A., Muhammad, N., Gilani, M.A. et al. (2017). Supported protic ionic liquid membrane based on 3-(trimethoxysilyl)propan-1-aminium acetate for the highly selective separation of CO_2. *Journal of Membrane Science* 543: 301–309.

108 U. Saeed, A.L. Khan, M.A. Gilani, M. Aslam, A.U. Khan, CO_2 separation by supported liquid membranes synthesized with natural deep eutectic solvents, (2020) 1-15.

109 Craveiro, R., Neves, L.A., Duarte, A.R.C., and Paiva, A. (2021). Supported liquid membranes based on deep eutectic solvents for gas separation processes. *Separation and Purification Technology* 254: 117593.

110 Carpenter, S.M. and Long, H.A. III, (2017). Integration of carbon capture in IGCC systems. In: *Integrated Gasification Combined Cycle (IGCC) Technologies* (eds. T. Wang and G. Stiegel), 445–463. Elsevier.

111 Gazzani, M., Macchi, E., and Manzolini, G. (2013). CO_2 capture in natural gas combined cycle with SEWGS. Part A: thermodynamic performances. *International Journal of Greenhouse Gas Control* 12: 493–501.

112 Fout, T., Shultz, T., Woods, M. et al. (2018). *Cost and Performance Baseline for Fossil Energy Plants, Volume 1: Bituminous Coal and Natural Gas to Electricity*, Revision 4. NETL.

113 Van Harmelen, T., Koornneef, J., van Horssen, A. et al. (2008). *The Impacts of CO_2 Capture Technologies on Transboundary Air Pollution in the Netherlands.* Universiteit Utrecht.

114 Ji, G. and Zhao, M. (2017). *Membrane Separation Technology in Carbon Capture*. IntechOpen.

115 Basu, S. and Debnath, A.K. (2019). Plant P&ID (process) discussions. In: *Power Plant Instrumentation and Control Handbook*, Chapter 3, 2e (eds. S. Basu and A.K. Debnath), 149–250. Boston, MA: Academic Press.

116 Al Juaied, M. and Whitmore, A. (2009). Realistic costs of carbon capture. In: *Energy Technology Innovation Policy Research Group*. United States Department of Energy.

117 David, J. and Herzog, H. (2000). The cost of carbon capture. *5th International Conference On Greenhouse Gas Control Technologies*, Cairns, Australia, pp. 13–16.

118 Rao, A., Rubin, E., and Berkenpas, M.B. (2004). An integrated modeling framework for carbon management technologies volume 1-technical documentation: amine-based CO_2 capture and storage systems for fossil fuel power plant. United States Department of Energy.

119 Barchas, R. and Davis, R. (1992). The Kerr-McGee/ABB Lummus Crest technology for the recovery of CO_2 from stack gases. *Energy Conversion and Management* 33: 333–340.

120 Abu-Zahra, M.R., Abbas, Z., Singh, P., and Feron, P.J.M. (2013). Carbon dioxide post-combustion capture: solvent technologies overview, status and future directions. 1: 923–934.

121 Iijima, M., Nagayasu, T., Okino, S., and Kajiya, Y. (2009). CO_2 recovery technology for coal-fired power plants. 46: 20.

122 GP Journal of the Energy Institute of London (2010). Design, Operation for Onshore Carbon Capture Installations and Onshore Pipelines.

123 Reddy, S., Scherffius, J., Freguia, S., and Roberts, C. (2003). Fluor's econamine FG plus SM technology. *Proceedings of the 2nd Annual Conference on Carbon Sequestration*, Citeseer, pp. 5–8.

124 Dooley, J.J., Davidson, C.L., and Dahowski, R.T. (2009). *An Assessment of The Commercial Availability of Carbon Dioxide Capture And Storage Technologies as of June 2009*. Richland, WA: Pacific Northwest National Lab. (PNNL).

125 Baron, H. (2010). *The Oil and Gas Engineering Guide*. Editions Technip.

126 Global CCS Institute (2017). *Projects Database: Large-Scale CCS Facilities*. Melbourne: Global CCS Institute.

127 Tatsumi, M., Yagi, Y., Kadono, K. et al. (2011). New energy efficient processes and improvements for flue gas CO_2 capture. *Energy Procedia* 4: 1347–1352.

128 Bandyopadhyay, A. (2014). *Carbon Capture and Storage: CO_2 Management Technologies*. CRC Press.

129 Hu, L., Pal, S., Nguyen, H. et al. (2020). Molecularly engineering polymeric membranes for H_2/CO_2 separation at 100–300 °C. *Journal of Polymer Science*: 1–15.

130 Rowe, B.W., Robeson, L.M., Freeman, B.D., and Paul, D.R. (2010). Influence of temperature on the upper bound: theoretical considerations and comparison with experimental results. 360: 58–69.

131 Merkel, T. (2011). *Novel Polymer Membrane Process for Pre-combustion CO_2 Capture from Coal-Fired Syngas*. Menlo Park, CA: Membrane Technology and Research, Incorporated.

132 Syngas, C.-F. (2010). *National Energy Technology Laboratory CO_2 Capture Technology Meeting*. United States Department of Energy.

133 Singh, R.P., Dahe, G.J., Dudeck, K.W. et al. (2014). High temperature polybenzimidazole hollow fiber membranes for hydrogen separation and carbon dioxide capture from synthesis gas. 63: 153–159.

134 Wang, S., Li, X., Wu, H. et al. (2016). Advances in high permeability polymer-based membrane materials for CO_2 separations. 9: 1863–1890.

135 Li, X., Singh, R.P., Dudeck, K.W. et al. (2014). Influence of polybenzimidazole main chain structure on H_2/CO_2 separation at elevated temperatures. 461: 59–68.

136 Zhu, L., Swihart, M.T., and Lin, H. (2018). Unprecedented size-sieving ability in polybenzimidazole doped with polyprotic acids for membrane H_2/CO_2 separation. 11: 94–100.

137 Carreon, M.A. *Membranes for Gas Separations*. World Scientific Publishing Company.

138 Ali, Z., Pacheco, F., Litwiller, E. et al. (2018). Ultra-selective defect-free interfacially polymerized molecular sieve thin-film composite membranes for H_2 purification. 6: 30–35.

139 Kim, D., Tzeng, P., Barnett, K.J. et al. (2014). Highly size-selective ionically crosslinked multilayer polymer films for light gas separation. 26: 746.

140 Illing, G., Hellgardt, K., Schonert, M. et al. (2005). Towards ultrathin polyaniline films for gas separation. *Journal of Membrane Science* 253: 199–208.

141 Moon, J.D., Bridge, A.T., D'Ambra, C. et al. (2019). Gas separation properties of polybenzimidazole/thermally-rearranged polymer blends. 582: 182–193.

142 Panapitiya, N.P., Wijenayake, S.N., Nguyen, D.D. et al. (2015). Interfaces, gas separation membranes derived from high-performance immiscible polymer blends compatibilized with small molecules. 7: 18618–18627.

143 Naderi, A., Tashvigh, A.A., Chung, T.-S. et al. (2018). Molecular design of double crosslinked sulfonated polyphenylsulfone/polybenzimidazole blend membranes for an efficient hydrogen purification. 563: 726–733.

144 Klaehn, J.R., Orme, C.J., and Peterson, E.S. (2016). Blended polybenzimidazole and melamine-co-formaldehyde thermosets. 515: 1–6.

145 Sánchez-Laínez, J., Zornoza, B., Carta, M. et al. (2018). Hydrogen separation at high temperature with dense and asymmetric membranes based on PIM-EA (H_2)-TB/PBI blends. 57: 16909–16916.

146 Giel, V., Morávková, Z., Peter, J., and Trchová, M. (2017). Thermally treated polyaniline/polybenzimidazole blend membranes: structural changes and gas transport properties. 537: 315–322.

147 Koros, W.J. and Zhang, C. (2017). Materials for next-generation molecularly selective synthetic membranes. 16: 289–297.

148 Riasat Harami, H., Riazi Fini, F., Rezakazemi, M., and Shirazian, S. (2019). Sorption in mixed matrix membranes: experimental and molecular dynamic simulation and grand canonical Monte Carlo method. *Journal of Molecular Liquids* 282: 566–576.

149 Rezakazemi, M., Shahidi, K., and Mohammadi, T. (2012). Sorption properties of hydrogen-selective PDMS/zeolite 4A mixed matrix membrane. *International Journal of Hydrogen Energy* 37: 17275–17284.

150 Rezakazemi, M., Shahidi, K., and Mohammadi, T. (2012). Hydrogen separation and purification using crosslinkable PDMS/zeolite A nanoparticles mixed matrix membranes. *International Journal of Hydrogen Energy* 37: 14576–14589.

151 Japip, S., Liao, K.S., and Chung, T.-S. (2017). Molecularly tuned free volume of vapor cross-linked 6FDA-durene/ZIF-71 MMMs for H_2/CO_2 separation at 150 °C. 29: 1603833.

152 Wijenayake, S.N., Panapitiya, N.P., Versteeg, S.H. et al. (2013). Surface cross-linking of ZIF-8/polyimide mixed matrix membranes (MMMs) for gas separation. 52: 6991–7001.

153 El Naga, A.O.A., Shaban, S.A., and El Kady, F.Y.A. (2018). Metal organic framework-derived nitrogen-doped nanoporous carbon as an efficient adsorbent for methyl orange removal from aqueous solution. 93: 363–373.

154 Japip, S., Liao, K.-S., Xiao, Y., and Chung, T.-S. (2016). Enhancement of molecular-sieving properties by constructing surface nano-metric layer via vapor cross-linking. *Journal of Membrane Science* 497: 248–258.

155 Kang, Z., Peng, Y., Qian, Y. et al. (2016). Mixed matrix membranes (MMMs) comprising exfoliated 2D covalent organic frameworks (COFs) for efficient CO_2 separation. 28: 1277–1285.

156 Yang, T., Shi, G.M., and Chung, T.-S. (2012). Symmetric and asymmetric zeolitic imidazolate frameworks (ZIFs)/polybenzimidazole (PBI) nanocomposite membranes for hydrogen purification at high temperatures. 2: 1358–1367.

157 Smith, Z.P., Tiwari, R.R., Murphy, T.M. et al. (2013). Hydrogen sorption in polymers for membrane applications. 54: 3026–3037.

158 Zhu, L., Yin, D., Qin, Y. et al. (2019). Sorption-enhanced mixed matrix membranes with facilitated hydrogen transport for hydrogen purification and CO_2 capture. 29: 1904357.

159 Kaldis, S.P., Pantoleontos, G.T., and Koutsonikolas, D.E. (2018). Membrane technology in igcc processes for precombustion CO_2 capture. In: *Current Trends and Future Developments on (Bio-) Membranes* (eds. A. Basile and E.P. Favvas), 329–357. Elsevier.

160 Huang, W., Jiang, X., He, G. et al. (2020). A novel process of H_2/CO_2 membrane separation of shifted syngas coupled with gasoil hydrogenation. 8: 590.

161 Li, B., He, G., Jiang, X. et al. (2016). Pressure swing adsorption/membrane hybrid processes for hydrogen purification with a high recovery. *Frontiers of Chemical Science and Engineering* 10: 264–255.

162 Aaron, D. and Tsouris, C. (2005). Separation of CO_2 from flue gas: a review. *Separation Science and Technology* 40: 321–348.

163 Sánchez-Laínez, J., Zornoza, B., Orsi, A.F. et al. (2018). Synthesis of ZIF-93/11 hybrid nanoparticles via post-synthetic modification of ZIF-93 and their use for H_2/CO_2 separation. *Chemistry – A European Journal*: 1–10.

164 Song, H., Zhao, S., Chen, J., and Qi, H. (2016). Hydrothermally stable Zr-doped organosilica membranes for H_2/CO_2 separation. 224: 277–284.

165 Hou, J., Wei, Y., Zhou, S. et al. (2018). Highly efficient H_2/CO_2 separation via an ultrathin metal-organic framework membrane. 182: 180–188.

166 D'Alessandro, D.M., Smit, B., and Long, J.R. (2010). Carbon dioxide capture: prospects for new materials. 49: 6058–6082.

167 Avci, G., Velioglu, S., and Keskin, S. (2018). High-throughput screening of MOF adsorbents and membranes for H_2 purification and CO_2 capture. 10: 33693–33706.

168 Tong, M., Yang, Q., and Zhong, C. (2015). Computational screening of covalent organic frameworks for CH_4/H_2, CO_2/H_2 and CO_2/CH_4 separations. 210: 142–148.

169 Avci, G., Erucar, I., and Keskin, S. (2020). Do new MOFs perform better for CO_2 capture and H_2 purification? Computational screening of the updated MOF database. *ACS Applied Materials Interfaces* 12: 41567–41579.

170 Farrukh, S. *Nanotechnology and the Generation of Sustainable Hydrogen*. Springer Nature.

9

Application of Membrane Contactor Technology for Post-combustion Carbon Dioxide (CO_2) Capture

Muhammad B. Wazir[1], Muhammad Daud[1], Mohammad Younas[1], and Mashallah Rezakazemi[2]

[1] *University of Engineering and Technology, Department of Chemical Engineering, 25120 Peshawar, Pakistan*
[2] *Shahrood University of Technology, Faculty of Chemical and Materials Engineering, Shahrood, Iran*

9.1 Introduction

CO_2 capture, storage, and/or subsequent conversion to useful products is considered a promising way to mitigate the effects of fossil fuel consumption on the environment [1, 2]. CO_2 capture can be carried out before the combustion process from syngas (pre-combustion) and after the complete combustion from flue gas (post-combustion). Both ways come tethered with their associated advantages and drawbacks. The pre-combustion CO_2 capture process intakes a mixture of CO_2/H_2 at high pressure of 15–20 bar, elevated temperature of 190–210 °C, and high concentration of CO_2 (~45 vol%). On the contrary, the post-combustion capture scheme sequesters CO_2 from a mixture of CO_2/N_2 at near ambient conditions and a relatively lower concentration of CO_2 (5–15 vol%). As the feed concentration of CO_2 is already high for pre-combustion, one can rightly perceive it as more efficient than the post-combustion process. However, harsh working conditions (high temperature and high pressure) and plants' existing configuration make it more complicated. Highly durable materials are required for the construction of membranes to withstand the harsh conditions. Moreover, additional constraints like inertness at elevated temperature and pressure, low volatility, and high CO_2 solubility at relatively higher temperature need to be considered to select suitable absorbents. Although the post-combustion CO_2 capture scheme suffers greatly from lower CO_2 concentration in the feed, it is still desirable due to the easy retrofitting of existing plants and relatively gentle working conditions. Figure 9.1 illustrates a typical post-combustion CO_2 capture process through membrane contactors.

Membrane Contactor Technology: Water Treatment, Food Processing, Gas Separation, and Carbon Capture, First Edition. Edited by Mohammad Younas and Mashallah Rezakazemi.

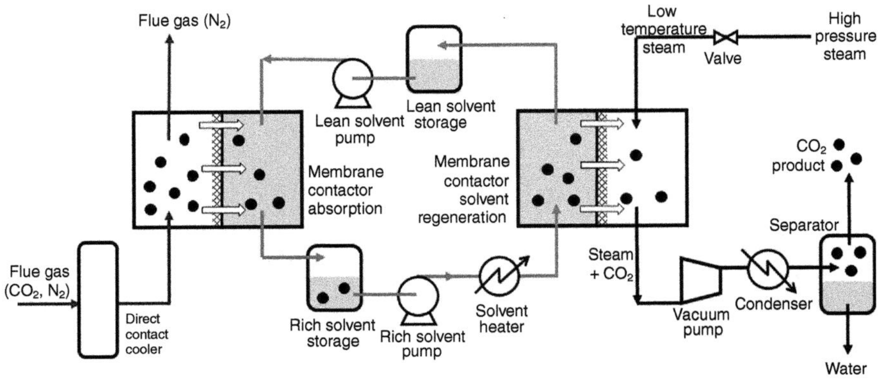

Figure 9.1 Schematic illustration of pilot-scale post-combustion CO_2 capture unit. Source: Scholes et al. [3].

The membrane absorption process is preferred due to easy upscaling, small footprint, operational flexibility, large liquid–gas interfacial area, and absence of foaming and flooding phenomenon associated with conventional absorption [4, 5]. However, several challenges are present on both the membrane side and the absorbent side. In Sections 9.3, 9.4, and 9.6, we will examine how contemporary research is trying to resolve the aforementioned issues with post-combustion CO_2 sequestration.

9.2 Membranes for Post-combustion CO_2 Capture

9.2.1 Membrane Types

The membranes employed in membrane contactors can be classified into porous membranes and nonporous composite membranes. The porous membranes are manufactured from hydrophobic materials like polytetrafluoroethylene (PTFE), polyvinylidenefluoride (PVDF), polyethylene (PE), and polypropylene (PP) [6]. The pores are ideally meant to be filled with gas to provide a high gas diffusion coefficient. However, with time, the solvent starts to seep into pores, increasing the overall mass transfer resistance. It was confirmed in pilot plant studies on CO_2 capture that, irrespective of membrane materials, pore wetting is problematic and unavoidable due to operational constraints [7, 8]. It has a huge detrimental effect on overall efficiency of CO_2 capture process as diffusivity in liquids is lower by order of 10^4 than in gases [9]. To counter the wetting process, nonporous composite membranes employ a thin dense layer to stop the solvent from entering the pores (Figure 9.2A) [8, 11, 13]. The thin film coating on porous part is illustrated in Figure 9.2B–D. The thin film coating also reduces the gas bubbling and solvent entrainment. Nonetheless, it adds to mass transfer resistance and reduces the CO_2 transport from flue gas to solvent. Therefore, materials with high permeability for CO_2 such as polymethylpentene (PMP), polymer of intrinsic

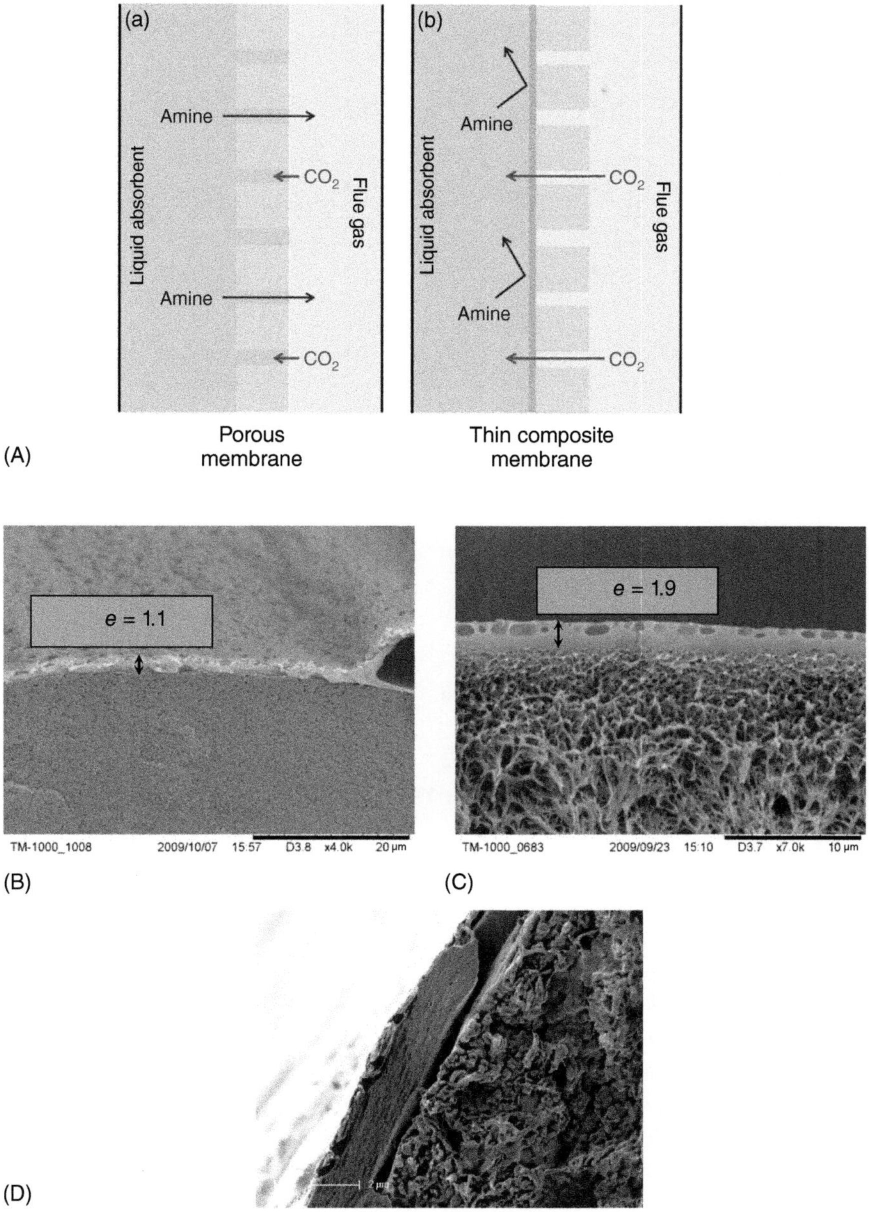

Figure 9.2 (A) Schematic illustration of membrane contactor employing a porous membrane (a) or thin film composite membrane (b) for the post-combustion CO_2 separation. Source: Ansaloni et al. [10]. (B–D) Scanning electron microscope (SEM) images of thin film composite membranes. (B) Teflon AF 2400 and (C) PTMSP. Source: Reproduced with permission from Nguyen et al. [11]. Copyright (2011) Elsevier. (D) PIM-1 on PP. Source: Reprinted with permission from Scholes et al. [12]. Copyright (2015) Elsevier. Comparison of the overall mass transfer coefficient (cm/s) for CO_2 separation by thin films of (E) PTMSP and (F) PIM-1 on PP membranes. Source: Scholes et al. [12]. (G) CO_2 capture ratio vs. gas velocity for hollow fiber contactors. Source: Nguyen et al. [11].

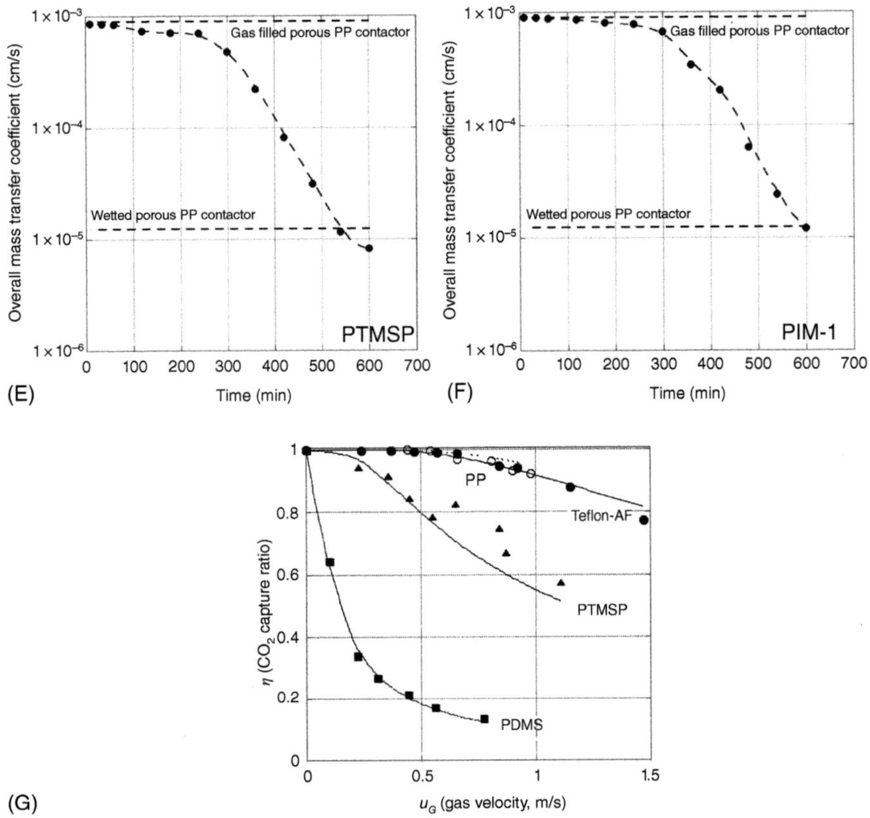

Figure 9.2 *(Continued)*

microporosity (PIM)-1, and poly(1-trimethylsilyl-1-propyne) (PTMSP) are often used for thin interfacial layer to minimize the additional resistance to overall mass transfer [12]. However, the aforementioned materials are not suitable for long-term operations because of drastic performance reduction due to the aging over time [14]. Although the CO_2 permeation of the aforementioned materials is high, the amount of water seeping through the nonporous thin layer of these materials results in pore wetting that in turn adds to the mass transfer resistance. This phenomenon is also depicted in Figure 9.2A. Consequently, their performance becomes comparable with that of wetted porous membranes approaches, as shown in Figure 9.2E,F. Fortunately, the perfluorinated Teflon AF 1600 thin film composite membranes are best suited for restricting the solvent from entering the nonporous and porous layers. Hence, fluorinated polymers exhibit better long-term performance for CO_2 capture, as shown in Figure 9.2G. This means that thin film-based membrane contactor and high permeance for CO_2 must exhibit low water permeability. To overcome detrimental aging, fluorinated polymers (e.g. Teflon AF 1600 and Teflon AF 2400) are used to construct nonporous thin film composite membranes [11, 12].

Table 9.1 Comparison of membrane contactors modules for post-combustion CO_2 capture.

Module type	Hollow fiber	Envelope	Spiral wound
Packing density (m^2/m^3)	<10 000	200–500	<1000
Cost (€/m^2)	2–8	40–150	8–37
Manufacturing	Cheap	Easy	Easy and cheap
Cleaning	Chemical washing or replaced	Medium	Hard
Pressure drop	High in the fibers	Moderate	High and longer permeate path

Source: Wang et al. [15].

9.2.2 Membrane Modules

In addition to the membrane type, the configuration of the membrane module also has a significant influence on the post-combustion CO_2 capture performance of membrane contactors. Three configuration, namely, hollow fiber, spiral wound, and envelope, are widely implemented [15]. The surface area per volume of membrane is a key indicator for better evaluation of these modules; it is reflected in the packing density of module. A comprehensive comparison is shown in Table 9.1.

9.3 Experimental Membrane Materials for Post-combustion CO_2 Sequestration

The contemporary researchers are giving considerable attention to developing and testing new membrane materials for post-combustion carbon capture. Kárászová et al. [16] article detailed the 27 membrane material listed in Table 9.2. These materials were utilized for CO_2 separation in mostly two configurations: flat sheet and hollow fibers. These membranes include liquid membranes (LMs), polymeric membranes (PMs) [44, 45], ceramic porous membranes (CMs) [46–49], and mixed matrix membranes (MMMs) [50–53]. The LMs exhibited superior diffusivity and can easily be modified for requisite application. Nevertheless, liquid leakage is very problematic and must be circumvented. In addition, the stability of these membranes also demands extra attention. The CMs showed high CO_2 capture efficiency of 90% from flue gas [54, 55]. The PMs also exhibit good permeability and selectivity; however, plasticization during the separation of CO_2 from flue gas is detrimental to overall performance. The MMMs incorporating metal–organic frameworks (MOFs) show a synergized performance due to the merger of separation and transport properties of the polymer matrix and nanoparticles' stabilization [29, 56]. However, in pristine form, MOFs suffer greatly in humid feeds, and the presence of impurities can also damage their framework. Therefore, careful modifications are often needed to stabilize these MOFs [1, 56].

Table 9.2 Experimental membrane materials for post-combustion CO_2 separation.

S. no.[a)]	Membrane type	Membrane	Permeability (barrer)	CO_2/N_2 selectivity	Note	References
1	SILM	[ATMPS][Ac] in polyimide 84	24	40		[17]
2	SILM	[emim][Tf_2N] and (Al(i-$C_3H_7O)_3$) + SAPO 34 in PSf	7.24	20.3		[18]
3	LM in composite	Pebax®1657/[emim][BF4]	280	42		[19]
4	Liquid membrane (LM) in MMM	GO-IL in Pebax 1657	143	71		[20]
5	LM in MMM	Polyionic liquid with free ionic liquid content on epoxide amine	900	140		[21]
6	MMM	CA + Tf_2N anion	8.9	26.8	Did not enhance the CA properties	[22]
7	MMM	Pebax membrane	194	44		[23]
8	MMM	Amino-functionalized polyhedral oligomeric silsesquioxane (amine POSS®) nanoparticles dispersed in polyvinyl alcohol (PVA)	22.2		Humidity effect: increase of CO_2 permeance	[24]
9	MMM	Pebax/polytetramethylene ether glycol (PTMEG)	350	60	Depends on PTMEG content	[25]
10	MMM	Polyurethane (PU) with NiO nanoparticles	350	68	Depending on nanoparticle content	[26]
11	MMM	Contained Fe(benzene 1,3,5 tricarboxylate [BTC]) incorporated in Matrimid	218	23		[27]
12	MOF	NH_2-MIL-53(Al) in CA	15.5	25		[28]
13	MOF	PU/UiO-66 (Zr)	75	34	Tamb and 4 bar	[29]
14	MOF	PU/MIL-101 (Cr)	83	42	Tamb and 4 bar	[29]
15	MOF	Zeolitic imidazolate framework (ZIF)/300/Pebax 1657	102.1	73.9	298 K, 12 bar dry feed CO_2:N_2 15 : 85 (v/v)	[30]

(Continued)

Table 9.2 (Continued)

S. no.[a)]	Membrane type	Membrane	Permeability (barrer)	CO_2/N_2 selectivity	Note	References
16	MOF	ZIF/300/Pebax 1657	466.2	59.2	298 K, 12 bar dry feed CO_2:N_2 15 : 85 (v/v) relative humidity 85%	[30]
17	MOF	PEDM/ZIF-8@GO-6–6	475	58.2	Tests at 25 °C under 1 bar	[31]
18	PIM	PIM-1	7500	19		[32]
19	PIM in MMM	PIM-1 – Schiff base network	7600	22.4		[33]
20	MMM in composite	Pebax/zeolite Y	795	35		[34]
21	Composite	Pebax without zeolite	1420	10		[34]
22	Composite	Pebax/polyethylene glycol (PEG) 200 polymer selective layer and Biomax PES with zeolite Y	745	25		[35]
23	Composite	Polyether imide (PEI) (PVC-g-POEM)	35			[36]
24	Hollow fiber	Asymmetric PES hollow fibers	50	45	Selectivity from numerous model	[37]
25	TRP	Polybenzoxazole		28		[38]
26	Fixed carriers	Polyvinylamine	222.2	300		[39]
27	Polymeric (modified PI)	Semi-alicyclic aromatic polyimide	8.6	28		[40]
28	Ceramic	$Ce_{0.8}Sm_{0.2}O_{1.9}$(SDC)/ Li_2CO_3-Na_2CO_3 ceramic carbonate dual-phase membrane		192	SO_2-poisoned CO_2 permeation, temperature of 550–750 °C	[41]
29	Ceramic	Amino-modified mesoporous ceramic membranes	30	22		[42]

a) For reference in Figure 9.3.
Source: Kárászová et al. [16].

To further explore the current dynamics in the field of post-combustion CO_2 sequestration through membrane-based separation, we will now proceed to the pilot-scale studies. These investigations will help us in evaluating the application of membrane materials in real-world scenarios. Let us begin with the pilot-scale study reported by Choi et al. [37]; the researchers investigated the CO_2 capture from flue gas of liquefied natural gas (LNG) through a membrane separation unit having hollow fiber configuration. The polyethersulfone (PES) was used as construction material for separation layers of membrane. After testing the unit with a model mixture of CO_2 and N_2, it was employed to separate flue gas containing CO_2, O_2, and N_2 in fractions of 10.8%, 2%, and 87.2%, respectively. The permeance of CO_2 was observed to be 40 GPU through the used membrane. The feed was supplied at 588 kPa with a flow rate of 500 l/min in the first stage of four-stage process. The permeate pressure was 19.6 kPa. They were able to maintain CO_2 purity at 99% and sequester 90% of CO_2. The experimental observations were also validated through simulation by Aspen Custom Modeler. However, the authors were unable to replicate the experimental results for unknown reasons that need further investigation. In another pilot-scale study, Eiberger et al. [42] investigated ceramic membranes with an asymmetrical structure coated with a mesoporous γ-Al_2O_3 or 8YSZ interlayer supported by macroporous α-Al_2O_3. A number of problems, including condensation of water, deposition of gypsum and fly ash complemented with the accumulation of sulfate crystals, and irreversible phase transformation of γ-Al_2O_3 interlayer, were observed to limit the extended use. Therefore, long-term operation would necessitate pretreatments of feed such as drying and eradication of ash particles.

9.4 Commercial Membranes for Post-combustion CO_2 Separation

A few commercial membranes have also been assessed for CO_2 separation from flue gas. The summary of commercial membranes tested for flue gas is presented in Table 9.3. We will briefly discuss their merits and limitations. A detailed discussion on these commercial membranes can be found in a recently published review [16].

Table 9.3 A summary of commercial membranes utilized CO_2 separation from flue gas.

S. no.[a)]	Membrane	Manufacturer	Permeance (GPU)	CO_2/N_2 selectivity	Polymer	References
A	Medal	Air liquid	—	50	PI	[57]
B	PRISM™	Air products	760	13	PSF	[58]
C	Polaris™ gen 1	MTR	1000	50	PE-PA copol.	[59]
D	Polaris gen 2	MTR	2000	49	PE-PA copol.	[59]
E	PolyActive	Helmholtz-Zentrum	1480	55	PEO-PBT	[60]

a) For reference in Figure 9.3.
Source: Kárászová et al. [16].

Scholes et al. [58] studied CO_2 capture performance of a commercial membrane PRISM™ manufactured by Air Products. The authors noticed that the water condensation on the membrane was detrimental to overall performance. In addition, plasticization of membrane resulted in drastic decrease in permeance and selectivity after few hours of operation that was further exacerbated due to the swelling of the membrane in the presence of water. Moreover, NO_x and SO_2 also permeated through the membrane. In short, the fluctuations in humidity aggravated by poor temperature control of the unit caused significant detrimental changes in transport and separation properties of the membrane [61]. The PolyActive™ membranes by Helmholtz–Zentrum are also actively examined for CO_2 capture from flue gas. However, at higher CO_2 fugacity, these membranes tend to swell, which decreases their selectivity for CO_2 [62]. To ameliorate these problems, Pohlmann et al. [63] suggested pretreat the flue gas for the elimination of condensates, water vapors, and dust particles. Other commercial membrane materials are listed in Table 9.3. Although a few commercial membranes have been tested for flue gas separation in particular settings, there is still a need for a comprehensive evaluation of all the membranes commercially available. The most pressing issue, for now, seems to be the durability of these commercial membranes. Table 9.3 lists the commercial membranes that are tested for post-combustion CO_2 sequestration.

9.5 Cost of Post-combustion CO_2 Capture in Membrane Contactors

As the lab-based studies discussed earlier are not reliable for predicting the economic impact of post-combustion CO_2 capture and large-scale pilot plants are too expensive, computational simulations are also considered cheap and reliable means to predict the economic performance of contactor-assisted CO_2 capture processes at large-scale operation. Zhai and Rubin [64] studied a computer-generated simulation of both single- and multistage carbon capture and separation plant utilizing the Integrated Environmental Control Model (IECM). It was observed that efforts to increase permeability could potentially yield more promising results than increasing CO_2/N_2 selectivity. In addition, Maas et al. [65] investigated a cascade membrane system operational at 25 °C for CO_2 capture, which achieved low energy consumption. However, they concluded that the cost allowance should surpass 37 €/ton CO_2 to make this technology economically viable.

Zhao et al. [15, 66] simulated a multistage membrane CO_2 separation process for a 600 MW power plant by using PRO/C. The system was tested for different CO_2 capture percentages: 50%, 70%, and 90%. It was witnessed that only 50% and 70% separation degrees can be deemed feasible due to energy requirements and cost constraints. It is noteworthy that exhaust gas only consisted CO_2 and N_2. Therefore, the effects of other gases such as O_2 and H_2O should be further explored as evidence of minuscule positive effect of water vapors on CO_2/N_2 separation is already present. Table 9.4 summarizes the economic evaluation of CO_2 sequestration from flue gas of power plants through membrane separation.

Table 9.4 Economic assessment of contactor-assisted post-combustion CO_2 sequestration.

Power output (MW)	Operating temperature (°C)	Permeance ($N m^3 m^{-2} h^{-1} bar^{-1}$)	CO_2 capture (%)	Efficiency loss	Membrane area	CO_2 avoidance cost (€/ton CO_2)	CO_2 capture cost (E/ton)	Cost of electricity	References
600	30	0.5	50	4.3	6.62/0.24				[66]
			70	6.8	13.92/0.34				
		3	70	6.4	2.44		31		
450	30	1000 GPU	90			47[a]	31[a]	89[a]	[64]
						72[b]	42[b]	106[b]	
600	25	3	90	9.6	0.40/0.07	45		91	[65]
	30	4.3		11.1	0.29/0.04				
	50	5		14.2	0.24/0.03				

a) With air sweep.
b) Without air sweep.
Source: Wang et al. [15].

The Robeson plot is widely used to compare membrane materials by contemporary researchers. Although the primary objective is to synthesis a membrane material that can cross the upper bound limit, the practical aspects such as requisite mechanical properties and a possibility to construct a homogenous defect-free thin film of sufficient area should also be considered. We have already established that the overall cost of the CO_2 capture process is a deciding factor. To make the membrane-based process competitive to traditional monoethanolamine (MEA)-based absorption process, a comprehensive cost analysis accounting all the contributive factors including cost of membrane materials and process cost such as cost of compression, cooling, evacuation, and other energy is needed. Fortunately, Kárászová et al. [16] compared the membrane materials listed in Tables 9.2 and 9.3 to MEA absorption process for post-combustion CO_2 sequestration on the Robeson chart. The authors also included the curves for percent savings when using the membrane-based process relative to MEA absorption (Figure 9.3). The authors suggested that CO_2 selectivity greater than 40 and permeance above 500 GPU could make membrane-based post-combustion CO_2 separation competitive to MEA-based separation process. Moreover, a permeability of higher than 200 barrer would be required at the same level of selectivity, given that a durable thin film 500 nm or less could be prepared. Increasing the permeability and selectivity of CO_2 can considerably reduce the operational costs. For instance, when the selectivity of CO_2 over N_2 was increased from 20 to 40 in PolyActive® membrane, the energy consumption was reduced by more than half of the original [67]. However, there is a trade-off relation between CO_2 selectivity and permeability for many membrane materials present today. This issue can also be circumvented by introducing a very thin (less than 500 nm) film of selective material upon porous support exhibiting superior permeability [68–70].

9.6 Absorbents for Post-combustion CO_2 Separation

The selection of absorbent is also among primary concerns for the application of membrane contactors in post-combustion CO_2 capture. Various factors need to be taken into consideration including selectivity of CO_2 over N_2 and other gases, chemical and thermal stabilities, corrosivity to membrane material and handling equipment, solubility of CO_2, etc. In addition, economic constraints such as regeneration and pumping cost are also of considerable interest. In general, absorbents exhibiting low volatility and corrosivity and high stability, selectivity, and solubility are preferred for post-combustion carbon capture [71, 72].

9.6.1 Amine-Based Absorbents

Different absorbents have been used for CO_2 capture [73, 74]. The different aspects of these solvents, such as thermodynamics, chemistry, kinetics, process modeling, etc., have been comprehensively investigated and described in the literature [75–77]. The amines comprise hydroxyl (–OH) and amino (–NH_2, –NR_2, and –NHR) groups

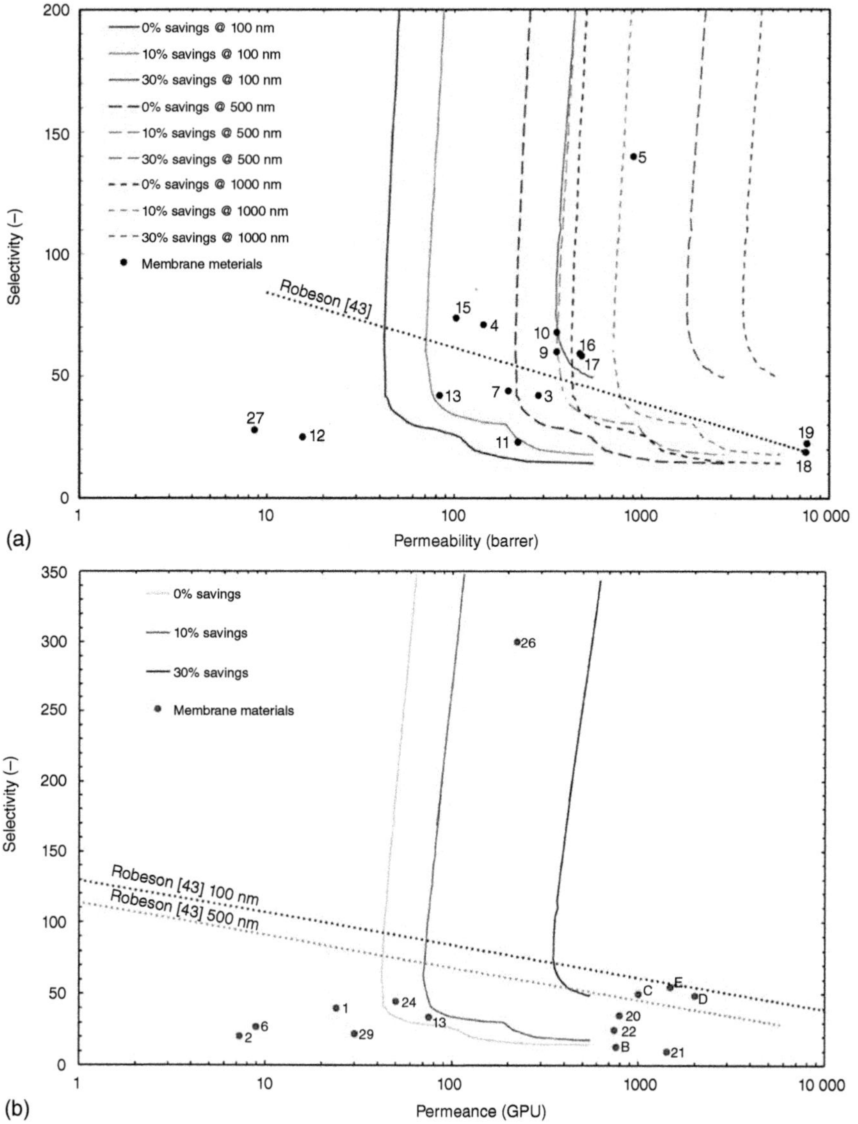

Figure 9.3 Comparison of cost-effectiveness of membrane materials of which (a) permeability and (b) permeance is known with traditional MEA based absorption. Source: Kárászová et al. [16].

tethered to an alkane chain. These can be classified into various types, including primary, secondary, tertiary, cyclic, and hindered amines. In the primary amines, an alkyl or aromatic substitutes a hydrogen atom in ammonia. Similarly, in secondary amines, there are two hydrogen replacements by aryl, alkyl, or both. A sterically hindered amine can be defined as a primary or secondary amine in which amine group is tethered to a tertiary carbon atom and/or secondary carbon atom [78]. However, in tertiary amines, all the hydrogen atoms are replaced by organic substituents, and their reaction mechanism for CO_2 capture is also different from other amines. The tertiary amines do not directly react with CO_2, they absorb the CO_2 through the production of hydroxyl radicals, and, hence, the presence of water is a prerequisite for CO_2 capture through tertiary amines [79, 80]. Among cyclic amines, pyrazine and morpholine are widely known and can be utilized for CO_2 capture [80–82]. Although pyrazine exhibits the same level of volatility as compared with MEA, it has an edge over it in terms of capacity and kinetics. In addition, it is also more stable and shows high resistance toward thermal or oxidative stimulus and can sustain up to 150 °C [83]. On the other hand, morpholine is being commercially utilized in the process named as Morphysorb®. It can capture various acidic gases from a wide range of feed sock with excellent selectivity [80].

A lot of new amines were introduced recently. The structural modifications are proving to be a way forward for these amines in CO_2 capture processes. The 4-diethylamino-2-butanol (DEAB), 1-diethylamino-2-propanol (1DEA2P), 2-(diethylamino)ethanol (DEEA), and 2-((2-aminoethyl)amino)ethanol (AEEA) are among the few new amines [84]. Among these, DEAB exhibits a high absorption capacity for CO_2 at a given CO_2 partial pressure. Benzylamine (BZA) is another class of amine-based solvents comprising a benzyl group tethered to the primary amine functional group. Richner et al. [85] reported the performance assessment and thermo-kinetic properties of BZA-based amines. The detailed studies on various amine-based absorbents for CO_2 capture can be found in the literature [86, 87].

9.6.2 Ammonia

The ammonia was first utilized for absorption of acidic gases in the 1950s. Up until now, a lot of researchers have studied its use for CO_2 absorption [88–91]. However, ammonia has various hazards associated and requires intense process control as it is toxic in nature and can instigate grave environmental issues even in small emissions. Makhloufi et al. [92] examined the feasibility of using ammonia as absorbent in a hollow fiber membrane contactor for CO_2 separation process. The authors utilized two dense thin film composite membranes: commercial poly(4-methyl-1-pentene) (TPXs) (PMP) and custom-made Teflon AF 2400. A large intensification factor for CO_2 mass transfer and reduced ammonia slip was observed as compared with column absorption process. Cui and deMontigny [93] investigated the aqueous ammonia for post-combustion CO_2 capture by using hollow fiber membrane contactors. The authors concluded that aqueous ammonia can achieve high CO_2 absorption performance comparable (98%) with MEA solvent at maximized liquid flow rate. However, PTFE membranes used in the contactor suffered greatly from wetting and fouling and resulted in performance degradation after one hour. This reduced performance can be attributed to the surface precipitation of ammonium salts, which, in addition to membrane fouling, can also reduce the hydrophobicity exacerbating the membrane wetting problem.

In another study, McLeod et al. [94] reported an interesting utilization of 2 M aqueous ammonia absorbent for CO_2 capture and simultaneous crystallization as ammonium bicarbonates on the shell side through surface tension-induced nucleation. However, limited ammonia concentration to control the surface tension might result in smaller CO_2 flux and lower reactivity. Nevertheless, there is a possibility for utilization of surface tension regulators and selection of better membranes to diminish the solvent seepage into pores. The kinetics for heterogeneous crystal growth in chemically reactive CO_2 absorption by ammonia is also detailed in a recent study by Bavarella et al. [95]. However, such utilization of ammonia as absorbent for post-combustion CO_2 capture through membrane contactors still needs to be explored in detail. Due to low mass transfer and low reactivity of ammonia-based CO_2 capture and simultaneous crystallization requiring large-scale operation, however, the pilot-scale studies are still nonexistent. Furthermore, the effect of crystallization, although, is deemed as of trivial importance in degrading the overall performance over time; it has yet to be confirmed at large-scale operation in practical conditions for post-combustion CO_2 capture.

9.6.3 Salt Solutions

The low cost of salt solutions like carbonate/bicarbonate and salts of amino acids make them preferable absorbents for CO_2 capture. The carbonates of sodium and potassium exhibit high CO_2 absorption capacity, high stability, and low toxicity. In addition, they are easy to regenerate, less corrosive in nature, and cost-effective chemical solvents. Due to the aforementioned merits, carbonate salt has been utilized more than 700 plants worldwide for H_2S and CO_2 absorption [96]. In a conventional process, an exothermic reaction of CO_2 with potassium carbonate is involved. The sodium carbonate-based process is also the same [97, 98]. The utilization of carbonate salts is suitable for high CO_2 concentrations. However, the reaction rate is slower, which can be addressed by adding organic and inorganic promoters [78, 96]. Moreover, the reaction rate can also be enhanced by elevating the operation temperature from ambient conditions to 75 °C [99]. However, at increased temperature, high loss of water occurs, which in turn requires a dehumidifier downstream to remove this extra water. Moreover, the scaling of these carbonates in membrane contactors is limiting the practical utilization of these solvents for post-combustion CO_2 sequestration. To facilitate these challenges, different amines such as piperazine, diethanolamine (DEA), and MEA are added as promoters to carbonate salt absorbents. This also results in lowering the operating temperature [100]. Moreover, mimic enzymes have also been utilized for enhancing post-combustion CO_2 capture in carbonate salts through membrane contactors [101].

Amino acids are composed of carboxyl and amine groups linked to a particular R chain. Amino acid salts (AASs) are obtained by reacting an inorganic base such as NH_3 [102] and KOH [103] with amino acids. Zhang et al. [104] reviewed the effectiveness of AAS for CO_2 absorption. The aqueous solutions of AAS are regarded as effective absorbents for membrane contactors because of their stable interaction with membrane materials, lower affinity for pore wetting, commercial availability, nontoxicity, thermal stability, and high reactive nature with CO_2 [105]. Moreover, they can also be easily regenerated and exhibit a zwitterionic reaction mechanism with CO_2 like primary and secondary amines [106, 107]. However, there is stern need for testing these absorbents at post-combustion condition in membrane contactor units at pilot scale.

9.6.4 Ionic Liquids

Ionic liquids (ILs) have recently amassed substantial research interest. These are organic salts composed of organic or inorganic anions and organic cations having bulky asymmetric structure [108–111]. ILs generally exhibit a melting point below 100 °C. Albo et al. [112] first applied the ILs in a membrane contactor for CO_2 separation from flue gas. They utilized commercially available PP membranes and [emim][$EtSO_4$] as an absorbent. Since then, a numerous IL-based absorbents were tested for CO_2 capture through membrane-based processes such as [bmim][DCA] [113], [apmim][BF_4], [bmim][BF_4] [114], and [emim][Ac] [115]. Dai et al. [116] summarized the integration of ILs absorbents in membrane-based CO_2 capture. Sohaib et al. [117] reported an efficient combination of [emim] cation-based ILs with PP contactors for post-combustion CO_2 sequestration. In another study, a coupled adsorption/desorption membrane contactor system employing imidazolium-based ILs as an absorbent with an absorption efficiency of 90% was designed for post-combustion CO_2 capture [118].

9.7 Conclusion and Future Perspective

Membrane-based CO_2 capture is becoming more and more competitive to traditional absorption-based technologies. Numerous experimental investigations and pilot-scale studies complemented by computational simulations focusing on material development and economic evaluation of membrane-based processes for CO_2 capture are available. However, many issues still remain in the commercialization of membrane contactors for post-combustion CO_2 capture at a large scale. The major issues arise in membrane durability and cost-competitiveness with traditional absorption-based processes. Various novel membrane materials are being developed and tested for post-combustion CO_2 capture. One such approach is the development of MMMs, which might result in a durable and cost-competitive membrane material. Besides, thin film composite membranes are also showing promising results.

The commercial membranes have also been tested for flue gas treatment. However, these greatly suffer from operation conditions, humidity, extraneous dust particles, and the presence of NO_x and SO_2 in the flue gas. Nevertheless, pilot-scale studies on commercial membrane materials have produced encouraging results.

Theoretically, any absorbent that is suitable to use in a conventional absorption column could be tried in membrane contactor-based post-combustion CO_2 capture. However, additional aspects like the pore-wetting property, stability at post-combustion conditions, and most importantly corrosivity toward membrane materials need to be considered before employing these absorbents in membrane contactors. Numerous absorbents are already available and have been tested for conventional CO_2 capture processes. However, detailed pilot-scale studies and comprehensive comparative techno-economic analysis of these absorbents for contactor-assisted post-combustion CO_2 capture are still needed for selecting the optimal solvent. In addition, the proposed investigations could also help in laying out the future directions for an ideal absorbent for post-combustion CO_2 capture. The comprehensive studies encompassing all the aspects, including interactions with membrane materials, stability, and cost of regeneration of commercially available and under development absorbents at post-combustion working conditions could potentially provide groundbreaking results and invigorate the research interest in post-combustion CO_2 capture using membrane contactors.

References

1 Younas, M., Rezakazemi, M., Daud, M. et al. (2020). Recent progress and remaining challenges in post-combustion CO_2 capture using metal-organic frameworks (MOFs). *Progress in Energy and Combustion Science* 80: 100849.

2 Dai, Z., Ansaloni, L., and Deng, L. (2016). Precombustion CO_2 capture in polymeric hollow fiber membrane contactors using ionic liquids: porous membrane versus nonporous composite membrane. *Industrial & Engineering Chemistry Research* 55: 5983–5992.

3 Scholes, C.A., Kentish, S.E., and Qader, A. (2020). Membrane gas-solvent contactor pilot plant trials for post-combustion CO_2 capture. *Separation and Purification Technology* 237: 116470.

4 Rezakazemi, M., Sadrzadeh, M., and Matsuura, T. (2018). Thermally stable polymers for advanced high-performance gas separation membranes. *Progress in Energy and Combustion Science* 66: 1–41.

5 Rezakazemi, M., Ebadi Amooghin, A., Montazer-Rahmati, M.M. et al. (2014). State-of-the-art membrane based CO_2 separation using mixed matrix membranes (MMMs): an overview on current status and future directions. *Progress in Polymer Science* 39: 817–861.

6 Mansourizadeh, A. and Ismail, A.F. (2009). Hollow fiber gas–liquid membrane contactors for acid gas capture: a review. *Journal of Hazardous Materials* 171: 38–53.

7 Scholes, C.A., Simioni, M., Qader, A. et al. (2012). Membrane gas–solvent contactor trials of CO_2 absorption from syngas. *Chemical Engineering Journal* 195–196: 188–197.

8 Scholes, C.A., Qader, A., Stevens, G.W., and Kentish, S.E. (2014). Membrane gas-solvent contactor pilot plant trials of CO_2 absorption from flue gas. *Separation Science and Technology* 49: 2449–2458.

9 Mosadegh-Sedghi, S., Rodrigue, D., Brisson, J., and Iliuta, M.C. (2014). Wetting phenomenon in membrane contactors – causes and prevention. *Journal of Membrane Science* 452: 332–353.

10 Ansaloni, L., Rennemo, R., Knuutila, H.K., and Deng, L. (2017). Development of membrane contactors using volatile amine-based absorbents for CO_2 capture: amine permeation through the membrane. *Journal of Membrane Science* 537: 272–282.

11 Nguyen, P.T., Lasseuguette, E., Medina-Gonzalez, Y. et al. (2011). A dense membrane contactor for intensified CO_2 gas/liquid absorption in post-combustion capture. *Journal of Membrane Science* 377: 261–272.

12 Scholes, C.A., Kentish, S.E., Stevens, G.W., and deMontigny, D. (2015). Comparison of thin film composite and microporous membrane contactors for CO_2 absorption into monoethanolamine. *International Journal of Greenhouse Gas Control* 42: 66–74.

13 Al-saffar, H.B., Ozturk, B., and Hughes, R. (1997). A comparison of porous and non-porous gas-liquid membrane contactors for gas separation. *Chemical Engineering Research and Design* 75: 685–692.

14 Chen, H.Z., Thong, Z., Li, P., and Chung, T.-S. (2014). High performance composite hollow fiber membranes for CO_2/H_2 and CO_2/N_2 separation. *International Journal of Hydrogen Energy* 39: 5043–5053.

15 Wang, Y., Zhao, L., Otto, A. et al. (2017). A review of post-combustion CO_2 capture technologies from coal-fired power plants. *Energy Procedia* 114: 650–665.

16 Kárászová, M., Zach, B., Petrusová, Z. et al. (2020). Post-combustion carbon capture by membrane separation, review. *Separation and Purification Technology* 238: 116448.

17 Ilyas, A., Muhammad, N., Gilani, M.A. et al. (2017). Supported protic ionic liquid membrane based on 3-(trimethoxysilyl)propan-1-aminium acetate for the highly selective separation of CO_2. *Journal of Membrane Science* 543: 301–309.

18 Ahmad, N.N.R., Leo, C.P., Mohammad, A.W., and Ahmad, A.L. (2017). Modification of gas selective SAPO zeolites using imidazolium ionic liquid to develop polysulfone mixed matrix membrane for CO_2 gas separation. *Microporous and Mesoporous Materials* 244: 21–30.

19 Fam, W., Mansouri, J., Li, H., and Chen, V. (2017). Improving CO_2 separation performance of thin film composite hollow fiber with Pebax®1657/ionic liquid gel membranes. *Journal of Membrane Science* 537: 54–68.

20 Huang, G., Isfahani, A.P., Muchtar, A. et al. (2018). Pebax/ionic liquid modified graphene oxide mixed matrix membranes for enhanced CO_2 capture. *Journal of Membrane Science* 565: 370–379.

21 Cowan, M.G., Gin, D.L., and Noble, R.D. (2016). Poly(ionic liquid)/ionic liquid ion-gels with high "free" ionic liquid content: platform membrane materials for CO_2/light gas separations. *Accounts of Chemical Research* 49: 724–732.

22 Nikolaeva, D., Azcune, I., Tanczyk, M. et al. (2018). The performance of affordable and stable cellulose-based poly-ionic membranes in CO_2/N_2 and CO_2/CH_4 gas separation. *Journal of Membrane Science* 564: 552–561.

23 Solimando, X., Lherbier, C., Babin, J. et al. (2016). Pseudopeptide bioconjugate additives for CO_2 separation membranes. *Polymer International* 65: 1464–1473.

24 Guerrero, G., Venturi, D., Peters, T. et al. (2017). Influence of functionalized nanoparticles on the CO_2/N_2 separation properties of PVA-based gas separation membranes. *Energy Procedia* 114: 627–635.

25 Rabiee, H., Ghadimi, A., Abbasi, S., and Mohammadi, T. (2015). CO_2 separation performance of poly(ether-b-amide6)/PTMEG blended membranes: permeation and sorption properties. *Chemical Engineering Research and Design* 98: 96–106.

26 Molki, B., Aframehr, W.M., Bagheri, R., and Salimi, J. (2018). Mixed matrix membranes of polyurethane with nickel oxide nanoparticles for CO_2 gas separation. *Journal of Membrane Science* 549: 588–601.

27 Nabais, A.R., Ribeiro, R.P.P.L., Mota, J.P.B. et al. (2018). CO_2/N_2 gas separation using Fe(BTC)-based mixed matrix membranes: a view on the adsorptive and filler properties of metal-organic frameworks. *Separation and Purification Technology* 202: 174–184.

28 Mubashir, M., Yeong, Y.F., Lau, K.K. et al. (2018). Efficient CO_2/N_2 and CO_2/CH_4 separation using NH_2-MIL-53(Al)/cellulose acetate (CA) mixed matrix membranes. *Separation and Purification Technology* 199: 140–151.

29 Rodrigues, M.A., de Souza Ribeiro, J., de Souza Costa, E. et al. (2018). Nanostructured membranes containing UiO-66 (Zr) and MIL-101 (Cr) for O_2/N_2 and CO_2/N_2 separation. *Separation and Purification Technology* 192: 491–500.

30 Ying, Y., Cheng, Y., Peh, S.B. et al. (2019). Plasticization resistance-enhanced CO_2 separation at elevated pressures by mixed matrix membranes containing flexible metal-organic framework fillers. *Journal of Membrane Science* 582: 103–110.

31 Chen, B., Wan, C., Kang, X. et al. (2019). Enhanced CO_2 separation of mixed matrix membranes with ZIF-8@GO composites as fillers: effect of reaction time of ZIF-8@GO. *Separation and Purification Technology* 223: 113–122.

32 Lasseuguette, E., Ferrari, M.-C., and Brandani, S. (2014). Humidity impact on the gas permeability of PIM-1 membrane for post-combustion application. *Energy Procedia* 63: 194–201.

33 Wu, X., Tian, Z., Wang, S. et al. (2017). Mixed matrix membranes comprising polymers of intrinsic microporosity and covalent organic framework for gas separation. *Journal of Membrane Science* 528: 273–283.

34 Chen, Y., Wang, B., Zhao, L. et al. (2015). New Pebax®/zeolite Y composite membranes for CO_2 capture from flue gas. *Journal of Membrane Science* 495: 415–423.

35 Zhao, L., Chen, Y., Wang, B. et al. (2016). Multilayer polymer/zeolite Y composite membrane structure for CO_2 capture from flue gas. *Journal of Membrane Science* 498: 1–13.

36 Kim, K., Hong, S., Kim, J., and Lee, H. (2014). Preparation and performance evaluation of composite hollow fiber membrane for SO_2 separation. *AICHE Journal* 60: 2298–2306.

37 Choi, S.-H., Kim, J.-H., and Lee, Y. (2013). Pilot-scale multistage membrane process for the separation of CO_2 from LNG-fired flue gas. *Separation and Purification Technology* 110: 170–180.

38 Cersosimo, M., Brunetti, A., Drioli, E. et al. (2015). Separation of CO_2 from humidified ternary gas mixtures using thermally rearranged polymeric membranes. *Journal of Membrane Science* 492: 257–262.

39 Sandru, M., Kim, T.-J., Capala, W. et al. (2013). Pilot scale testing of polymeric membranes for CO_2 capture from coal fired power plants. *Energy Procedia* 37: 6473–6480.

40 Park, C.-Y., Kim, E.-H., Kim, J.H. et al. (2018). Novel semi-alicyclic polyimide membranes: synthesis, characterization, and gas separation properties. *Polymer* 151: 325–333.

41 Chen, T., Yu, B., Zhao, Y. et al. (2017). Carbon dioxide permeation through ceramic-carbonate dual-phase membrane-effects of sulfur dioxide. *Journal of Membrane Science* 540: 477–484.

42 Eiberger, J., Wilkner, K., Reetz, C. et al. (2015). Influence of coal power plant exhaust gas on the structure and performance of ceramic nanostructured gas separation membranes. *International Journal of Greenhouse Gas Control* 43: 46–56.

43 Robeson, L.M. (2008). The upper bound revisited. *Journal of Membrane Science* 320 (1–2): 390–400.

44 Rezakazemi, M. and Shirazian, S. (2018). Gas permeation prediction through polymeric membranes using compressible regular solution theory. *International Journal of Hydrogen Energy* 43: 22357–22364.

45 Rezakazemi, M. and Shirazian, S. (2018). Computational simulation of mass transfer in molecular separation using microporous polymeric membranes. *Chemical Engineering & Technology* 41: 1975–1981.

46 Sheikh, M., Pazirofteh, M., Dehghani, M. et al. (2020). Application of ZnO nanostructures in ceramic and polymeric membranes for water and wastewater technologies: a review. *Chemical Engineering Journal* 391: 123475.

47 Rezakazemi, M. and Shirazian, S. (2018). Separation performance of nanostructured ceramic membranes: analytical model development. *Journal of Non-Equilibrium Thermodynamics* 43: 245.

48 Rezakazemi, M., Maghami, M., and Mohammadi, T. (2018). High loaded synthetic hazardous wastewater treatment using lab-scale submerged ceramic membrane bioreactor. *Periodica Polytechnica Chemical Engineering* 62: 299–304.

49 Rezakazemi, M., Mirzaei, S., Asghari, M., and Ivakpour, J. (2017). Aluminum oxide nanoparticles for highly efficient asphaltene separation from crude oil using ceramic membrane technology, oil & gas Science and technology - rev. *IFP Energies nouvelles* 72: 34.

50 Mirqasemi, M.S., Homayoonfal, M., and Rezakazemi, M. (2020). Zeolitic imidazolate framework membranes for gas and water purification. *Environmental Chemistry Letters* 18: 1–52.

51 Shahverdi, M., Baheri, B., Rezakazemi, M. et al. (2012). Pervaporation study of ethylene glycol dehydration through synthesized (PVA–4A)/polypropylene mixed matrix composite membranes. *Polymer Engineering & Science* 53: 1487–1493.

52 Rezakazemi, M., Shahidi, K., and Mohammadi, T. (2012). Sorption properties of hydrogen-selective PDMS/zeolite 4A mixed matrix membrane. *International Journal of Hydrogen Energy* 37: 17275–17284.

53 Rezakazemi, M., Shahidi, K., and Mohammadi, T. (2012). Hydrogen separation and purification using crosslinkable PDMS/zeolite a nanoparticles mixed matrix membranes. *International Journal of Hydrogen Energy* 37: 14576–14589.

54 Cui, Q., Liu, S., Xu, L. et al. (2020). Modification of rich-split carbon capture process using ceramic membrane for reducing the reboiler duty: effect of membrane arrangements. *Separation and Purification Technology* 235: 116148.

55 Lee, H.J., Park, Y.G., Kim, M.K. et al. (2019). Study on CO_2 absorption performance of lab-scale ceramic hollow fiber membrane contactor by gas/liquid flow direction and module design. *Separation and Purification Technology* 220: 189–196.

56 Trickett, C.A., Helal, A., Al-Maythalony, B.A. et al. (2017). The chemistry of metal–organic frameworks for CO_2 capture, regeneration and conversion. *Nature Reviews Materials* 2: 17045.

57 Hasse, D., Ma, J., Kulkarni, S. et al. (2014). CO_2 capture by cold membrane operation. *Energy Procedia* 63: 186–193.

58 Scholes, C.A., Qader, A., Stevens, G.W., and Kentish, S.E. (2015). Membrane pilot plant trials of CO_2 separation from flue gas. *Greenhouse Gases: Science and Technology* 5: 229–237.

59 White, L.S., Amo, K.D., Wu, T., and Merkel, T.C. (2017). Extended field trials of Polaris sweep modules for carbon capture. *Journal of Membrane Science* 542: 217–225.

60 Brinkmann, T., Lillepärg, J., Notzke, H. et al. (2017). Development of CO_2 selective poly(ethylene oxide)-based membranes: from laboratory to pilot plant scale. *Engineering* 3: 485–493.

61 Pfister, M., Belaissaoui, B., and Favre, E. (2017). Membrane gas separation processes from wet postcombustion flue gases for carbon capture and use: a critical reassessment. *Industrial & Engineering Chemistry Research* 56: 591–602.

62 Schuldt, K., Pohlmann, J., Shishatskiy, S., and Brinkmann, T. (2018). Applicability of PolyActive™ thin film composite membranes for CO_2 separation from C_2H_4 containing multi-component gas mixtures at pressures up to 30 Bar. *Membranes* 8: 27.

63 Pohlmann, J., Bram, M., Wilkner, K., and Brinkmann, T. (2016). Pilot scale separation of CO_2 from power plant flue gases by membrane technology. *International Journal of Greenhouse Gas Control* 53: 56–64.

64 Zhai, H. and Rubin, E.S. (2013). Techno-economic assessment of polymer membrane systems for postcombustion carbon capture at coal-fired power plants. *Environmental Science & Technology* 47: 3006–3014.

65 Maas, P., Nauels, N., Zhao, L. et al. (2016). Energetic and economic evaluation of membrane-based carbon capture routes for power plant processes. *International Journal of Greenhouse Gas Control* 44: 124–139.

66 Zhao, L., Riensche, E., Blum, L., and Stolten, D. (2010). Multistage gas separation membrane processes used in post-combustion capture: energetic and economic analyses. *Journal of Membrane Science* 359: 160–172.

67 Zhao, L., Riensche, E., Blum, L., and Stolten, D. (2011). How gas separation membrane competes with chemical absorption in postcombustion capture. *Energy Procedia* 4: 629–636.

68 Park, H.B., Kamcev, J., Robeson, L.M. et al. (2017). Maximizing the right stuff: the trade-off between membrane permeability and selectivity. *Science* 356: eaab0530.

69 Koros, W.J. and Zhang, C. (2017). Materials for next-generation molecularly selective synthetic membranes. *Nature Materials* 16: 289–297.

70 Favre, E. (2007). Carbon dioxide recovery from post-combustion processes: can gas permeation membranes compete with absorption? *Journal of Membrane Science* 294: 50–59.

71 Dindore, V.Y., Brilman, D.W.F., Geuzebroek, F.H., and Versteeg, G.F. (2004). Membrane–solvent selection for CO_2 removal using membrane gas–liquid contactors. *Separation and Purification Technology* 40: 133–145.

72 Constantinou, A., Barrass, S., and Gavriilidis, A. (2014). CO_2 absorption in polytetrafluoroethylene membrane microstructured contactor using aqueous solutions of amines. *Industrial & Engineering Chemistry Research* 53: 9236–9242.

73 Soroush, E., Shahsavari, S., Mesbah, M. et al. (2018). A robust predictive tool for estimating CO_2 solubility in potassium based amino acid salt solutions. *Chinese Journal of Chemical Engineering* 26: 740–746.

74 Mesbah, M., Shahsavari, S., Soroush, E. et al. (2018). Accurate prediction of miscibility of CO_2 and supercritical CO_2 in ionic liquids using machine learning. *Journal of CO_2 Utilization* 25: 99–107.

75 Sharifzadeh, M. and Shah, N. (2019). MEA-based CO_2 capture integrated with natural gas combined cycle or pulverized coal power plants: operability and controllability through integrated design and control. *Journal of Cleaner Production* 207: 271–283.

76 Zhao, B., Tao, W., Zhong, M. et al. (2016). Process, performance and modeling of CO_2 capture by chemical absorption using high gravity: a review. *Renewable and Sustainable Energy Reviews* 65: 44–56.

77 Wang, M., Lawal, A., Stephenson, P. et al. (2011). Post-combustion CO_2 capture with chemical absorption: a state-of-the-art review. *Chemical Engineering Research and Design* 89: 1609–1624.

78 Sartori, G. and Savage, D.W. (1983). Sterically hindered amines for carbon dioxide removal from gases. *Industrial & Engineering Chemistry Fundamentals* 22: 239–249.

79 Borhani, T.N.G., Afkhamipour, M., Azarpour, A. et al. (2016). Modeling study on CO_2 and H_2S simultaneous removal using MDEA solution. *Journal of Industrial and Engineering Chemistry* 34: 344–355.

80 Borhani, T.N. and Wang, M. (2019). Role of solvents in CO_2 capture processes: the review of selection and design methods. *Renewable and Sustainable Energy Reviews* 114: 109299.

81 Dashti, A., Raji, M., Razmi, A. et al. (2019). Efficient hybrid modeling of CO_2 absorption in aqueous solution of piperazine: applications to energy and environment. *Chemical Engineering Research and Design* 144: 405–417.

82 Freeman, S.A. and Rochelle, G.T. (2012). Thermal degradation of aqueous piperazine for CO_2 capture. 1. Effect of process conditions and comparison of thermal stability of CO_2 capture amines. *Industrial & Engineering Chemistry Research* 51: 7719–7725.

83 Plaza, J.M. and Rochelle, G.T. (2011). Modeling pilot plant results for CO_2 capture by aqueous piperazine. *Energy Procedia* 4: 1593–1600.

84 Afkhamipour, M. and Mofarahi, M. (2018). A modeling-optimization framework for assessment of CO_2 absorption capacity by novel amine solutions: 1DMA2P, 1DEA2P, DEEA, and DEAB. *Journal of Cleaner Production* 171: 234–249.

85 Richner, G., Puxty, G., Carnal, A. et al. (2015). Thermokinetic properties and performance evaluation of benzylamine-based solvents for CO_2 capture. *Chemical Engineering Journal* 264: 230–240.

86 Liang, Z., Rongwong, W., Liu, H. et al. (2015). Recent progress and new developments in post-combustion carbon-capture technology with amine based solvents. *International Journal of Greenhouse Gas Control* 40: 26–54.

87 Bernhardsen, I.M. and Knuutila, H.K. (2017). A review of potential amine solvents for CO_2 absorption process: absorption capacity, cyclic capacity and pKa. *International Journal of Greenhouse Gas Control* 61: 27–48.

88 Wu, X., Yu, Y., Qin, Z., and Zhang, Z. (2014). The advances of post-combustion CO_2 capture with chemical solvents: review and guidelines. *Energy Procedia* 63: 1339–1346.

89 Zhao, S., Feron, P.H.M., Deng, L. et al. (2016). Status and progress of membrane contactors in post-combustion carbon capture: a state-of-the-art review of new developments. *Journal of Membrane Science* 511: 180–206.

90 Yeh, A.C. and Bai, H. (1999). Comparison of ammonia and monoethanolamine solvents to reduce CO_2 greenhouse gas emissions. *Science of the Total Environment* 228: 121–133.

91 Bai, H. and Yeh, A.C. (1997). Removal of CO_2 greenhouse gas by ammonia scrubbing. *Industrial & Engineering Chemistry Research* 36: 2490–2493.

92 Makhloufi, C., Lasseuguette, E., Remigy, J.C. et al. (2014). Ammonia based CO_2 capture process using hollow fiber membrane contactors. *Journal of Membrane Science* 455: 236–246.

93 Cui, Z. and deMontigny, D. (2017). Experimental study of carbon dioxide absorption into aqueous ammonia with a hollow fiber membrane contactor. *Journal of Membrane Science* 540: 297–306.

94 McLeod, A., Buzatu, P., Autin, O. et al. (2015). Controlling shell-side crystal nucleation in a gas–liquid membrane contactor for simultaneous ammonium bicarbonate recovery and biogas upgrading. *Journal of Membrane Science* 473: 146–156.

95 Bavarella, S., Brookes, A., Moore, A. et al. (2020). Chemically reactive membrane crystallisation reactor for CO_2–NH_3 absorption and ammonium bicarbonate crystallisation: kinetics of heterogeneous crystal growth. *Journal of Membrane Science* 599: 117682.

96 Borhani, T.N.G., Azarpour, A., Akbari, V. et al. (2015). CO_2 capture with potassium carbonate solutions: a state-of-the-art review. *International Journal of Greenhouse Gas Control* 41: 142–162.

97 Knuutila, H., Hessen, E.T., Kim, I. et al. (2010). Vapor–liquid equilibrium in the sodium carbonate–sodium bicarbonate–water–CO_2-system. *Chemical Engineering Science* 65: 2218–2226.

98 Park, S.W., Sung, D.H., Choi, B.S. et al. (2006). Sorption of carbon dioxide onto sodium carbonate. *Separation Science and Technology* 41: 2665–2684.

99 Williamson, R.V. and Mathews, J.H. (1924). Rate of absorption and equilibrium of carbon dioxide in alkaline solutions. *Industrial & Engineering Chemistry* 16: 1157–1161.

100 Tseng, P.C., Ho, W.S., and Savage, D.W. (1988). Carbon dioxide absorption into promoted carbonate solutions. *AICHE Journal* 34: 922–931.

101 Saeed, M. and Deng, L. (2016). Post-combustion CO_2 membrane absorption promoted by mimic enzyme. *Journal of Membrane Science* 499: 36–46.

102 Yang, N., Yu, H., Xu, D.y. et al. (2014). Amino acids/NH3 mixtures for CO_2 capture: effect of neutralization methods on CO_2 mass transfer and NH3 vapour loss. *Energy Procedia* 63: 773–780.

103 Song, H.-J., Park, S., Kim, H. et al. (2012). Carbon dioxide absorption characteristics of aqueous amino acid salt solutions. *International Journal of Greenhouse Gas Control* 11: 64–72.

104 Zhang, Z., Li, Y., Zhang, W. et al. (2018). Effectiveness of amino acid salt solutions in capturing CO_2: a review. *Renewable and Sustainable Energy Reviews* 98: 179–188.

105 Yan, S., He, Q., Zhao, S. et al. (2014). Biogas upgrading by CO_2 removal with a highly selective natural amino acid salt in gas–liquid membrane contactor. *Chemical Engineering and Processing: Process Intensification* 85: 125–135.

106 Lu, J.-G., Hua, A.-C., Bao, L.-L. et al. (2011). Performance evaluation on complex absorbents for CO_2 capture. *Separation and Purification Technology* 82: 87–92.

107 Lu, J.-G., Zhang, H., Ji, Y. et al. (2010). Amino acid-salt-based complex absorbents for CO_2 capture in a membrane contactor. *Energy & Fuels* 24: 4617–4626.

108 Sohaib, Q., Muhammad, A., Younas, M. et al. (2021). Rigorous non-isothermal modeling approach for mass and energy transport during CO_2 absorption into aqueous solution of amino acid ionic liquids in hollow fiber membrane contactors. *Separation and Purification Technology* 254: 117644.

109 Swati, I.K., Sohaib, Q., Cao, S. et al. (2020). Protic/aprotic ionic liquids for effective CO_2 separation using supported ionic liquid membrane. *Chemosphere* 267: 128894.

110 Sohaib, Q., Muhammad, A., Younas, M., and Rezakazemi, M. (2020). Modeling pre-combustion CO_2 capture with tubular membrane contactor using ionic liquids at elevated temperatures. *Separation and Purification Technology* 241: 116677.

111 Pishnamazi, M., Nakhjiri, A.T., Marjani, A. et al. (2020). Computational study on SO_2 molecular separation applying novel EMISE ionic liquid and DMA aromatic amine solution inside microporous membranes. *Journal of Molecular Liquids* 313: 113531.

112 Albo, J., Luis, P., and Irabien, A. (2011). Absorption of coal combustion flue gases in ionic liquids using different membrane contactors. *Desalination and Water Treatment* 27: 54–59.

113 Mulukutla, T., Chau, J., Singh, D. et al. (2015). Novel membrane contactor for CO_2 removal from flue gas by temperature swing absorption. *Journal of Membrane Science* 493: 321–328.

114 Lu, J.-G., Lu, C.-T., Chen, Y. et al. (2014). CO_2 capture by membrane absorption coupling process: application of ionic liquids. *Applied Energy* 115: 573–581.

115 Gómez-Coma, L., Garea, A., and Irabien, A. (2014). Non-dispersive absorption of CO_2 in [emim][$EtSO_4$] and [emim][Ac]: temperature influence. *Separation and Purification Technology* 132: 120–125.

116 Dai, Z., Noble, R.D., Gin, D.L. et al. (2016). Combination of ionic liquids with membrane technology: a new approach for CO_2 separation. *Journal of Membrane Science* 497: 1–20.

117 Sohaib, Q., Vadillo, J.M., Gómez-Coma, L. et al. (2020). Post-combustion CO_2 capture by coupling [emim] cation based ionic liquids with a membrane contactor; pseudo-steady-state approach. *International Journal of Greenhouse Gas Control* 99: 103076.

118 Sohaib, Q., Vadillo, J.M., Gómez-Coma, L. et al. (2020). CO_2 capture with room temperature ionic liquids; coupled absorption/desorption and single module absorption in membrane contactor. *Chemical Engineering Science* 223: 115719.

10

Market Prospects of Membrane Contactors

Zahra Pezeshki[1], Mohammad Younas[2], and Mashallah Rezakazemi[3]

[1]*Shahrood University of Technology, Faculty of Electrical and Robotic Engineering, Shahrood, Semnan, Iran*
[2]*University of Engineering and Technology, Department of Chemical Engineering, Peshawar 25120, Pakistan*
[3]*Shahrood University of Technology, Faculty of Chemical and Materials Engineering, Shahrood, Semnan, Iran*

10.1 Membrane Contactor Market Dynamics

Today, major technologies have evolved into mature technologies because of the new discoveries that membrane technology is one of them. However, the traditional membrane separation processes present validated better management for waste, pollution, and cleaning in the long run on an industrial scale [1]. Currently, engineers and scientists pay more attention to the technological sustainability of membrane technology [2, 3]. Table 10.1 shows some membrane contactor technology used for different applications.

Table 10.1 Some membrane contactor used for different applications.

Membrane name	Discovered by	References
Artificial semipermeable	Traube	[4]
Developing artificial semipermeable	Pfeffer	[5]
High-quality nitrocellulose	Bechhold	[5, 6]
Asymmetric reverse osmosis	Loeb and Sourirajan	[6]
Gas separation and distillation	Henis and Tripodi	[7]

Source: Selvamuthukumaran [3].

The membrane technology was developed and applied in various applications that attracted increasing attention to the membrane market [8–10]. During these years, estimates show that 75% of the money allocated for this purpose has been spent on industrial use and the rest on laboratory use. Membrane technologies have many applications in water and wastewater treatment [11–14], food and beverage processing [3], medical and pharmaceutical processing, and industrial processing [15]. The benefits of membrane technology consist of separation and concentration of a wide variety of components from mixtures, creating high-quality resulting products,

Membrane Contactor Technology: Water Treatment, Food Processing, Gas Separation, and Carbon Capture, First Edition. Edited by Mohammad Younas and Mashallah Rezakazemi.

operating without the addition of any additives, having low operating temperatures and energy consumption, having increasingly efficient and high separation coefficient, being environmentally friendly and highly versatile, and having less investment and operation costs [3, 16]. The common membrane separation processes and their applications in the water/wastewater treatment and food/beverage processing have been shown in Table 10.2.

Among the membrane separation processes, membrane contactor technology is one of the latest separation technologies with more than two decades applied for various applications using microporous membranes. They are often in hollow fiber form, but they can have flat or spiral wound form. This hollow fiber form of membrane contactors is usually sold at a higher price than the other two forms; typically, US\$ 300–700 per m^2 membrane area. The flat form of membrane contactors is sold at approximately US\$ 130–180 per m^2, and the spiral wound form is sold at US\$ 120 per m^2. There are three types of membrane contactors including fluid-phase membrane contactor, solid–fluid-phase membrane contactor, and multiple-phase-interface-based membrane contactor [6, 25]. Table 10.3 shows the range of using this kind of membrane.

10.2 Market Overview

There are currently a wide range of membrane contactors that some companies around the world are selling [19]. As our focus is on water and wastewater treatment, food and beverage processing, and carbon capture, the names of the key players in this market have been given in Table 10.4.

The membrane contactors are different in their transport properties, separation mechanisms, and morphology based on their fabrication technologies and material properties. Their market is highly capital intensive [16, 19]. This means that the main costs of this market are capital cost and investment cost. Perhaps one of the advantages of this market is the large difference in energy costs due to its much lower operating pressure and production cost. According to the report [16], the market for membrane contactors has grown to US\$ 257 million in 2020, and it is projected to reach US\$ 340 million by 2025 at compound annual growth rate (CAGR) of 5.8% from 2020 to 2025. Thus, this market for water and wastewater treatment and food and beverage processing has grown to US\$ 136.21 million in 2020, and it is projected to reach US\$ 180.2 million by 2025 at CAGR of 4.8% from 2020 to 2025 (Figure 10.1). From these two industries, the water and wastewater treatment segment dominated the demand for membrane contactor.

The companies typically develop membrane contactors for specific applications such as removal of microorganisms, acids, and heavy metals, disinfection, recovery, desalination, separation, concentration, and absorption in water and wastewater treatment and the applications including treatment, separation, concentration, absorption, humidity control, and reduction of the alcohol content in food and beverage processing. Among these, the removal of microorganisms, acids, and heavy metals in water and wastewater treatment and treatment in food and beverage

Table 10.2 The application of the membrane for water and wastewater treatment and food and beverage processing and their markets.

Markets		Water and wastewater treatment [17–21]	Food and beverage processing [3, 22–24]
Process	**Driving force**	**Application**	
Microfiltration (MF)	Pressure difference	– Removal of microorganisms – Disinfection – Treatment – Recovery	– Removal of microorganisms – Separation – Fractionation – Clarification and stabilization – Dewaxing and decolorization – Treatment
Ultrafiltration (UF)		– Removal of microorganisms	– Separation – Fractionation – Concentration – Clarification and stabilization – Degumming – Production
Nanofiltration (NF)		– Removal of microorganisms – Desalination	– Partial demineralization – Concentration – Removal of microorganisms – Control – Reduction of the alcohol content – Fractionation – Desalination – Degumming – Treatment
Reverse osmosis (RO)		– Removal of microorganisms – Recovery	– Concentration – Demineralization – Clarification and dealcoholization – Desalination
Forward osmosis (FO)	Concentration difference	– Removal of microorganisms – Recovery	– Concentration – Desalination
Diffusion dialysis and dialysis		– Recovery – Removal of acids and heavy metals – Separation	– Recovery – Removal of acids – Dealcoholization
Gas separation		– Separation – Treatment	– Production – Separation
Pervaporation		– Removal of microorganisms – Separation	– Recovery – Dealcoholization – Dehydration and removal

(Continued)

Table 10.2 (Continued)

Markets		Water and wastewater treatment [17–21]	Food and beverage processing [3, 22–24]
Liquid extraction		– Concentration – Removal of microorganisms	– Separation – Removal of microorganisms and heavy metals
Electrodialysis	Electrical potential difference	– Removal of microorganisms and heavy metals – Recovery – Desalination	– Demineralization – Desalination – Removal of microorganisms – Deacidification – Production of oxygen-enriched air – Recovery and concentration – Treatment
Membrane distillation	Temperature difference	– Separation – Recovery – Desalination – Treatment	– Concentration – Recovery – Dealcoholization – Treatment
Membrane contactor	– Pressure difference – Temperature difference	– Removal of microorganisms, acids, and heavy metals – Pertraction – Recovery – Desalination – Separation – Concentration – Absorption	– Treatment – Separation – Concentration – Absorption – Ozonation – Humidity control – Dehydration – Reduction of the alcohol content

Table 10.3 The range of the membrane contactor application domain.

Type of membrane contactor	Range of use
Fluid-phase membrane contactor	– Gas–liquid – Liquid–liquid – Vapor–liquid – Supercritical fluid–fluid
Solid–fluid-phase membrane contactor	– Liquid–solid – Gas/vapor–solid – Supercritical fluid–solid
Multiple-phase-interface-based membrane contactor	– Gas–liquid (1) and liquid–gas (2) – Liquid–liquid (1) and Liquid–liquid (2)

Source: Based on Uragami [6].

Table 10.4 The names of the key players in the membrane contactor markets related to water and wastewater treatment and food and beverage processing industries.

Name of company	Country	Website	Commercial name
3M Company	United States	https://www.3m.com	3M™ Liqui-Cel™
Cobetter Filtration Equipment Pvt. Ltd.	China	https://www.cobetterfiltration.com	Arrayforce™ gas–liquid
EUROWATER	Denmark	https://www.eurowater.com	Membrane degassing units (MDUs)
JU.CLA.S Srl	Italy	https://www.juclas.it	MFTS and MASTERMIND® REMOVE
Veolia Environnement SA	France	https://www.veolia.com	BERKEFELD PurBev® UF
PTI Pacific Pty. Ltd.	Australia	https://ptipacific.com.au	Satroclear® depth filter sheets, Satrocell® lenticular depth filter modules
Kværner ASA	Norway	https://www.akersolutions.com	Kværner membrane contactor
Lenntech B.V.	Netherlands	https://www.lenntech.com	Lenntech B.V. membrane contactor
Pure Water Group	Netherlands	https://purewatergroup.com	Membrane degassing
TNO Company	Netherlands	https://www.tno.nl/en	TNO absorption membranes
Euwa H. H. Eumann GmbH (Euwa)	Germany	https://www.euwa.com	CALMIX and EUWA
Hydro-Elektrik GmbH	Germany	https://www.hydrogroup.de	Membrane degassing
KH TEC GmbH	Germany	https://www.kh-tec.com	KH TEC GmbH membrane contactor
Romfil GmbH	Germany	https://www.romfil.com	Romfil membrane contactor

Source: Membrane Contactor Market [16] and Klaassen et al. [19].

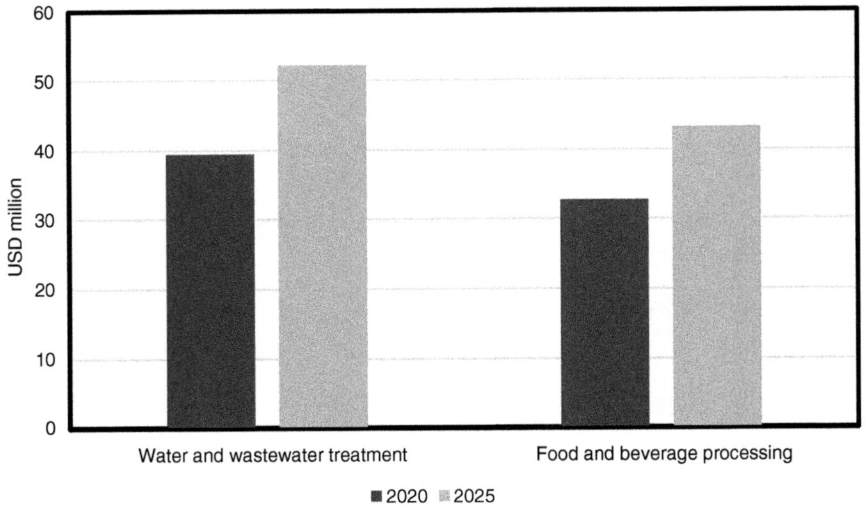

Figure 10.1 The growth estimation of the water and wastewater treatment and food and beverage processing applications in USD million.

processing is more visible. It is expected that these applications drive the membrane contactors market in the future. By water and wastewater treatment industry, the removal of microorganisms, acids, and heavy metals is estimated to dominate the membrane contactors market worldwide with market size of US$ 11.46 million in 2020.

Both industries – i.e. water and wastewater treatment and food and beverage processing – use carbon capture in their applications. In this regard, water and wastewater treatment use carbon capture for the removal of microorganisms, acids, and heavy metals, disinfection, recovery, and desalination. Food and beverage processing use this process for treatment. Figure 10.2 shows the carbon capture market share in these two industries. Also, Figure 10.3 shows the growth estimation of this market in USD million. As you see, the demand for the carbon capture for the water and wastewater treatment segment is higher.

Table 10.5 shows the drivers, restraints, opportunities, and challenges of this market. These factors lead companies to focus more on their research and development (R&D). In addition, Table 10.6 shows the focuses of the companies on developing market technology.

The value chain of membrane contactor includes suppliers, manufacturers (Table 10.4), system integrators, and end users. In fact, suppliers, system integrators, and end users are the investors of this chain. This chain may be different worldwide (Figures 10.4–10.6) depending on the membrane contactor physical and chemical properties, but manufacturers are the main manufacturers shown in Table 10.4.

According to Figure 10.5, Cobetter Filtration Equipment Pvt. Ltd. is on top of the list of the most number of suppliers, service integrators, and end users in this value chain, followed by 3M Company having 10 suppliers. Then, they are succeeded by Pure Water Group, Kværner ASA, PTI Pacific Pty. Ltd., and JU.CLA.S Srl with 6, 4,

Table 10.5 The drivers, restraints, opportunities, and challenges in the membrane contactor markets related to water and wastewater treatment and food and beverage processing industries and definition of these characters for carbon capture as a strong stimulus for the mentioned industries (these data obtained from the website of companies depicted in Table 10.4).

Market overview	Water and wastewater treatment	Food and beverage processing	CO_2 capture
Drivers	– Optimal and long-term solutions for drinking water and process water – Rapid separation of microorganisms indigenous to the must – Allow production in harsh environments and remote locations	– Rapid and precise in-line dissolved gas control in a compact design – Technical experience and deep understanding of sterilization process	– Human developments and improving in living standards – Reduction of population growth – Elimination of inequalities in low-income countries – Empowering communities and citizens – Locally capture
Restraints	– Reliability – High cost because of the devaluation of the current currency of the country – COVID-19 – Restricted markets because of sanction – Emerging black markets		– High costs – Decrease of crude oil and fossil fuel prices
Opportunities	– Allow seawater treatment for injection – Allow subsea gas/liquid separation and boosting – Enable long-distance subsea tiebacks – Increase hydrocarbon recovery – Pleasant and skin-friendly wellness water	– Fully automatic O_2 adjustment and remote-control display	– Employment with job creation – Achieving large GHG emissions reductions – Creating cheaper facilities – Strengthening networks among different partners (locally and globally) – Reducing energy imports and free government resources
Challenges	– Increased reliability – Protection and lowering overall costs of consumables – Turning innovative water technology into sustainable economic growth	– Protecting the original taste, color, and texture of food products – Increasing or extending hydrocarbon recovery – Production from low-pressure reservoirs – Subsea water injection – Long-distance subsea tiebacks – Increasing or extending hydrocarbon recovery	– Lack of markets for these technologies – Decrease of costs – No long-term contracts – Lack of government or private incentives – Lack of regulatory pressures

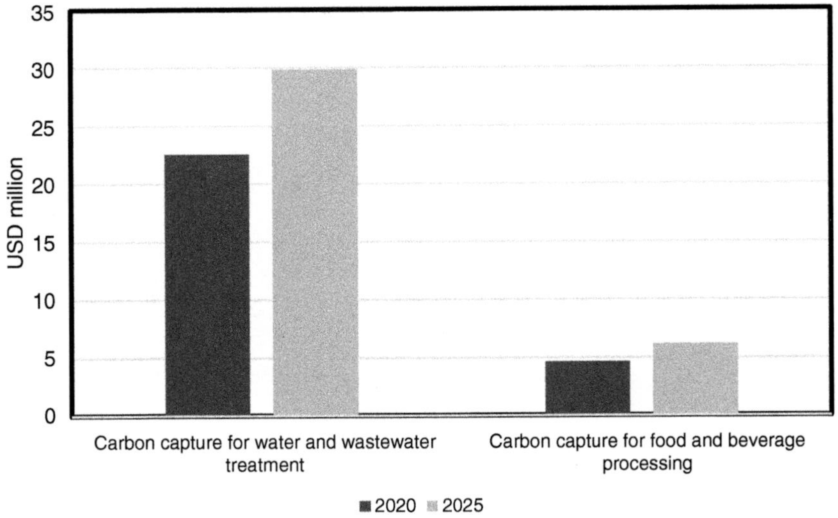

Figure 10.2 The share of carbon capture market for water and wastewater treatment and food and beverage processing applications.

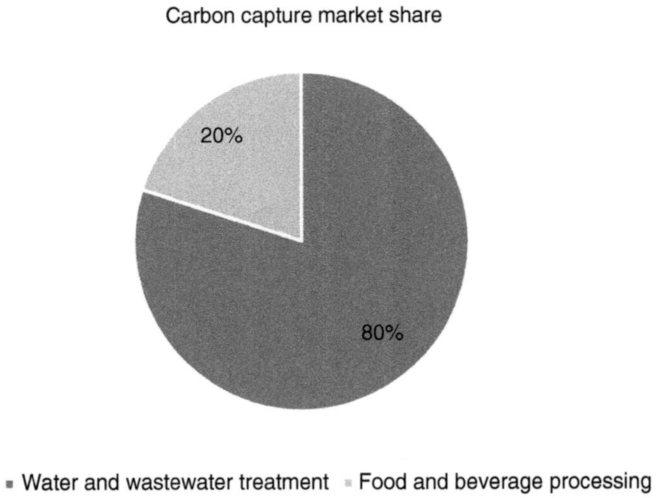

Figure 10.3 The growth estimation of carbon capture for water and wastewater treatment and food and beverage processing applications in USD million.

3, and 2 agencies worldwide, respectively. Also, PTI Pacific Pty. Ltd., Hydro-Elektrik GmbH, JU.CLA.S Srl, Lenntech B.V., and KH TEC GmbH having 6, 3, and 2 agencies are the most system integrators. Finally, Pure Water Group, Kværner ASA, PTI Pacific Pty. Ltd., Kværner ASA, and Veolia Environnement SA having 6, 4, 3, and 2 agencies are the most end users.

According to Figure 10.6, EUROWATER and TNO Company are the most end users in this value chain.

Table 10.6 Market focus on application based in module manufacturer companies.

Name of company	Development focus
3M Company	Beverage processing
Cobetter Filtration Equipment Pvt. Ltd.	Food and beverage processing
EUROWATER	Water treatment
JU.CLA.S Srl	Water and wastewater treatment, beverage processing
Veolia Environnement SA	Water and wastewater treatment, food and beverage processing
PTI Pacific Pty. Ltd.	Water treatment, food and beverage processing
Kværner ASA	Water treatment
Lenntech B.V.	Water and wastewater treatment, food and beverage processing
Pure Water Group	Water and wastewater treatment
TNO Company	Water and wastewater treatment
Euwa H. H. Eumann GmbH (Euwa)	Water and wastewater treatment, beverage processing
Hydro-Elektrik GmbH	Water and wastewater treatment
KH TEC GmbH	Food and beverage processing
Romfil GmbH	Water and wastewater treatment, beverage processing

Now, the market overview is analyzed and discussed based on application, membrane, and regions in water and wastewater treatment and food and beverage processing.

10.3 Membrane Contactor Market by Application

10.3.1 Water and Wastewater Treatment Market

In water and wastewater treatment, the market of membrane contactors is projected to grow at the highest CAGR of 4.8% between 2020 and 2025. This industry offers the removal of microorganisms, acids, heavy metals, disinfection, recovery, desalination, separation, concentration, and absorption. Today, water contamination and scarcity are the important issues that treat the humans' lives around the world. So, this market is considered to be a great benefit for countries with huge populations due to providing a solution for water and wastewater treatment. Table 10.7 shows the investors in this market while Figure 10.7 shows the investment analysis in this market.

According to this research, EUROWATER is the largest investor in this market. It is projected the removal of microorganisms, acids, and heavy metals and desalination in this market attributed to the presence of the different key investors (see Table 10.7)

Table 10.7 The investors of the membrane contractor for water and wastewater treatment (these data obtained from the companies' website depicted in Table 10.4).

	Application						
Investor name	**Removal of microorganisms, acids, and heavy metals**	**Disinfection**	**Recovery**	**Desalination**	**Separation**	**Concentration**	**Absorption**
EUROWATER	×	×	×		×	×	
JU.CLA.S Srl					×		
Veolia Environnement SA	×	×	×	×		×	
PTI Pacific Pty. Ltd.	×		×		×		
Kværner ASA			×		×		
Lenntech B.V.	×	×		×		×	
Pure Water Group	×	×		×			
TNO Company	×			×			
Euwa H. H. Eumann GmbH (Euwa)	×	×					
Hydro-Elektrik GmbH	×	×		×	×	×	×
KH TEC GmbH	×						
Romfil GmbH	×						

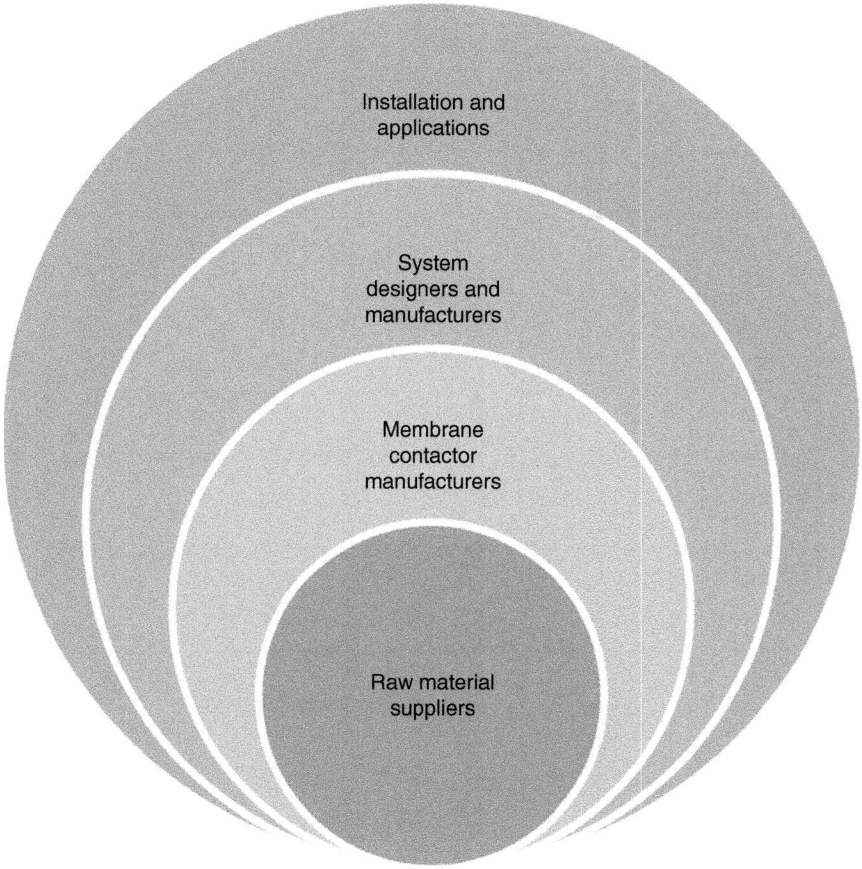

Figure 10.4 Value chain of membrane contactors.

to grow at the highest CAGR between 2020 and 2025. So, according to this estimation, the prospect for water and wastewater treatment is expected to increase in the future.

Regardless of the benefits of this market, there are always limitations in this market due to unreliability and high cost. Researchers believe that these restrictions are more relevant to the devaluation of the current currency of the country, restricted markets because of sanction, and emerging black markets.

Figure 10.8 shows the graph of service period of companies in membrane contactor markets for water and wastewater treatment where Veolia Environnement SA and Kværner ASA are the oldest companies. Table 10.8 depicts the service period of the companies that work in this market.

10.3.2 Food Processing Market

In food and beverage processing, the market of membrane contactors is projected to grow at the highest CAGR of 4.8% between 2020 and 2025. This industry offers

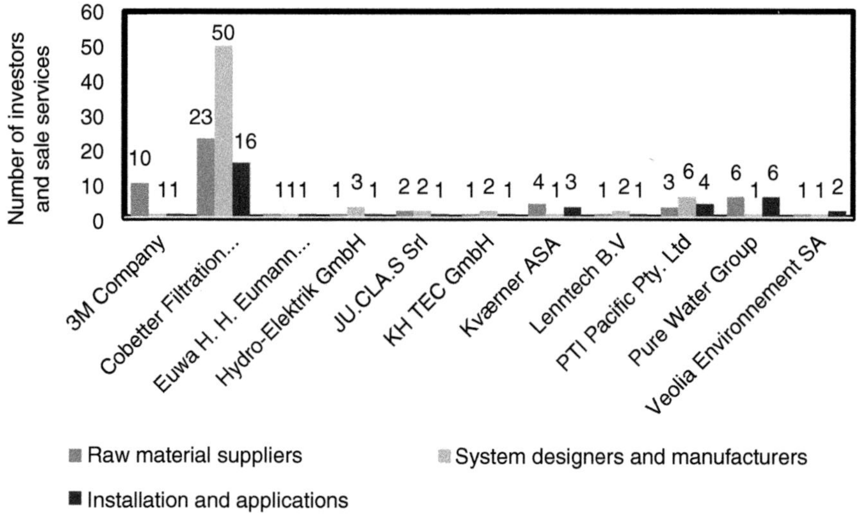

Figure 10.5 Value chain analysis of membrane contactors in the companies that have both investors and sale services.

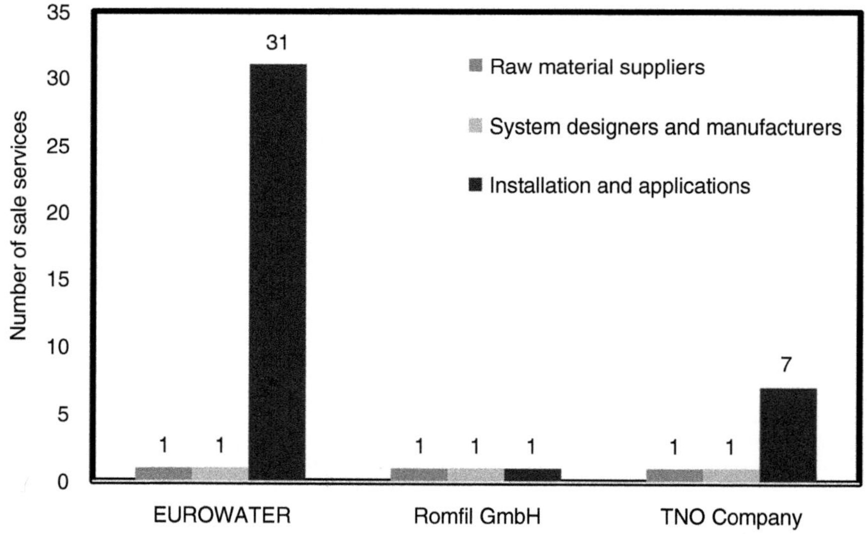

Figure 10.6 Value chain analysis of membrane contactors in the companies that have only sale services.

treatment, separation, concentration, absorption, humidity control, and reduction of the alcohol content. Today, the food and beverage industry are required to have an adequate supply of drinking water (i.e. potable water) available for use in food production to ensure foods are not contaminated. So, this market is considered as a great benefit for countries with huge populations due to providing a solution for food and beverage processing. Table 10.9 shows the investors in this market. Figure 10.9 shows the investment analysis in this market.

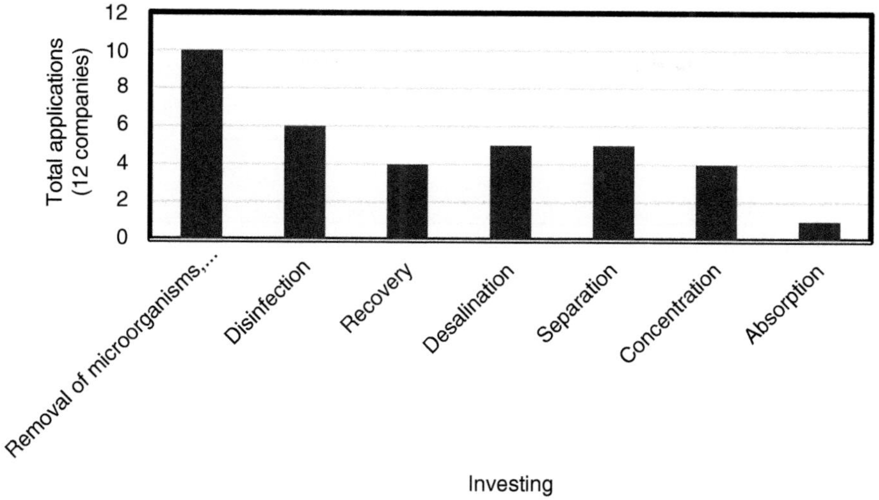

Figure 10.7 The investment analysis of the membrane contractor for water and wastewater treatment.

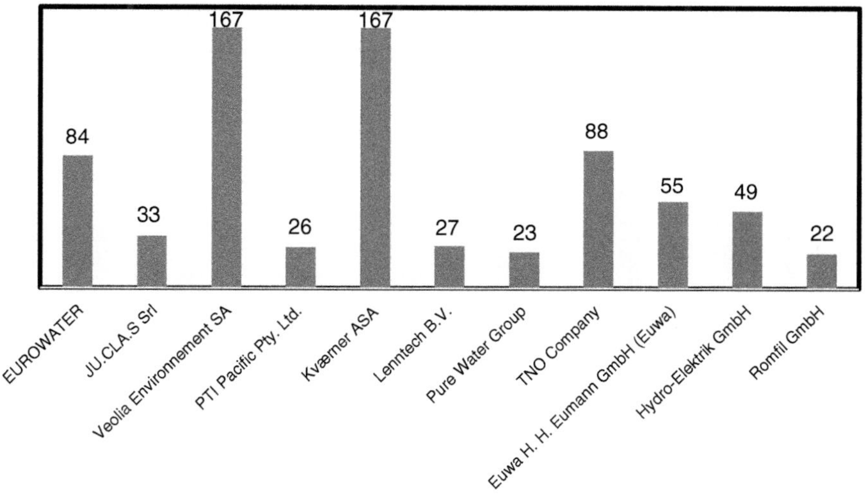

Figure 10.8 The service period (in years) of companies in membrane contactor markets related to water and wastewater treatment.

According to this research, Cobetter Filtration Equipment Pvt. Ltd. is the largest investor in this market. It is projected the treatment in this market attributed to the presence of the different key investors (Table 10.9) to grow at the highest CAGR between 2020 and 2025. So, according to this estimation, the prospect for food and beverage processing is expected to increase in the future.

Regardless of the benefits of this market, there are some restraints such as reliability, high cost because of the devaluation of the current currency of the country, COVID-19, restricted markets because of sanction, and emerging black markets.

Figure 10.10 shows the graph of service period of companies in membrane contactor markets for food and beverage processing where Veolia Environnement SA and

Table 10.8 The service period of the companies that work for water and wastewater treatment.

Name of company	Service period in year
EUROWATER	84
JU.CLA.S Srl	33
Veolia Environnement SA	167
PTI Pacific Pty. Ltd.	26
Kværner ASA	167
Lenntech B.V.	27
Pure Water Group	23
TNO Company	88
Euwa H. H. Eumann GmbH (Euwa)	55
Hydro-Elektrik GmbH	49
Romfil GmbH	22

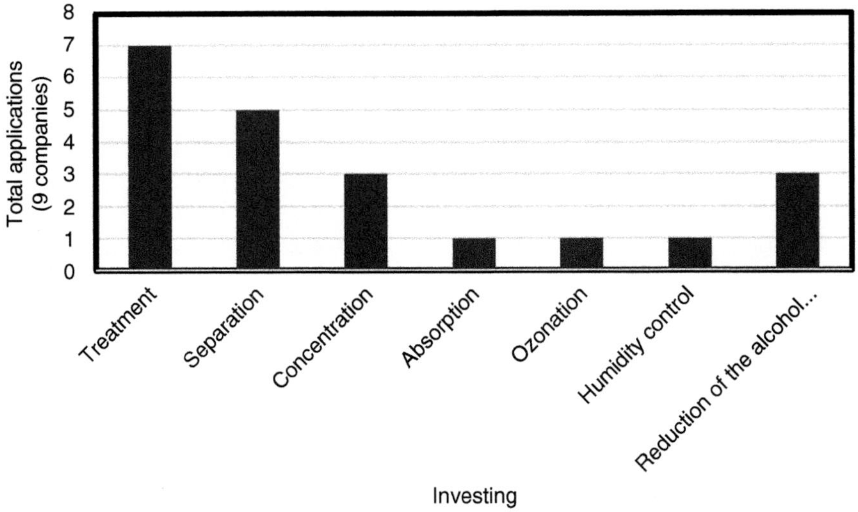

Figure 10.9 The investment analysis of the membrane contractor for food and beverage processing.

3M are the oldest companies. Table 10.10 depicts the service period of the companies that work in this market.

10.3.3 Gas Separation Market

The market for gas separation applications used in membrane contactors is rising due to increasing demand especially for biogas, because of the growing focus

Table 10.9 The investors of the membrane contractor for food and beverage processing.

	Application						
Investor name	**Treatment**	**Separation**	**Concentration**	**Absorption**	**Ozonation**	**Humidity control**	**Reduction of alcohol content**
3M Company		×	×			×	
Cobetter Filtration Equipment Pvt. Ltd.	×	×					
JU.CLA.S Srl		×					
Veolia Environnement SA	×	×	×				×
PTI Pacific Pty. Ltd.	×						
Lenntech B.V.	×	×	×		×		×
Euwa H. H. Eumann GmbH (Euwa)	×						
KH TEC GmbH	×						×
Romfil GmbH	×			×			

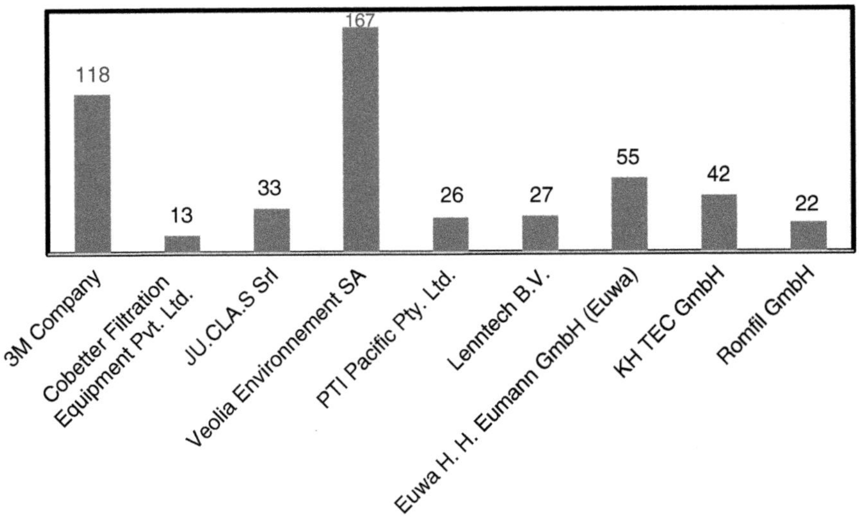

Figure 10.10 The service period (in years) of companies in membrane contactor markets related to food and beverage processing.

Table 10.10 The service period of the companies that work for food and beverage processing.

Name of company	Service period in year
3M Company	118
Cobetter Filtration Equipment Pvt. Ltd.	13
JU.CLA.S Srl	33
Veolia Environnement SA	167
PTI Pacific Pty. Ltd.	26
Lenntech B.V.	27
Euwa H. H. Eumann GmbH (Euwa)	55
KH TEC GmbH	42
Romfil GmbH	22

on renewable energies and increasing concerns over Greenhouse Gas (GHG) emissions. The gas separation technology is used for the separation of hydrogen, nitrogen, hydrocarbons, light hydrocarbons, carbon dioxide, traces of hydrogen sulfide, ammonia, carbon monoxide, oxygen, and methane [26–28]. But, because of the COVID-19, it is seen a slowdown in business activities of this market. According to Table 10.4, all companies except TNO Company, Euwa H. H. Eumann GmbH (Euwa), and KH TEC GmbH can play an essential role in this market. It means that Europe is on top of the list and has the most companies driving this market. Figure 10.11 shows this market based on the regions.

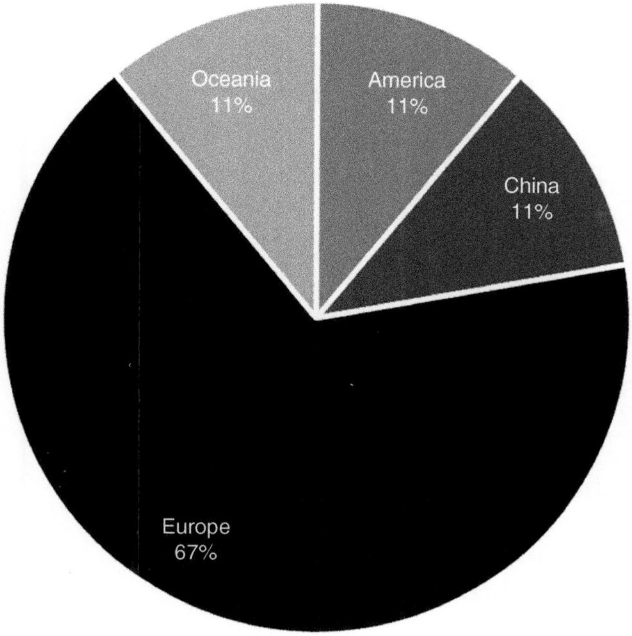

Figure 10.11 The markets for gas separation applications used in membrane contactors based on the region in 2020.

10.3.4 Carbon Capture Market

Carbon capture applications are divided into pre-combustion and post-combustion carbon capture [29–31]. The market for carbon capture applications used in membrane contactors is smaller than gas separation, but the demand for this market is growing due to the presence of renewable energies. These energies can bring about new opportunities for investing. Due to the dependence on fossil fuels and impact of COVID-19, this market can offer novel design and technologies for clean/green energy, low-carbon/low-emission energy, and emission-neutral energy. So, market players should focus more on the growth prospects in this regard. According to this research (Table 10.4), eight companies including 3M Company, EUROWATER, JU.CLA.S Srl, Veolia Environnement SA, PTI Pacific Pty. Ltd., Kværner ASA, Lenntech B.V., and TNO Company are the key players in this market. Here, Europe is on top of the list and has the most companies driving this market. Figure 10.12 shows this market based on the regions.

10.4 Membrane Contactor Market, by Membrane

As mentioned before, membrane contactors are mostly in hollow fiber form, but they can have flat or spiral wound form. Based on their form, they can make from different polymeric materials (Table 10.11). These polymeric materials are prepared

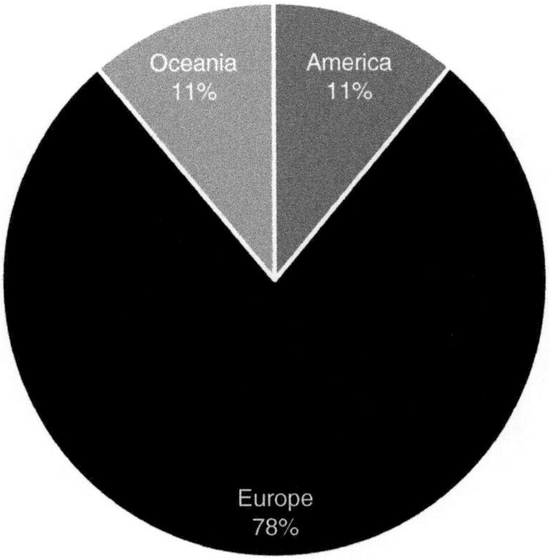

Figure 10.12 The markets for carbon capture applications used in membrane contactors based on the region in 2020.

Table 10.11 The polymer materials of membrane contactors.

Name	References	Name	References
Polysulfone (PSF)	[32]	Polymethylpentene (PMP)	[33]
Polyethersulfone (PES)	[34]	Polydimethyl siloxane (PDMS)	[35]
Polyether ether ketone (PEEK)	[36]	Polycarbonate (PC)	[37]
Polyetherimide (PEI)	[38]	Polyvinyl alcohol (PVA)	[37]
Polyethylene (PE)	[37, 39]	Polyacrylonitrile (PAN)	[37]
Polypropylene (PP)	[37, 40]	Polyphenylene sulfide (PPS)	[37]
Polyvinylidene fluoride (PVDF)	[37, 41]	Teflon (T)	[37]
Polytetrafluoroethylene (PTFE)	[42]	Cellulose acetate (CA)	[37]
Polyphenyl sulfone (PPSU)			

Source: Based on Bazhenov et al. [22].

by different techniques such as melting extrusion, temperature-induced phase separation, and so on [12, 43, 44]. These materials must create the special operating conditions and features for membrane contactors [22] to act as an engine for the membrane contactors.

According to this research, all companies (Table 10.4), except Pure Water Group and Romfil GmbH, active such suppliers and distributors of the polymeric materials. Table 10.12 depicts the existed polymeric materials of these companies.

In this market, Lenntech B.V. and TNO Company are the largest investors due to distributing and supplying the most polymeric materials, followed by 3M Company, shown in Figure 10.13. Also, according to this research, polypropylene (PP) is the most supplied material in this market, followed by polytetrafluoroethylene (PTFE),

Table 10.12 The polymer materials supported by the membrane contactors companies.

Name	Polymeric material supported																
	PSF	PES	PEEK	PEI	PE	PP	PVDF	PTFE	PMP	PDMS	PC	PVA	PAN	PPS	T	CA	PPSU
3M Company	×	×		×	×	×	×	×	×		×				×		×
Cobetter Filtration Equipment Pvt. Ltd.		×			×	×	×	×						×	×		
EUROWATER					×	×									×		
JU.CLA.S Srl						×	×	×									
Veolia Environnement SA					×	×									×		×
PTI Pacific Pty. Ltd.		×				×											
Kværner ASA						×											
Lenntech B.V.	×	×	×		×	×	×	×			×		×	×	×	×	
TNO Company	×	×	×	×	×	×	×	×		×	×		×		×		
Euwa H. H. Eumann GmbH (Euwa)		×		×			×	×			×				×		
Hydro-Elektrik GmbH						×											
KH TEC GmbH						×		×				×				×	

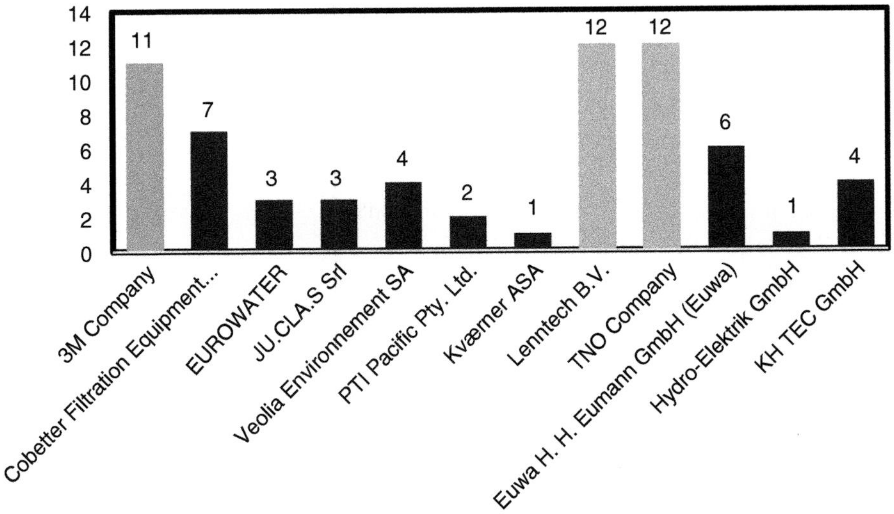

Figure 10.13 The investment analysis for polymeric materials of membrane contactors.

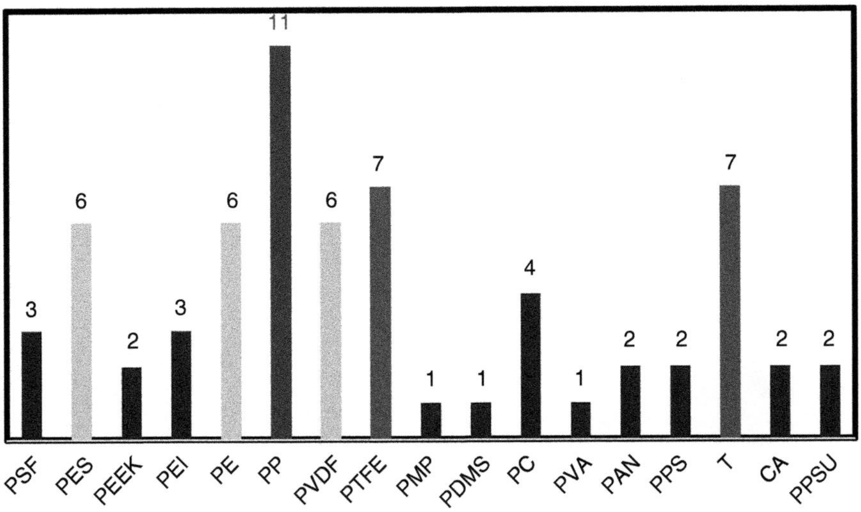

Figure 10.14 The market allocated to polymeric materials of membrane contactors.

teflon (T), polyvinylidene fluoride (PVDF), polyethylene (PE), and polyethersulfone (PES) (Figure 10.14).

The polymeric materials can be used such single or multi-layer polymeric materials called polymeric composite membranes [45, 46]. Table 10.13 shows these membranes used in membrane contactors.

Table 10.14 depicts the companies including 3M Company, PTI Pacific Pty. Ltd., Lenntech B.V., TNO Company, and Euwa H. H. Eumann GmbH (Euwa) involved in distributing and supplying the most polymeric composite membranes. Among

Table 10.13 The polymeric composite membranes used in membrane contactors.

Type	Materials for polymeric composite membranes			
	Substrate	**First layer**	**Second layer**	**Third layer**
Single layer	PAN	Silicone rubber or PDMS	—	—
	PSF	Silicone rubber or PDMS	—	—
	PSF	PEBA[a)]	—	—
	PSF	CA	—	—
	PES	PDMS	—	—
	PVDF	PDMS	—	—
	PEI	PDMS	—	—
	PEI	PEBA	—	—
Multilayer	PSF	PAS[b)]	PPO[c)]	—
	PSF	PVP[d)]	Silicone rubber	—
	PSF	CN[e)]	Silicone rubber	—
	PAN	PDMS	PEBA	—
	PAN	PDMS	PIM[f)]	—
	PEI	PDMS	PEBA	PDMS
	PAN	PTMSP[g)]	Matrimid	PDMS
	PES	PEG[h)]hybrid	PEG-hybrid	PEG-hybrid

a) PEBA: polyether block amide.
b) PAS: polyaminosiloxane.
c) PPO: polydimethylphenylene oxide.
d) PVP: polyvinylpyridine.
e) CN: cellulose nitrate.
f) PIM: polymer of intrinsic microscopy.
g) PTMSP: polytrimethylsilylpropyne.
h) PEG: polyethylene glycol.
Source: Liang et al. [45].

these companies, TNO Company is the largest investor in this market, followed by Lenntech B.V. and 3M Company (Figure 10.15).

Also, according to this research, the most supplied material is "PES + PEG hybrid + PEG hybrid + PEG hybrid" that is a multilayer polymeric composite membrane (Figure 10.16).

10.5 Membrane Contactor Market, by Region

In this section, we classify the regions based on the value chain (Figures 10.3–10.5) of membrane contactor global markets. So, Asia and Europe are estimated to be the largest global market, followed by Africa and America. Figure 10.17 shows this market based on the regions.

Table 10.14 The polymeric composite membranes supported by the membrane contactors companies.

Type	Polymeric composite membranes	3M Company	PTI Pacific Pty. Ltd.	Lenntech B.V.	TNO Company	Euwa H. H. Eumann GmbH (Euwa)
Single layer	PAN + silicone rubber			×		
	PAN + PDMS				×	
	PSF + silicone rubber	×		×		×
	PSF + PDMS				×	
	PSF + PEBA	×			×	
	PSF + CA			×		
	PES + PDMS				×	
	PVDF + PDMS				×	
	PEI + PDMS				×	
	PEI + PEBA	×			×	
Multilayer	PSF + PAS + PPO				×	
	PSF + PVP + silicone rubber	×		×		
	PSF + CN + silicone rubber			×		×
	PAN + PDMS + PEBA				×	
	PAN + PDMS + PIM				×	
	PEI + PDMS + PEBA + PDMS				×	
	PAN + PTMSP + Matrimid + PDMS				×	
	PES + PEG hybrid + PEG hybrid + PEG hybrid	×	×	×	×	×

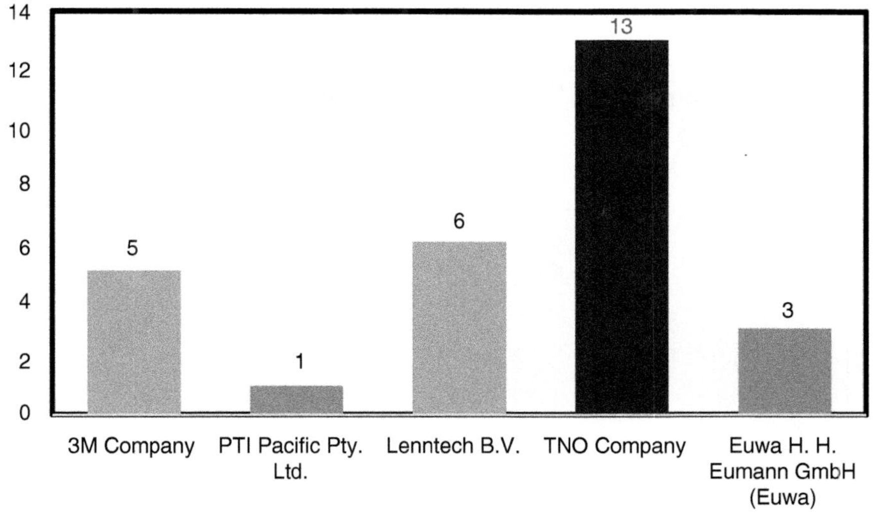

Figure 10.15 The investment analysis for polymeric composite membranes used in membrane contactors.

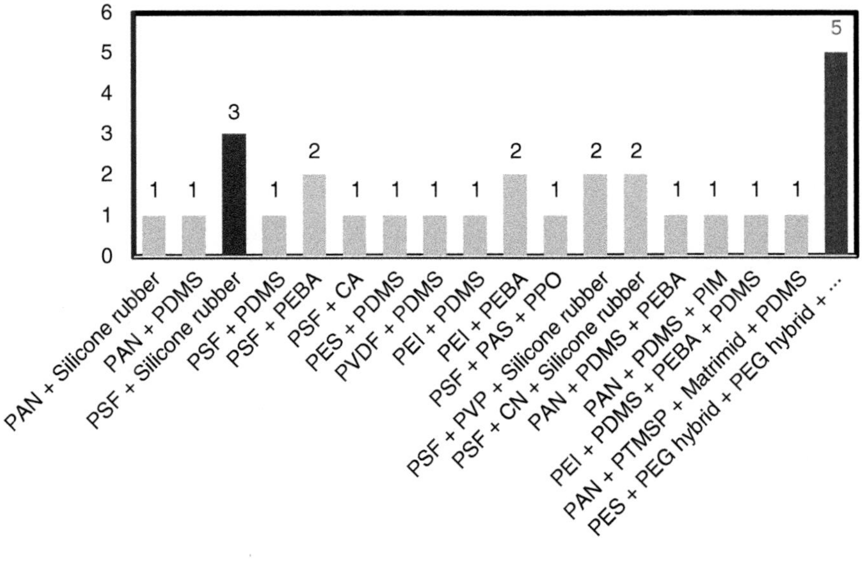

Figure 10.16 The market allocated to polymeric composite membranes used in membrane contactors.

According to this research, the membrane contactor market has a positive outlook in water and wastewater treatment and food and beverage processing as this technology can widely employ on a large scale for these industries. Also, it is projected to grow, attract more attention, witness substantial demands, and create more agencies worldwide. This would help manufacturers increase their profit margins and add value to their products.

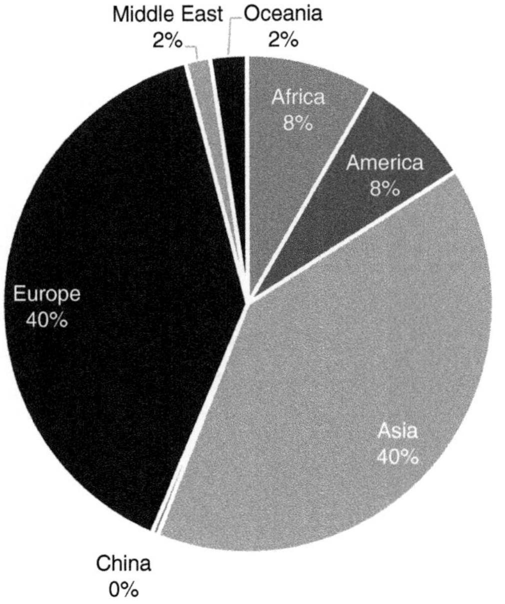

Figure 10.17 The membrane contactor markets based on the region (these data obtained from the website of companies depicted in Table 10.4).

10.6 Recent Developments of Membrane Contactor Companies

In this section, we follow the recent development done by the companies (see Table 10.6). All these developments have been given in Sections 10.6.1–10.6.14 as follows.

10.6.1 3M Company

3M Company is one of the membrane contactor companies that follow science to improve lives and help solve the world's toughest challenges. During recent years, this company has carried out many activities for the growth and development of the membrane contactors, which are listed as follows:

- Manufacturing and developing a collection of unique materials that provide differentiated performances to additives and coatings.
- Defining metamaterials smaller than the wavelengths that can exhibit properties not observed in the constituent materials to control and manipulate the stimuli wavelengths.
- Exploring a new world at the nanometer level where materials exhibit unique physical, chemical, and biological properties.
- Finding new ways to refine and apply nonwoven materials by developing new nonwoven materials for using in products.
- Developing high-performance materials by combining unique materials such as ceramics [47], flouromaterials, nanomaterials, advance silicone materials, and specialty monomers and polymers [48].

10.6.2 Cobetter Filtration Equipment Pvt. Ltd.

Cobetter Filtration Equipment Pvt. Ltd. is another membrane contactor company that has carried out many activities for the growth and development of membrane contactors during recent years. Some of their activities are as follows:

- Developing and manufacturing nano-PTFE microfiltration (MF) membrane for protective facial masks with bacteria filtration efficiency of more than 99%.
- Developing regenerated cellulose (RC) filter media with ultralow non-specific binding characteristic.
- Developing and manufacturing polyethersulfone ultrafiltration (UF) membranes for fluid purification or clarification variety of applications.
- Developing and manufacturing Consieve United Electronic Technology (UET) Cassette membranes with high retention efficiency.

10.6.3 EUROWATER

EUROWATER has supplied the complete water treatment plant and developed the local area plan to pay due respect to nature conservation and the unique surrounding environment as follows:

- Developing and producing ultrapure water using hydrochloric acid and sodium hydroxide.
- Launching two new plants for producing ultrapure water.

10.6.4 JU.CLA.S Srl

JU.CLA.S Srl has always been concerned about the correct management and application of new technologies in the production of wines and has developed an innovative system designed to adjust continuously and precisely the concentration of gases present in wines so as to control or reduce alcohol as follows:

- Smart management of wine gases by controlling the gases including all reduction and impregnation treatments of gases.

10.6.5 Veolia Environnement SA

Veolia has developed technological, contractual, social, and managerial innovations on a daily basis to be able to offer high added value services to its municipal, industrial, and tertiary customers. Some of these activities are as follows:

- Water and water cycle management.
- Water treatment process improvements.
- Wastewater treatment and recovery.
- Waste management, recycling, and recovery.
- Recycled raw materials.
- Energy optimization in buildings and installations.
- Smart management of cities and their infrastructure and services.

10.6.6 PTI Pacific Pty. Ltd.

PTI Pacific Pty. Ltd. has developed an innovative machine based on the MemCell technology and service program for the removal of non-condensable gases from steam sterilization processes.

10.6.7 Kværner ASA

This company has made a new technology by subsea gas separation. This technology in combination with the subsea gas compression technology could make offshore handling of CO_2 for enhanced oil recovery (EOR) technically and economically attractive.

10.6.8 Lenntech B.V.

Lenntech supplies any type of water treatment in a containerized version from desalination plants to emergency units. Some of its activity related to developing membrane contactor are as follows:

- Manufacturing new replacement seals for membranes.

10.6.9 Pure Water Group

This company has developed membrane degassing in many high-purity and ultrapure water systems.

10.6.10 TNO Company

TNO Company is a major player in a growing international network composed of leading scientific institutes, companies with ambitious development profiles, universities, and other partners in knowledge. Some of its activities related to membrane contactors are as follows:

- Creating high-pressure membrane contactors to remove CO_2 from pressurized CO_2/H_2 mixtures from reformer/shift reactors [19, 49].

10.6.11 Euwa H. H. Eumann GmbH (Euwa)

The innovative treatment plants of this company guarantee maximum brewing and beverage water quality. They are used around the world and set the benchmark for efficiency, quality, and performance. Core competence of EUWA is the water treatment technology for the beverage industry. This company with more than 50 years of experience creates innovative water treatment plants individually.

10.6.12 Hydro-Elektrik GmbH

Hydro-Elektrik GmbH has developed and manufactured high-quality products and innovative system solutions. This company is an industry leader, especially with oxidative methods using ozone, optimization of classic multilayer filtration, and ozone biofiltration. Some of its activities related to membrane contactors are as follows:

- Developing membrane contactors for membrane degassing system.

10.6.13 KH TEC GmbH

KH TEC GmbH is another company that is active in the membrane contactor field. This company is in line with continuous changes and the expansion of the product range. Some of its activities during recent years has been given as follows:

- Developing systems for gas management in liquids and SO_2 reduction.
- Developing systems for reducing alcohol content in liquids.

10.6.14 Romfil GmbH

Romfil GmbH Technology has dealt with the application of membranes since 1975. In close collaboration with customers and oenologists of renowned national and international institutions, it constantly develops and improves its systems [50].

10.7 Future Directions

Future direction sets out the vision for approaches that provide more support to help people. 3M Company intends to develop a dynamic approach based on new technologies with an efficient harvest and clarification process to facilitate robust integration of multiple steps. This company wants to pursue these approaches primarily in the pharmaceutical industry. It believes that single-use centrifugation, sonication, depth filtration, and flocculants or body feed-based unit operations can reduce the contaminant load by further removal and help generate higher-purity harvest. These approaches can reduce operating costs too. EUROWATER follows the reduction of pollution by developing the standard technology for removing CO_2 and O_2 from water in many high-purity and industrial applications especially membrane degassing units in water and wastewater treatment, oil and gas generation, and pharmaceutical industries. These approaches can protect capital investment, extend life time of equipment, and decrease operating costs by improving the process efficiency. JU.CLA.S Srl is more interested in developing import and export activities especially in the United States by tracking its competitors and learning about the factories overseas. This company more tends to enforce exclusivity agreements there. Veolia Environnement SA and PTI Pacific Pty. Ltd. want to develop their water technologies and beverage, pharmaceutical and medical degassing, and gas management systems. The Kværner ASA is oriented toward sustainable energy production and a zero-emission future. In fact, this company has followed these approaches since the late 1980s [51]. The future orientations of Lenntech B.V. are toward maximum respect for the environment, maximum energy savings, and maximum respect for health. Meanwhile, TNO Company seems more concerned about climate change than other companies. It believes that this problem is a result of the increased atmospheric concentration of GHGs such as CO_2 closely linked to the use of fossil energy, e.g. for the generation of electricity. This company wants to follow all of the CO_2 capture options for power plants. Euwa H. H. Eumann GmbH (Euwa), Hydro-Elektrik GmbH, KH TEC GmbH, and Romfil GmbH are the companies that want to expand their product range. Finally, Cobetter Filtration Equipment Pvt. Ltd. and Pure Water Group have not provided an approach for the future. Overall, future directions are underpinned by nine strategic priorities:

- More support to help people.
- Developing a dynamic approach based on new technologies.
- Reducing the contaminant load.
- Reducing the operating costs.
- Developing import and export activities.
- Sustainable energy production and zero-emission future.
- Respect for the environment.
- Maximum energy savings.
- Expanding the product range.

10.8 Conclusion

Membrane contactor is a new technology with a broad range of applications. This chapter discussed the key drivers influencing market growth by these applications and analyzed this market and its impact on present and future developments.

References

1 Rezakazemi, M., Khajeh, A., and Mesbah, M. (2018). Membrane filtration of wastewater from gas and oil production. *Environmental Chemistry Letters* 16: 367–388.

2 Kumar Purkait, M., Singh, R., Mondal, P., and Haldar, D. (2020). Chapter 8: Membrane contactors. In: *Thermal Induced Membrane Separation Processes* (eds. M. Kumar Purkait, R. Singh, P. Mondal and D. Haldar), 143–162. Elsevier.

3 Selvamuthukumaran, M. (2020). *Applications of Membrane Technology for Food Processing Industries*. CRC Press.

4 Strathmann, H. (1981). Membrane separation processes. *Journal of Membrane Science* 9: 121–189.

5 Hendricks, D.W. (2018). *Water Treatment Unit Processes: Physical and Chemical*. CRC Press.

6 Uragami, T. (2017). *Science and Technology of Separation Membranes*. Wiley.

7 Hägg, M.-B. (2016). Gas permeation: permeability, permeance, and separation factor. In: *Encyclopedia of Membranes* (eds. E. Drioli and L. Giorno), 1–4. Berlin, Heidelberg: Springer Berlin Heidelberg.
8 Mohammad, A.W., Teow, Y.H., Ho, K.C., and Rosnan, N.A. (2019). Chapter 5: Recent developments in nanofiltration for food applications. In: *Nanomaterials for Food Applications* (eds. A. López Rubio, M.J. Fabra Rovira, M. martínez Sanz and L.G. Gómez-Mascaraque), 101–120. Elsevier.
9 Saffarionpour, S. and Ottens, M. (2018). Recent advances in techniques for flavor recovery in liquid food processing. *Food Engineering Reviews* 10: 81–94.
10 Rosalinda, M., Emma, P., Abaynesh Yihdego, G., and Lidietta, G. (2017). Membrane bioreactors in food, pharmaceutical and biofuel applications: state of the art, progresses and perspectives. *Current Organic Chemistry* 21: 1671–1701.
11 Sheikh, M., Pazirofteh, M., Dehghani, M. et al. (2020). Application of ZnO nanostructures in ceramic and polymeric membranes for water and wastewater technologies: a review. *Chemical Engineering Journal* 391: 123475.
12 Hajilary, N., Rezakazemi, M., and Shirazian, S. (2019). Biofuel types and membrane separation. *Environmental Chemistry Letters* 17: 1–18.
13 Rezakazemi, M., Maghami, M., and Mohammadi, T. (2018). High loaded synthetic hazardous wastewater treatment using lab-scale submerged ceramic membrane bioreactor. *Periodica Polytechnica, Chemical Engineering* 62: 299–304.
14 Rezakazemi, M., Maghami, M., and Mohammadi, T. (2018). Wastewaters treatment containing phenol and ammonium using aerobic submerged membrane bioreactor. *Chemistry Central Journal* 12: 79.
15 Pezeshki, Z. (2020). Classification, modeling, and requirements for separators in rechargeable batteries. In: *Rechargeable Batteries* (eds. R. Boddula, Inamuddin, R. Pothu and A.M. Asiri), 265–314. https://doi.org/10.1002/9781119714774.ch13.
16 Membrane Contactor Market - Global Forecast to 2025, in, 2020. https://www.asdreports.com/ASDR-563578, pp. 132.
17 Obotey Ezugbe, E. and Rathilal, S. (2020). Membrane technologies in wastewater treatment: a review. *Membranes* 10 https://doi.org/10.1080/19443994.2014.974216.
18 Licon, E., Reig, M., Villanova, P. et al. (2015). Ammonium removal by liquid–liquid membrane contactors in water purification process for hydrogen production. *Desalination and Water Treatment* 56: 3607–3616.
19 Klaassen, R., Feron, P.H.M., and Jansen, A.E. (2005). Membrane contactors in industrial applications. *Chemical Engineering Research and Design* 83: 234–246.
20 Susanto, H. and Wenten, I.G. (2020). Fresh water production in coastal and remote areas by solar powered liquid–liquid membrane contactor. *Journal of Coastal Development* 6: 135–144.

21 Vecino, X., Reig, M., Gibert, O. et al. (2020). Integration of liquid–liquid membrane contactors and electrodialysis for ammonium recovery and concentration as a liquid fertilizer. *Chemosphere* 245: 125606.

22 Bazhenov, S.D., Bildyukevich, A.V., and Volkov, A.V. (2018). Gas–liquid hollow fiber membrane contactors for different applications. *Fibers* 6 https://doi.org/10.3390/fib6040076.

23 Valdés, H., Unda, K., and Saavedra, A. (2019). Numerical simulation on supercritical CO_2 fluid dynamics in a hollow fiber membrane contactor. *Computation* 7 https://doi.org/10.3390/computation7010008.

24 Roizard, D. (2016). Gas–liquid membrane contactor. In: *Encyclopedia of Membranes* (eds. E. Drioli and L. Giorno), 1–4. Berlin, Heidelberg: Springer Berlin Heidelberg.

25 Kovvali, A.S. and Sirkar, K.K. (2003). Chapter 7: Membrane contactors: recent developments. In: *Membrane Science and Technology* (eds. D. Bhattacharyya and D.A. Butterfield), 147–164. Elsevier.

26 Rezakazemi, M., Sadrzadeh, M., and Matsuura, T. (2018). Thermally stable polymers for advanced high-performance gas separation membranes. *Progress in Energy and Combustion Science* 66: 1–41.

27 Rezakazemi, M., Ebadi Amooghin, A., Montazer-Rahmati, M.M. et al. (2014). State-of-the-art membrane based CO_2 separation using mixed matrix membranes (MMMs): an overview on current status and future directions. *Progress in Polymer Science* 39: 817–861.

28 Rezakazemi, M., Vatani, A., and Mohammadi, T. (2015). Synergistic interactions between POSS and fumed silica and their effect on the properties of crosslinked PDMS nanocomposite membranes. *RSC Advances* 5: 82460–82470.

29 Sohaib, Q., Muhammad, A., Younas, M., and Rezakazemi, M. (2020). Modeling pre-combustion CO_2 capture with tubular membrane contactor using ionic liquids at elevated temperatures. *Separation and Purification Technology* 241: 116677.

30 Younas, M., Tahir, T., Wu, C. et al. (2020). Post-combustion CO_2 capture with sweep gas in thin film composite (TFC) hollow fiber membrane (HFM) contactor. *Journal of CO_2 Utilization* 40: 101266.

31 Younas, M., Rezakazemi, M., Daud, M. et al. (2020). Recent progress and remaining challenges in post-combustion CO_2 capture using metal-organic frameworks (MOFs). *Progress in Energy and Combustion Science* 80: 100849.

32 Vladisavljević, G.T. (1999). Use of polysulfone hollow fibers for bubbleless membrane oxygenation/deoxygenation of water. *Separation and Purification Technology* 17: 1–10.

33 Huang, X., Wang, W., Zheng, Z. et al. (2016). Surface monofunctionalized polymethyl pentene hollow fiber membranes by plasma treatment and hemocompatibility modification for membrane oxygenators. *Applied Surface Science* 362: 355–363.

34 Bakeri, G., Ismail, A.F., Rahimnejad, M., and Matsuura, T. (2014). Porous polyethersulfone hollow fiber membrane in gas–liquid contacting processes. *Chemical Engineering Research and Design* 92: 1381–1390.

35 Tilahun, E., Bayrakdar, A., Sahinkaya, E., and Çalli, B. (2017). Performance of polydimethylsiloxane membrane contactor process for selective hydrogen sulfide removal from biogas. *Waste Management* 61: 250–257.

36 Nymeijer, K., Visser, T., Assen, R., and Wessling, M. (2004). Super selective membranes in gas–liquid membrane contactors for olefin/paraffin separation. *Journal of Membrane Science* 232: 107–114.

37 Singh, R. (2015). Chapter 1: Introduction to membrane technology. In: *Membrane Technology and Engineering for Water Purification*, 2e (ed. R. Singh), 1–80. Oxford: Butterworth-Heinemann.

38 Kneifel, K., Nowak, S., Albrecht, W. et al. (2006). Hollow fiber membrane contactor for air humidity control: modules and membranes. *Journal of Membrane Science* 276: 241–251.

39 Leiknes, T. and Semmens, M.J. (2001). Vacuum degassing using microporous hollow fiber membranes. *Separation and Purification Technology* 22–23: 287–294.

40 Lv, Y., Yu, X., Tu, S.-T. et al. (2012). Experimental studies on simultaneous removal of CO_2 and SO_2 in a polypropylene hollow fiber membrane contactor. *Applied Energy* 97: 283–288.

41 Su, J. and Wei, Y. (2019). Novel tri-bore PVDF hollow fiber membranes for the control of dissolved oxygen in aquaculture water. *Journal of Water Process Engineering* 30: 100584.

42 Merle, T., Pronk, W., and von Gunten, U. (2017). MEMBRO3X, a novel combination of a membrane contactor with advanced oxidation (O_3/H_2O_2) for simultaneous micropollutant abatement and bromate minimization. *Environmental Science & Technology Letters* 4: 180–185.

43 Rezakazemi, M., Dashti, A., Riasat Harami, H. et al. (2018). Fouling-resistant membranes for water reuse. *Environmental Chemistry Letters* 16: 715–763. https://doi.org/10.1007/s10311-018-0717-8.

44 Mirqasemi, M.S., Homayoonfal, M., and Rezakazemi, M. (2020). Zeolitic imidazolate framework membranes for gas and water purification. *Environmental Chemistry Letters* 18: 1–52.

45 Liang, C.Z., Chung, T.-S., and Lai, J.-Y. (2019). A review of polymeric composite membranes for gas separation and energy production. *Progress in Polymer Science* 97: 101141.

46 Rezakazemi, M., Shahidi, K., and Mohammadi, T. (2015). Synthetic PDMS composite membranes for pervaporation dehydration of ethanol. *Desalination and Water Treatment* 54: 1542–1549.

47 Koonaphapdeelert, S. and Li, K. (2006). The development of ceramic hollow fibre membranes for a membrane contactor. *Desalination* 200: 581–583.

48 Kattan, O., Ebbers, K., Koolaard, A. et al. (2018). Membrane contactors: an alternative for de-aeration of salt solutions? *Separation and Purification Technology* 205: 231–240.

49 Ansaloni, L., Arif, A., Ciftja, A.F. et al. (2016). Development of membrane contactors using phase change solvents for CO_2 capture: material compatibility study. *Industrial & Engineering Chemistry Research* 55: 13102–13113.

50 Chuah, C.Y., Kim, K., Lee, J. et al. (2020). CO_2 absorption using membrane contactors: recent progress and future perspective. *Industrial & Engineering Chemistry Research* 59: 6773–6794.

51 O. Falk-Pedersen, H. Dannstroem, Methods for Removing Carbon Dioxide from Gases, in, 1998, PCT/NO1997/000190.

11

Conclusions and Perspective

Mohammad Younas[1] *and Mashallah Rezakazemi*[2]

[1] *University of Engineering and Technology, Department of Chemical Engineering, Peshawar 25120, Pakistan*
[2] *Shahrood University of Technology, Faculty of Chemical and Materials Engineering, Shahrood, Iran*

Membrane technology has become an important part of our life. Among different membrane separation processes and techniques, membrane contactor technology is an emerging technology and has become increasingly important in water treatment, gas separation, and food and beverage production. Membrane contactor technology undergone a detailed theoretical study followed by laboratory analysis before the realization of its industrial application in water, food, and environment. Today, membrane contactor technology is widely accepted as the best available sustainable technology for the applications mentioned above. This technology offers a viable alternative to conventional pressure-driven membrane separations and purifications in water treatment, food and beverage processing, energy production, and gas separations. Many successful large-scale real applications have brought membrane contactor technology to the forefront of emerging modern technologies to impart great benefits for better human life. As in the twenty-first century, the world is facing great challenges in water, energy, food, and environment. Membrane contactor technology has greatly enhanced the development of new membrane materials and efficient design of membrane modules to restructure production processes and protect the environment and public health and, therefore, found its place in the best available sustainable technologies. The scope of the applications of membrane contactor technology is still extending, stimulated by the developments of the novel and improved membrane materials, integrated processes, and new efficient modules with improved species permeability, selectivity, and cost-effectiveness.

An overview of membrane technology was presented in **Chapter 1**. Traditional membrane separation processes were described with the help of schematics. Besides that, a brief historical perspective was also presented. The need to develop novel membranes, innovative processes, and compact modular design was highlighted. Though the conventional membrane separation processes are being used successfully in the industries for the last century, still there remained a gap to develop the membrane technology with higher productivity, lower cost of production, and increased development speed. This gap can be bridged by introducing the new

Membrane Contactor Technology: Water Treatment, Food Processing, Gas Separation, and Carbon Capture, First Edition. Edited by Mohammad Younas and Mashallah Rezakazemi.

materials, processes, and compact design in membrane contactor technology. The best choice of membrane materials, modules, and specific processes makes the membrane contactors the best available technology for the sustainable development and growth at the nexus of water, food, energy, and environment.

In **Chapter 2**, an introduction to membrane contactor technology was detailed. Theoretical aspects of membrane contactor separation processes were described with the help of schematics and mathematical equations. The current state of the knowledge indicates that the independent use of membrane contactors is still looking for its commercial use at an industrial scale and has great potential in developing various successful applications. However, membrane contactors are sometimes driven with the organic solvents that create the driving force. Therefore, there is a need to develop less toxic and environment-friendly organic solvents (receiving phase), making the membrane contactor separation processes a green process. Similarly, it was pointed out that a deeper understanding of mass transfer phenomena across the membrane interface in multiphase systems in diverse applications using separation science principles is required to increase the long-term operation of membrane contactor separation systems.

Chapter 3 was focused on the transport theory of species in membrane contactor operation. All the theoretical aspects of this technology were discussed in detail. It includes the film theory, interface analysis at pore mouths, and species transport models (algebraic, analytical, and numerical). Phase equilibria have also been presented. All the theoretical phenomena can be presented for liquid–liquid (L-L), gas–liquid (G-L), and gas–gas (G-G) interfacial systems. The guidelines presented in **Chapter 3** could be used to model the membrane contactor module for a specific operation in L-L, G-L, and G-G applications.

The development in membrane contactor technology also inspired scientists to develop the new membrane materials and innovative module geometries that were discussed in **Chapter 4**. The overview and performance of membrane modules based on their geometry and flow, membrane materials, and membrane synthesis methods were presented.

Various modes in which membrane contactor technology can be used and is operated have been discussed with schematics and theoretical explanation in **Chapter 5**. These include the dispersion-free L-L, G-L, and G-G extractions, membrane distillation, osmotic evaporation, pressure-retarded osmosis, membrane reactor contactors, membrane crystallization, etc. All processes were described with the mathematical models as well as the progress in the experimental aspect of all modes of operation. The membrane contactor technology has been proved to be very advantageous in all these modes of operation.

The applications of membrane contactor technology in water, food, and carbon capture were presented in **Chapters 6–9**, respectively. Nowadays, the treatment of wastewater is considered as one of the major challenges. Membrane contactor technology is reliable and easy to scale up and shows potential for growth in wastewater treatment. In **Chapter 6**, the application of membrane contactor in wastewater is highlighted. **Chapter 6** describes how membrane contactor technology works in treating wastewater. Membrane contactor technology has been used for the removal of toxic organic pollutants present in wastewater effluent. In wastewater treatment, membrane contactor is a highly efficient technology that requires less footprint and a small amount of organic extractant. A stable membrane is needed that immediately controls its swelling in the presence of an organic extractant. On the other hand, recent work on membrane synthesis is to prepare the ultrathin membrane with high surface area into compact modules. Therefore, further improvement and advancement in extractants, polymeric materials, and membrane structure enhance separation performance with prolonged life of membrane contactor.

Membrane contactor that combined extraction/stripping process has been widely used in the resource's extraction from wastewater because of its less footprint and small amount of organic extractant usage. Membrane contactors have been successfully employed in the wastewater treatment to recover heavy metals as in dispersion-free liquid–liquid extraction, water desalination as in membrane distillation, and brackish water treatment merged with power production as in pressure-retarded osmosis.

Applications of membrane contactor technology in food and beverage industry were briefly discussed in **Chapter 7**. Mostly membrane distillation (MD) and osmotic membrane distillation (OMD) were described in detail for juice clarification and concentration. A variety of fruit juices, dairy products, and olive mill water were described for processing through MD/OMD in membrane contactors. The transmembrane flux, membrane wetting, and fouling were presented schematically, and quantitative results were tabulated.

The membrane contactors have been discussed for pre- and post-combustion capture of CO_2 in **Chapters 8 and 9**, respectively. The membrane materials for use in selective separation in H_2 and CO_2 were explored in **Chapter 8**. The recent developments and the future prospects were presented for utilizing membrane contactor technology for selective separation of H_2 and CO_2 that can enable the proper utilization of hydrogen fuel produced by steam reforming of natural gas or by coal or biomass gasification. **Chapter 8** outlines that the membrane contactors, primarily supported liquid membranes (SLMs), are very helpful in improving separation performance and hence have a high potential to be used in future membrane gas separation technology.

Chapter 9 delineates various experimental and pilot-scale studies complemented by computational simulations to make membrane-based CO_2 capture competitive to traditional absorption-based technologies. The main focus is of these studies is material development and economic evaluation, to point out and address the major issues that arise in membrane durability and cost-competitiveness with traditional absorption-based processes. Among others, one seminal approach is the development of mixed matrix membrane (MMMs), which might result in a durable and cost-competitive membrane material. Moreover, thin film composite membranes are also showing promising results. Significant progress has also been made in the arena of absorbents for post-combustion CO_2 sequestration. However, there is still need of comprehensive studies encompassing all the aspects, including membrane/absorbent interactions, stability, and cost of regeneration at post-combustion working conditions that could potentially provide groundbreaking results and invigorate the research interest in post-combustion CO_2 capture using membrane contactors.

Chapter 10 presents a thorough overview for membrane contactor market and future trends in this market in two industries including water and wastewater treatment and food and beverage processing. **Chapter 10** explores the gas separation and carbon capture applications in this market too. The major goal is to show how these two industries can affect the supply chain and market size growth. This analysis is done with the help of the leading market players of these industries.

11.1 Future Directions

The major challenges involved in large-scale applications of membrane contactors are the membrane swelling induced by the organic extractant and high mass transfer resistance. The promising direction of research is to fabricate a high-permeable and solvent-resistant membrane to overcome these limitations. A thermally stable and mild green solvent, e.g. ionic liquid, can be employed as an extractant. Meanwhile, further research is also required to design and fabricate membrane modules that offer improved flow characteristics to control concentration polarization and easy module cleaning in membrane fouling.

Future aspects of this technology include its further integration into the existing system, both as a standalone alternative to conventional technologies and a hybrid in combination with other typical technologies. Whatever framework is used, one thing is certain that using membrane technologies to any extent can improve the performance of the system. The issues that need to be worked out are cost optimization and proper design of the system to avoid losses.

Most of the efforts are focused on improving membrane contactors' efficiency in terms of flux, selectivity, solute rejection, and antifouling properties due to membrane materials and synthesis techniques. It is now recommended to focus on using the cutting-edge artificial intelligence technology and process modeling and simulation to develop more robust modules with improved process efficiency. Environmental regulations, energy consumption, sustainable development, nanotechnology, and waste minimization are the key drivers for membrane contactor technology to be promoted at industrial level. The technology would rise exponentially if the green, nontoxic fluids are used as contacting fluid and membrane stability is increased. Further sustained transmembrane flux is also the paramount challenging factor in membrane synthesis and process design. Other challenges are the efficient design of modules to enhance the mass transfer coefficient and development of low-cost membrane materials that can simultaneously provide the appreciable selectivity and flux.

There is still considerable work to be done to implement the membrane contactor technology on an industrial scale. There is immense potential for progress in membrane contactor technology because of the numerous advantages that have been discussed in the book. Therefore, it is worthwhile to further work on this technology and explore its many facets for product quality improvement within the industry. In fact, all these technologies have made quite an impact and are already being used in conjunction with other technologies to give better results than conventional technologies. Soon, these hybrid technologies may be replaced completely by these newer technologies as a standalone option. Until then, work needs to be done to remove or reduce the effects of all the disadvantages of these processes, which are hindering their use on a much larger scale. Membrane contactor technology is envisaged to play a significant role in addressing these issues through extensive research on membrane materials, integrated process improvement, and efficient module design.

Index

Membrane Contactor Technology: Water Treatment, Food Processing, Gas Separation, and Carbon Capture, First Edition. Edited by Mohammad Younas and Mashallah Rezakazemi.

n

o